An Introduction to Medicinal Chemistry

An Introduction to

Medicinal Chemistry

GRAHAM L. PATRICK

Department of Chemistry,
Paisley University

Oxford   New York   Tokyo
OXFORD UNIVERSITY PRESS

Oxford University Press, Walton Street, Oxford OX2 6DP

Oxford   New York
Athens   Auckland   Bangkok   Bombay
Calcutta   Cape Town   Dar es Salaam   Delhi
Florence   Hong Kong   Istanbul   Karachi
Kuala Lumpur   Madras   Madrid   Melbourne
Mexico City   Nairobi   Paris   Singapore
Taipei   Tokyo   Toronto
and associated companies in
Berlin   Ibadan

Oxford is a trade mark of Oxford University Press

Published in the United States
by Oxford University Press Inc., New York

© *Graham L. Patrick, 1995*
First published in 1995
Reprinted 1995

A catalogue record for this book is available from the British Library

Library of Congress Cataloging in Publication Data
Patrick, Graham L.
An introduction to medicinal chemistry / Graham L. Patrick. — 1st ed.
Includes bibliographical references and index.
1. Pharmaceutical chemistry.   I. Title.
RS403.P38   1995   615/.19—dc20   94–24704
ISBN 0 19 855872 4 (Hbk)
ISBN 0 19 855871 6 (Pbk)

Printed in Great Britain on acid-free paper by
the Bath Press

Preface

This text is aimed at undergraduates who have a basic grounding in chemistry and are interested in a future career in the pharmaceutical industry. It attempts to convey something of the fascination of working in a field which overlaps the disciplines of chemistry, biochemistry, cell biology, and pharmacology.

No previous knowledge of biology is assumed and the first six chapters cover the basics of cell structure, proteins, and nucleic acids as applied to drug design.

Chapters 7, 8, and 9 describe the general tactics employed in developing an effective drug and also the difficulties faced by the medicinal chemist in this task.

Chapters 10, 11, and 12 cover three particular areas of medicinal chemistry and are representative of the classifications which prevail in medicinal chemistry. By doing this, it is hoped that the advantages and disadvantages of the various classification schemes are exemplified.

The three areas of medicinal chemistry described in Chapters 10, 11, and 12 have long histories and much of the early development of these drugs relied heavily on random variations of lead compounds on a trial and error basis. This approach is wasteful and the future of medicinal chemistry lies in the rational design of drugs based on a firm understanding of their biology and chemistry. The development of the antiulcer drug cimetidine represents one of the best examples of the rational approach to medicinal chemistry and is covered in Chapter 13.

Paisley   G.L.P.
June 1994

Acknowledgements

Figure 7.2 and all the figures in Chapter 1 were designed and drawn by Mr G. Leerie, to whom the author is indebted. The author also wishes to express his gratitude to Professor J. Mann for many helpful and constructive comments on the text, and also to Dr D. Marrs for helping with proofreading; however, any errors in the text are the author's responsibility alone.

Contents

Classification of drugs

The way drugs are classified or grouped can be confusing. Different textbooks group drugs in different ways.

1 *By pharmacological effect*

Drugs are grouped depending on the biological effect they have, e.g. analgesics, antipsychotics, antihypertensives, antiasthmatics, antibiotics, etc.

This is useful if one wishes to know the full scope of drugs available for a certain ailment. However, it should be emphasized that such groupings contain a large and extremely varied assortment of drugs. This is because there is very rarely one single way of dealing with a problem such as pain or heart disease. There are many biological mechanisms which the medicinal chemist can target to get the desired results, and so to expect all painkillers to look alike or to have some common thread running through them is not realistic.

A further point is that many drugs do not fit purely into one category or another and some drugs may have more uses than just one. For example, a sedative might also have uses as an anticonvulsant. To pigeon-hole a drug into one particular field and ignore other possible fields of action is folly.

The antibacterial agents (Chapter 10) are a group of drugs that are classified according to their pharmacological effect.

2 *By chemical structure*

Many drugs which have a common skeleton are grouped together, e.g. penicillins, barbiturates, opiates, steroids, catecholamines, etc.

In some cases (e.g. penicillins) this is a useful classification since the biological activity (e.g. antibiotic activity for penicillins) is the same. However, there is a danger that one could be confused into thinking that all compounds of a certain chemical group have the same biological action. For example, barbiturates may look much alike and yet have completely different uses in medicine. The same holds true for steroids.

It is also important to consider that most drugs can act at several sites of the body and have several pharmacological effects.

The opiates (Chapter 12) are a group of drugs with similar chemical structures.

3 *By target system*

These are compounds which are classed according to whether they affect a certain target system in the body—usually involving a neurotransmitter, e.g. antihistamines, cholinergics, etc.

This classification is a bit more specific than the first, since it is identifying a system with which the drugs interact. It is, however, still a system with several stages and so the same point can be made as before—for example, one would not expect all antihistamines to be similar compounds since the system by which histamine is synthesized, released, interacts with its receptor, and is finally removed, can be attacked at all these stages.

4 *By site of action*

These are compounds which are grouped according to the enzyme or receptor with which they interact. For example, anticholinesterases (Chapter 11) are a group of drugs which act through inhibition of the enzyme acetylcholinesterase.

This is a more specific classification of drugs since we have now identified the precise target at which the drugs act. In this situation, we might expect some common ground between the agents included, since a common mechanism of action is a reasonable though not inviolable assumption.

It is easy, however, to lose the wood for the trees and to lose sight of why it is useful to have drugs which switch off a particular enzyme or receptor site. For example, it is not intuitively obvious why an anticholinergic agent should paralyse muscle and why that should be useful.

1 ▪ Drugs and the medicinal chemist

In medicinal chemistry, the chemist attempts to design and synthesize a medicine or a pharmaceutical agent which will benefit humanity. Such a compound could also be called a 'drug', but this is a word which many scientists dislike since society views the term with suspicion.

With media headlines such as 'Drugs Menace', or 'Drug Addiction Sweeps City Streets', this is hardly surprising. However, it suggests that a distinction can be drawn between drugs which are used in medicine and drugs which are abused. Is this really true?

Can we draw a neat line between 'good drugs' like penicillin and 'bad drugs' like heroin? If so, how do we define what is meant by a good or a bad drug in the first place? Where would we place a so-called social drug like cannabis in this divide? What about nicotine, or alcohol?

The answers we would get would almost certainly depend on who we were to ask. As far as the law is concerned, the dividing line is defined in black and white. As far as the party-going teenager is concerned, the law is an ass.

As far as we are concerned, the questions are irrelevant. To try and divide drugs into two categories—safe or unsafe, good or bad, is futile and could even be dangerous.

First of all, let us consider the so-called 'good' drugs—the medicines. How 'good' are they? If a medicine is to be truly 'good' it would have to satisfy the following criteria. It would have to do what it is meant to do, have no side-effects, be totally safe, and be easy to take.

How many medicines fit these criteria?

The short answer is 'none'. There is no pharmaceutical compound on the market today which can completely satisfy all these conditions. Admittedly, some come quite close to the ideal. Penicillin, for example, has been one of the most effective antibacterial agents ever discovered and has also been one of the safest. Yet it too has drawbacks. It has never been able to kill *all* known bacteria and as the years have gone by, more and more bacterial strains have become resistant.

Nor is penicillin totally safe. There are many examples of patients who show an allergic reaction to penicillin and are required to take alternative antibacterial agents.

Whilst penicillin is a relatively safe drug, there are some medicines which are distinctly dangerous. Morphine is one such example. It is an excellent analgesic, yet it suffers from the serious side-effects of tolerance, respiratory depression, and addiction. It can even kill if taken in excess.

Barbiturates are also known to be dangerous. At Pearl Harbor, American casualties undergoing surgery were given barbiturates as general anaesthetics. However, because of a poor understanding about how barbiturates are stored in the body, many patients received sudden and fatal overdoses. In fact, it is reputed that more casualties died at the hands of the anaesthetists at Pearl Harbor than died of their wounds.

To conclude, the 'good' drugs are not as perfect as one might think.

What about the 'bad' drugs then? Is there anything good that can be said about them? Surely there is nothing we can say in defence of the highly addictive drug heroin?

Well, let us look at the facts about heroin. Heroin is one of the best painkillers known to man. In fact, it was named heroin at the end of the nineteenth century because it was thought to be the 'heroic' drug which would banish pain for good. The drug went on the market in 1898, but five years later the true nature of heroin's addictive properties became evident and the drug was speedily withdrawn from general distribution.

However, heroin is still used in medicine today—under strict control of course.

The drug is called diamorphine and it is the drug of choice when treating patients dying of cancer. Not only does diamorphine reduce pain to acceptable levels, it also produces a euphoric effect which helps to counter the depression faced by patients close to death. Can we really condemn a drug which can do that as being all 'bad'?

By now it should be evident that the division between good drugs and bad drugs is a woolly one and is not really relevant to our discussion of medicinal chemistry. All drugs have their good points and their bad points. Some have more good points than bad and vice versa, but like people, they all have their own individual characteristics. So how are we to define a drug in general?

One definition could be to classify drugs as 'compounds which interact with a biological system to produce a biological response'.

This definition covers all the drugs we have discussed so far, but it goes further. There are chemicals which we take every day and which have a biological effect on us. What are these everyday drugs?

One is contained in all the cups of tea, coffee, and cocoa which we consume. All of these beverages contain the stimulant caffeine. Whenever you take a cup of coffee, you are a drug user. We could go further. Whenever you crave a cup of coffee, you are a drug addict. Even kids are not immune. They get their caffeine 'shot' from coke or pepsi. Whether you like it or not, caffeine is a drug. When you take it, you experience a change of mood or feeling.

So too, if you are a worshipper of the 'nicotine stick'. The biological effect is

Caffeine

different. In this case you crave sedation or a calming influence, and it is the nicotine in the cigarette smoke which induces that effect.

There can be little doubt that alcohol is a drug and as such causes society more problems than all other drugs put together. One only has to study road accident statistics to appreciate that fact. It has been stated that if alcohol was discovered today, it would be restricted in exactly the same way as cocaine or cannabis. If one considers alcohol in a purely scientific way, it is a most unsatisfactory drug. As many will testify, it is notoriously difficult to judge the correct dose of alcohol required to gain the beneficial effect of 'happiness' without drifting into the higher dose levels which produce unwanted side-effects. Alcohol is also unpredictable in its biological effects. Happiness *or* depression may result depending on the user's state of mind. On a more serious note, addiction and tolerance in certain individuals have ruined the lives of addicts and relatives alike.

Even food can be a drug. Junk foods and fizzy drinks have been blamed for causing hyperactivity in children. It is believed that junk foods have high concentrations of certain amino acids which can be converted in the body to neurotransmitters. These are chemicals which pass messages between nerves. If an excess of these chemical messengers should accumulate, then too many messages are being transmitted in the brain, leading to the disruptive behaviour observed in susceptible individuals. Allergies due to food additives and preservatives are also well recorded.

Our definition of a drug can also be used to include compounds which we might not at first sight consider to be drugs.

Consider how the following examples fit our definition.

- Morphine—reacts with the body to bring pain relief.
- Snake venom—reacts with the body and may cause death!
- Strychnine—reacts with the body and may cause death!
- LSD—reacts with the body to produce hallucinations.
- Coffee—reacts with the body to waken you up.
- Penicillin—reacts with bacterial cells and kills them.
- Sugar—reacts with the tongue to produce a sense of taste.

All of these compounds fit our definition of drugs. It may seem strange to include poisons and snake venoms as drugs, but they too react with a biological system and produce a biological response—a bit extreme perhaps, but a response all the same.

The idea of poisons and venoms acting as drugs may not appear so strange if we consider penicillin. We have no problem in thinking of penicillin as a drug, but if we were to look closely at how penicillin works, then it is really a poison. It interacts with bacteria (the biological system) and kills them (the biological response). Fortunately for us, penicillin has no effect on human cells.

Even those medicinal drugs which do not act as poisons have the potential to become poisons—usually if they are taken in excess. We have already seen this with morphine. At low doses it is a painkiller. At high doses, it is a poison which kills by suffocation. Therefore, it is important that we treat all medicines as potential poisons and keep them well protected from children searching the house for concealed sweets.

There is a term used in medicinal chemistry known as the therapeutic index which indicates how safe a particular drug is. The therapeutic index is a measure of the drug's beneficial effects at a low dose versus its harmful effects at a high dose. A high therapeutic index means that there is a large safety margin between beneficial and toxic doses. The values for cannabis and alcohol are 1000 and 10 respectively.

If useful drugs can be poisons at high doses, does the opposite hold true? Can a poison be a medicine at low doses? In certain cases, this is found to be the case.

Arsenic is well known as a poison, but arsenic-based compounds were used at the beginning of the century as antiprotozoal agents.

Curare is a deadly poison which was used by the Incas to tip their arrows such that a minor arrow wound would be fatal, yet compounds based on the tubocurarine structure (the active principle of curare) have been used in surgical operations to relax

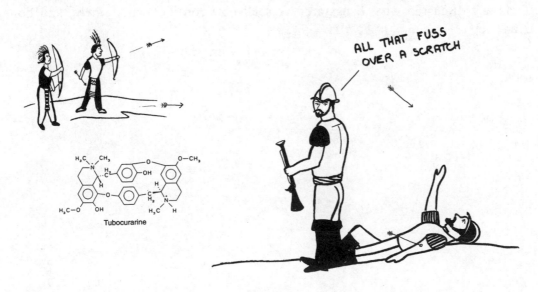

Tubocurarine

ALL THAT FUSS OVER A SCRATCH

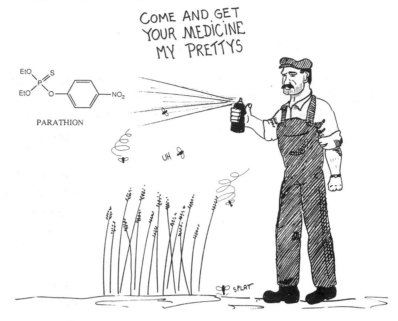

muscles. Under proper control and in the correct dosage, a lethal poison may well have an important medical role.

Since our definition covers any chemical which interacts with any biological system, we could include all pesticides and herbicides as drugs. They interact with bacteria, fungi, and insects, kill them and thus protect plants.

Sugar (or any sweet item for that matter) can be classed as a drug. It reacts with a biological system (the taste buds on the tongue) to produce a biological response (sense of sweetness).

Having discussed what drugs are, we shall now consider why, where, and how drugs act.

2 ▪ The why and the wherefore

2.1 *Why should drugs work?*

Why indeed? We take it for granted that they do, but why should chemicals, some of which have remarkably simple structures, have such an important effect on such a complicated and large structure as a human being? The answer lies in the way that the human body operates. If we could see inside our bodies to the molecular level, we would no doubt get a nasty shock, but we would also see a magnificent array of chemical reactions taking place, keeping the body healthy and functioning.

Drugs may be mere chemicals but they are entering a world of chemical reactions with which they can interact. Therefore, there should be nothing odd in the fact that they can have an effect. The surprise might be that they can have such specific effects. This is more a result of *where* they react in the body.

2.2 *Where do drugs work?*

Since life is made up of cells, then quite clearly drugs must act on cells. The structure of a typical cell is shown in Fig. 2.1.

All cells in the human body contain a boundary wall called the cell membrane. This encloses the contents of the cell—the cytoplasm.

The cell membrane seen under the electron microscope consists of two identifiable layers. Each layer is made up of an ordered row of phosphoglyceride molecules such as phosphatidylcholine (lecithin).[1] Each phosphoglyceride molecule consists of a small polar head-group, and two long hydrophobic chains (Fig. 2.2).

In the cell membrane, the two layers of phospholipids are arranged such that the hydrophobic tails point to each other and form a fatty, hydrophobic centre, while the

[1] The outer layer of the membrane is made up of phosphatidylcholine whereas the inner layer is made up of phosphatidylethanolamine, phosphatidylserine, and phosphatidylinositol.

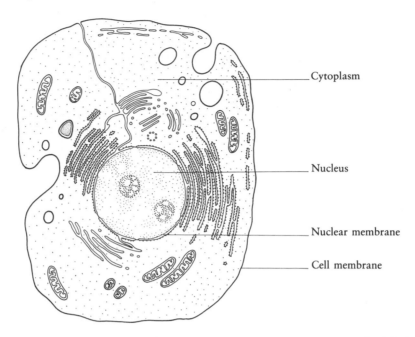

Fig. 2.1 A typical cell. Taken from J. Mann, *Murder, magic, and medicine*, Oxford University Press (1992), with permission.

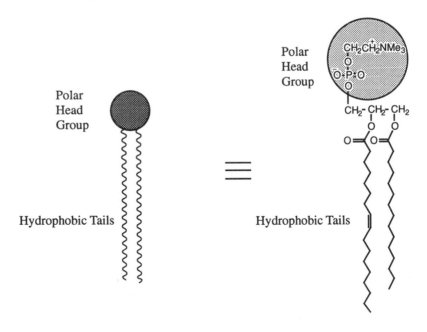

Fig. 2.2 Phosphoglyceride structure.

ionic head-groups are placed at the inner and outer surfaces of the cell membrane (Fig. 2.3). This is a stable structure since the ionic, hydrophilic head-groups can interact with the aqueous media inside and outside the cell, while the hydrophobic tails maximize van der Waals bonding with each other and are kept away from the aqueous environments. The overall result of this structure is to construct a fatty barrier between the cell's interior and its surroundings.

The membrane is not just made up of phospholipids, however. There are a large variety of proteins situated in the cell membrane (Fig. 2.4). Some proteins lie on the surface of the membrane. Others are embedded in it with part of their structure exposed to one surface of the membrane or the other. Other proteins traverse the whole membrane and have areas exposed both to the outside and the inside of the cell. The extent to which these proteins are embedded within the cell membrane structure depends on the type of amino acid present. Portions of protein which are embedded in the cell membrane will have a large number of hydrophobic amino acids, whereas those portions which stick out on the surface will have a large number of hydrophilic

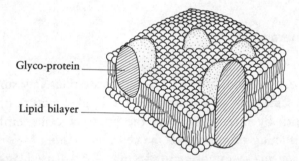

Fig. 2.3 Cell membrane. Taken from J. Mann, *Murder, magic, and medicine*, Oxford University Press (1992), with permission.

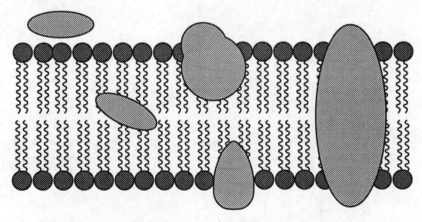

Fig. 2.4 The position of proteins associated with the cell membrane.

amino acids. Many surface proteins also have short chains of carbohydrates attached to them and are thus classed as glycoproteins. These carbohydrate segments are thought to be important towards cell recognition.

Within the cytoplasm there are several structures, one of which is the nucleus. This acts as the 'control centre' for the cell. The nucleus contains the genetic code—the DNA—and contains the blueprints for the construction of all the cell's enzymes.

There are many other structures within a cell, such as the mitochondria, the golgi apparatus, and the endoplasmic reticulum, but it is not the purpose of this book to look at the structure and function of these organelles. Suffice it to say that different drugs act at different locations in the cell and there is no one target site which we could pinpoint as **the** spot where drugs act. Nor would we get any closer to understanding how drugs work by cataloguing which drug acts at which particular cell component.

We need to magnify the picture, move down to the molecular level, and find out what types of molecules in the cell are affected by drugs. When we do that, we find that there are three main molecular targets:

(1) lipids
(2) proteins (glycoproteins)
(3) nucleic acids

The number of drugs which interact with lipids are relatively small and, in general, they all act in the same way—by disrupting the lipid structure of cell membranes.

Anaesthetics work by interacting with the lipids of cell membranes to alter the structure and conducting properties of the cell membrane.

The antifungal agent—amphotericin B (Fig. 2.5)—(used against athletes foot) interacts with the lipids of fungal cell membranes to build 'tunnels' through the membrane. Once in place, the contents of the cell are drained away and the cell is killed.

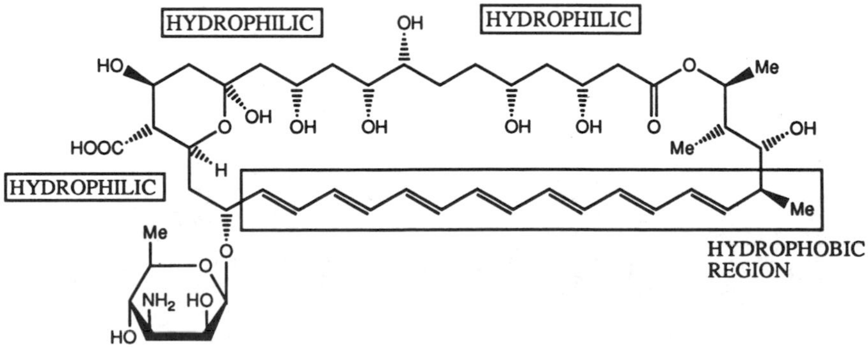

Fig. 2.5 Amphotericin B.

Amphotericin is a fascinating molecule in that one half of the structure is made up of double bonds and is hydrophobic, while the other half contains a series of hydroxyl groups and is hydrophilic. It is a molecule of extremes and as such is ideally suited to act on the cell membrane in the way that it does. Several amphotericin molecules cluster together such that the alkene chains are to the exterior and interact favourably with the hydrophobic centre of the cell membrane. The tunnel resulting from this cluster is lined with the hydroxyl groups and so is hydrophilic, allowing the polar contents of the cell to escape (Fig. 2.6).

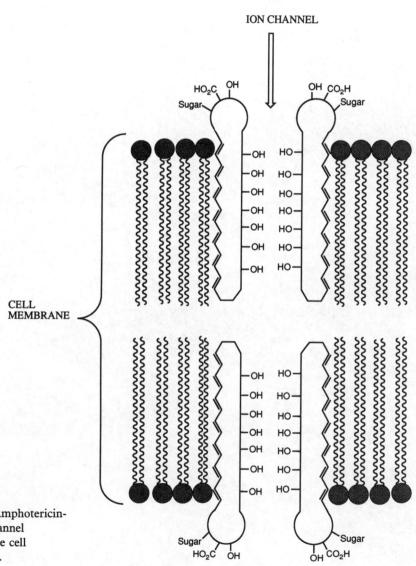

Fig. 2.6 Amphotericin-formed channel through the cell membrane.

The antibiotics valinomycin and gramicidin A operate by acting within the cell membrane as ion carriers and ion channels respectively (see Chapter 10).

These drugs apart, the vast majority of drugs interact with proteins or nucleic acids, and in particular with proteins. We shall therefore concentrate our attention in the next three chapters on proteins, then consider nucleic acids in Chapter 6.

3 ▪ Protein structure

In order to understand how drugs interact with proteins, it is necessary to understand their structure.

Proteins have four levels of structure—primary, secondary, tertiary, and quaternary.

3.1 *The primary structure of proteins*

The primary structure is quite simply the order in which the individual amino acids making up the protein are linked together through peptide bonds (Fig. 3.1).

The 20 common amino acids found in man are listed below and the structures are shown in Appendix 1.

Synthesized in the human body

Alanine	(Ala)
Arginine	(Arg)
Asparagine	(Asn)
Aspartic acid	(Asp)
Cysteine	(Cys)
Glutamic acid	(Glu)
Glutamine	(Gln)
Glycine	(Gly)
Proline	(Pro)
Serine	(Ser)
Tyrosine	(Tyr)

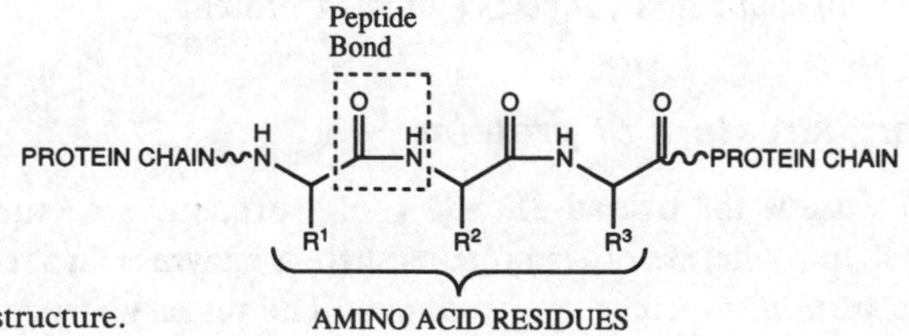

Essential to the diet

Histidine	(His)
Isoleucine	(Ile)
Leucine	(Leu)
Lysine	(Lys)
Methionine	(Met)
Phenylalanine	(Phe)
Threonine	(Thr)
Tryptophan	(Trp)
Valine	(Val)

Fig. 3.1 Primary structure.

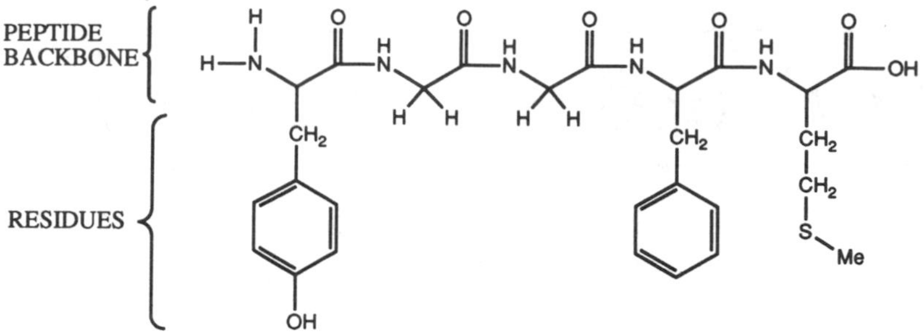

Fig. 3.2 Met-enkephalin.   SHORTHAND NOTATION: H-TYR-GLY-GLY-PHE-MET-OH

The primary structure of Met-enkephalin (one of the body's own painkillers) is shown in Fig. 3.2.

3.2 *The secondary structure of proteins*

The secondary structure of proteins consists of regions of ordered structures taken up by the protein chain. There are two main structures—the alpha helix and the beta-pleated sheet.

3.2.1 The alpha helix

The alpha helix results from coiling of the protein chain such that the peptide bonds making up the backbone are able to form hydrogen bonds between each other. These hydrogen bonds are directed along the axis of the helix as shown in Fig. 3.3. The residues of the component amino acids 'stick out' at right angles from the helix, thus minimizing steric interactions and further stabilizing the structure.

3.2.2 The beta-pleated sheet

The beta-pleated sheet is a layering of protein chains, one on top of another as shown in Fig. 3.4. Here too, the structure is held together by hydrogen bonds between the peptide links. The residues are situated at right angles to the sheets, once again to reduce steric interactions.

In structural proteins such as wool and silk, secondary structures are extensive and determine the overall shape and properties of such proteins.

3.3 *The tertiary structure of proteins*

The tertiary structure is the overall 3D shape of a protein. Structural proteins are quite ordered in shape, whereas other proteins such as enzymes and receptors fold up on themselves to form more complex structures. The tertiary structure of enzymes

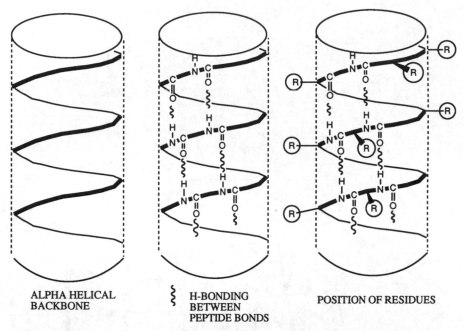

ALPHA HELICAL
BACKBONE

H-BONDING
BETWEEN
PEPTIDE BONDS

POSITION OF RESIDUES

Fig. 3.3 The alpha helix.

and receptors is crucial to their function and also to their interaction with drugs. Therefore, it is important to appreciate the forces which control their tertiary structure.

Enzymes and receptors have a great variety of amino acids arranged in what appears to be a random fashion—the primary structure of the protein. Regions of secondary structure may also be present, but the extent varies from protein to protein. For example, myoglobin (a protein which stores oxygen molecules) has extensive regions of alpha-helical secondary structure (Fig. 3.5), whereas the digestive enzyme chymotrypsin has very little secondary structure. Nevertheless, the protein chains in both myoglobin and chymotrypsin fold up upon themselves to form a complex globular shape such as that shown for myoglobin (Fig. 3.5).

How does this come about?

At first sight the 3D structure of myoglobin looks like a ball of string after the cat has been at it. In fact, the structure shown is a very precise shape which is taken up by every molecule of myoglobin synthesized in the body. There is no outside control or directing force making the protein take up this shape. It occurs spontaneously and is a consequence of the protein's primary structure[1]—that is, the amino acids making up the protein and the order in which they are linked. This automatic folding of proteins takes place even as the protein is being synthesized in the cell (Fig. 3.6).

[1] Some proteins contain species known as cofactors (e.g. metal ions or small organic molecules) which also have an effect on tertiary structure.

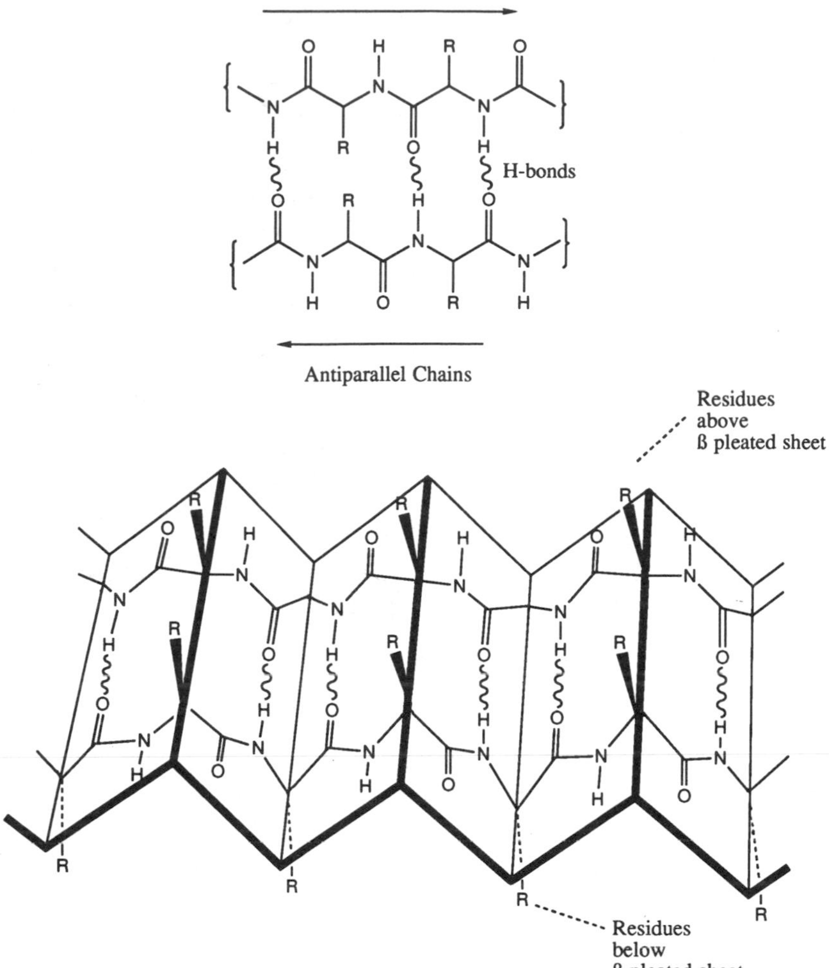

Fig. 3.4 β-pleated sheet.

Proteins are synthesized within cells on nucleic acid bodies called ribosomes. The ribosomes move along 'ticker-tape' shaped molecules of another nucleic acid called messenger RNA. This messenger RNA contains the code for the protein and is called a messenger since it has carried the message from the cell's DNA. The mechanism by which this takes place need not concern us here. We need only note that the ribosome holds on to the growing protein chain as the amino acids are added on one by one. As the chain grows, it automatically folds into its 3D shape such that by the time the protein is fully synthesized and released from the ribosome, the 3D shape is already adopted. The 3D shape is identical for every molecule of the particular protein being synthesized.

This poses a problem. Why should a chain of amino acids take up such a precise 3D shape? At first sight, it does not make sense. If we place a length of string on the table,

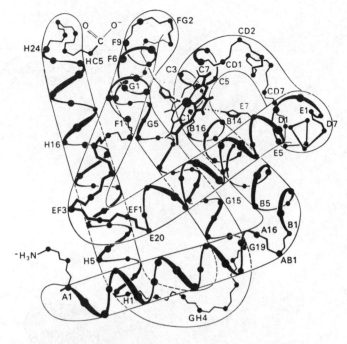

Fig. 3.5 Myoglobin. Taken from Stryer, *Biochemistry*, 3rd edn, Freeman (1988), with permission.

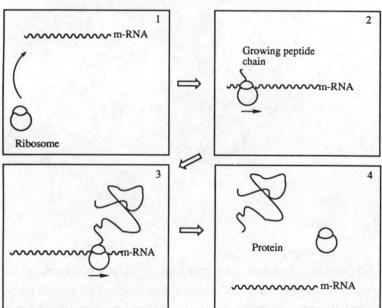

Fig. 3.6 Protein synthesis.

it does not fold itself up into a precise complex shape. So why should a 'string' of amino acids do such a thing?

The answer lies in the fact that a protein is not just a bland length of string. It has a whole range of chemical functional groups attached along the length of its chain.

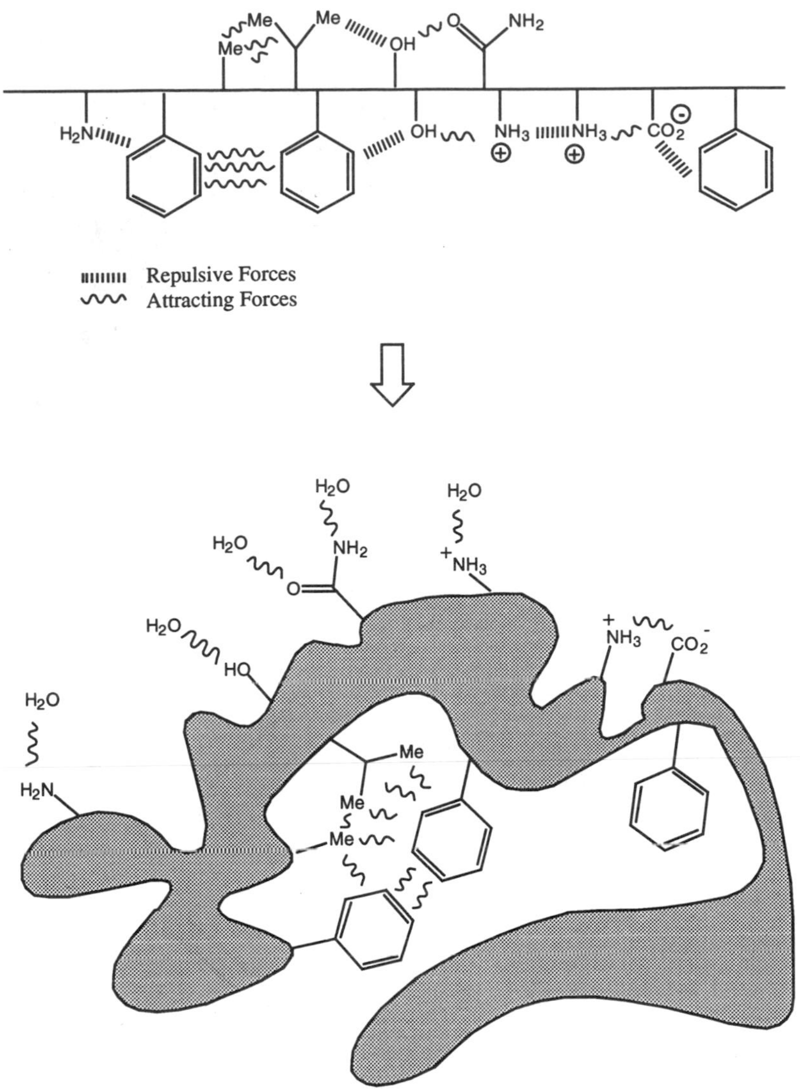

ıııııı Repulsive Forces
ᨃᨃ Attracting Forces

Fig. 3.7 Forces producing tertiary structure.

These are, of course, the residues of each amino acid making up the chain. These residues can interact with each other. Some will attract each other. Others will repel. Thus the protein will twist and turn to minimize the unfavourable interactions and to maximize the favourable ones until the most favourable shape (conformation) is found—the tertiary structure (Fig. 3.7).

What then are these important forces? Let us consider the attracting or binding forces first of all. There are four to consider: covalent bonds; ionic bonds; hydrogen bonds; and van der Waals bonds.

3.3.1 Covalent bonds

Covalent bonds are the strongest bonding force available

$$\text{Bond strength (S–S)} = 250 \text{ kJ mol}^{-1}$$

When two cysteine residues are close together, a covalent disulfide bond can be formed between them as a result of oxidation. A covalent bonded bridge is thus formed between two different parts of the protein chain (Fig. 3.8).

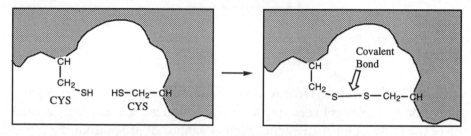

Fig. 3.8 Covalent bond.

3.3.2 Ionic bonds

Ionic bonds are a strong bonding force between groups having opposite charges

$$\text{Bond strength} = 20 \text{ kJ mol}^{-1}$$

An ionic bond can be formed between the carboxylate ion of an acidic residue such as aspartic acid and the ammonium ion of a basic residue such as lysine (Fig. 3.9).

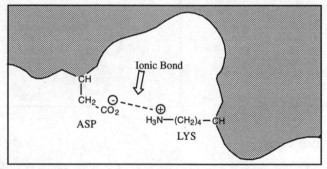

Fig. 3.9 Ionic bond.

3.3.3 Hydrogen bonds

$$\text{Bond strength} = 7\text{–}40 \text{ kJ mol}^{-1}$$

Hydrogen bonds are formed between electronegative atoms, such as oxygen, and protons attached to electronegative atoms (Fig. 3.10).

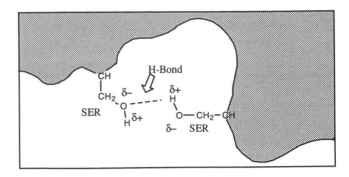

Fig. 3.10 Hydrogen bond.

3.3.4 Van der Waals bonds

Bond strength = 1.9 kJ mol^{-1}

The van der Waals bonding force results from an interaction between hydrophobic molecules (for example, between two aromatic residues such as phenylalanine (Fig. 3.11) or between two aliphatic residues such as valine). It arises from the fact that the electronic distribution in these neutral, non-polar residues is never totally even or symmetrical. As a result, there are always transient areas of high electron density and low electron density, such that an area of high electron density on one residue can have an attraction for an area of low electron density on another molecule.

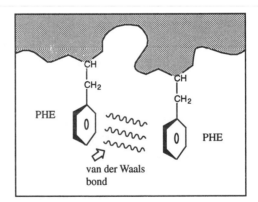

Fig. 3.11 Van der Waals bond.

3.3.5 Repulsive forces

Repulsive forces would arise if a hydrophilic group such as an amino function was too close to a hydrophobic group such as an aromatic ring. Alternatively, two charged groups of identical charge would be repelled.

3.3.6 Relative importance of binding forces

Based on the strengths of the four types of bonds above, we might expect the relative importance of the bonding forces to follow the same order as their strengths, that is, covalent, ionic, hydrogen bonding, and finally van der Waals. In fact, the opposite is usually true. In most proteins, the most important binding forces in tertiary structure are those due to van der Waals interactions and hydrogen bonding, while the least important forces are those due to covalent and ionic bonding.

There are two reasons for this.

First of all, in most proteins there are far more van der Waals and hydrogen bonding interactions possible, compared to covalent or ionic bonding. We only need to consider the number and types of amino acids in any typical globular protein to see why. The only covalent bond which can contribute to tertiary structure is a disulfide bond. Only one amino acid out of our list can form such a bond—cysteine. However, there are eight amino acids which can interact with each other through van der Waals bonding: Gly, Ala, Val, Leu, Ile, Phe, Pro and Met.

There *are* examples of proteins with a large number of disulfide bridges, where the relative importance of the covalent link to tertiary structure *is* more significant. Disulfide links are also more significant in small polypeptides such as the peptide hormones vasopressin (Fig. 3.12) and oxytocin (Fig. 3.13). However, in the majority of proteins, disulfide links play a minor role in controlling tertiary structure.

As far as ionic bonding is concerned, only four amino acids (Asp, Glu, Lys, Arg) are involved, whereas eight amino acids can interact through hydrogen bonding (Ser, Thr, Cys, Asn, Gln, His, Tyr, Trp). Clearly, the number of possible ionic and covalent bonds is greatly outnumbered by the number of hydrogen bonds or van der Waals interactions.

There is a second reason why van der Waals interactions, in particular, can be the most important form of bonding in tertiary structure. Proteins do not exist in a vacuum. The body is mostly water and as a result all proteins will be surrounded by this medium. Therefore, amino acid residues at the surface of proteins must interact with water molecules. Water is a highly polar compound which forms strong hydrogen bonds. Thus, water would be expected to form strong hydrogen bonds to the hydrogen bonding amino acids previously mentioned.

Water can also accept a proton to become positively charged and can form ionic bonds to aspartic and glutamic acids (Fig. 3.14).

Therefore, water is capable of forming hydrogen bonds or ionic bonds to the following amino acids: Ser, Thr, Cys, Asn, Gln, His, Tyr, Trp, Asp, Glu, Lys, Arg. These polar amino acids are termed hydrophilic. The remaining amino acids (Gly, Ala, Val, Leu, Ile, Phe, Pro, Met) are all non-polar amino acids which are hydrophobic or lipophilic. As a result, they are repelled by water.

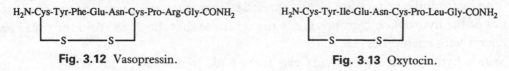

Fig. 3.12 Vasopressin.   **Fig. 3.13** Oxytocin.

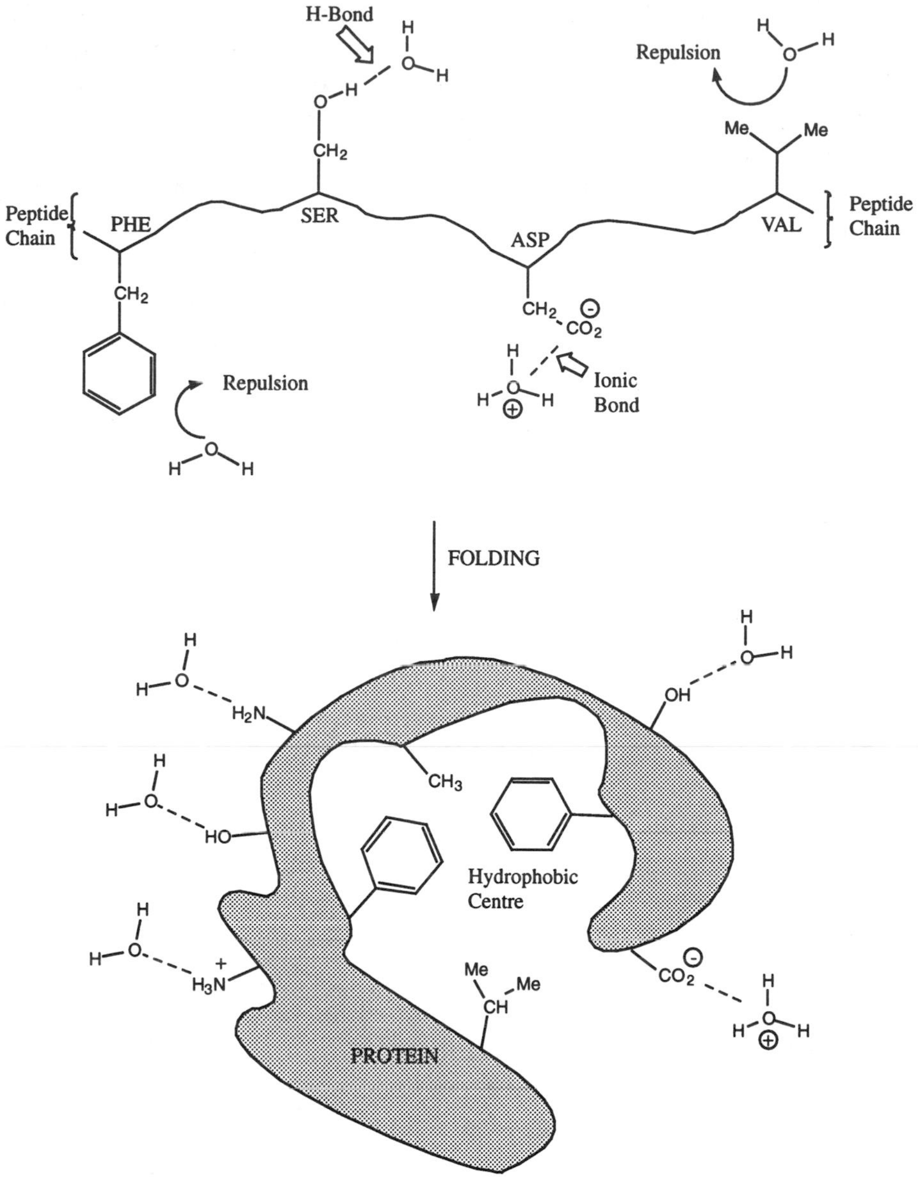

Fig. 3.14 Bonding interactions with water.

Therefore, the most stable tertiary structure of a protein will be the one where most of the hydrophilic groups are on the surface so that they can interact with water, and most of the hydrophobic groups are in the centre so that they can avoid the water and interact with each other.

Since hydrophilic amino acids can form ionic/hydrogen bonds with water, the number of ionic and hydrogen bonds contributing to the tertiary structure is reduced. The hydrophobic amino acids in the centre have no choice in the matter and must interact with each other. Thus, the hydrophobic interactions within the structure outweigh the hydrophilic interactions and control the shape taken up by the protein.

One important feature of this tertiary structure is that the centre of the proteins is hydrophobic and non-polar. This has important consequences as far as the action of enzymes is concerned and helps to explain why reactions which should be impossible in an aqueous environment can take place in the presence of enzymes. The enzyme can provide a non-aqueous environment for the reaction to take place.

3.4 *The quaternary structure of proteins*

Quaternary structure is confined to those proteins which are made up of a number of protein subunits (Fig. 3.15). For example, haemoglobin is made up of four protein molecules—two identical alpha subunits and two identical beta subunits (not to be confused with the alpha and beta terminology used in secondary structure). The quaternary structure of haemoglobin is the way in which these four protein units associate with each other.

Since this must inevitably involve interactions between the exterior surfaces of proteins, ionic bonding can be more important to quaternary structure than it is to tertiary structure. Nevertheless, hydrophobic (van der Waals) interactions have a role to play. It is not possible for a protein to fold up such that all hydrophobic groups are placed to the centre. Some such groups may be stranded on the surface. If they form a small hydrophobic area on the protein surface, there would be a distinct advantage for two protein molecules to form a dimer such that the two hydrophobic areas face each other rather than be exposed to an aqueous environment.

3.5 Conclusion

We have now discussed the four types of structure. The tertiary structure is the most important feature as far as drug action is concerned, although it must be emphasized

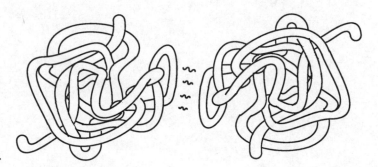

Fig. 3.15 Quaternary structure.

again that tertiary structure is a consequence of primary structure. Tertiary structure is a result of interactions between different amino acid residues and interactions between amino acids and water.

We are now ready to discuss the two types of protein with which drugs interact—enzymes and receptors.

4 · Drug action at enzymes

4.1 *Enzymes as catalysts*

Enzymes are the body's catalysts. Without them, the cell's chemical reactions would be too slow to be useful, and many would not occur at all. A catalyst is an agent which speeds up a chemical reaction without being changed itself.

An example of a catalyst used frequently in the laboratory is palladium on activated charcoal (Fig. 4.1)

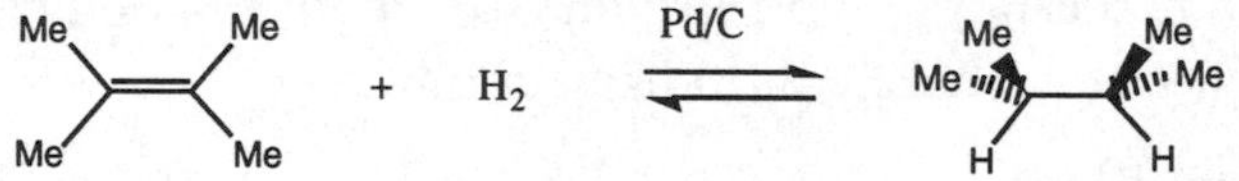

Fig. 4.1 An example of a catalyst.

Note that the above reaction is shown as an equilibrium. It is therefore more correct to describe a catalyst as an agent which can speed up the approach to equilibrium. In an equilibrium reaction, a catalyst will speed up the reverse reaction just as efficiently as the forward reaction. Consequently, the final equilibrium concentrations of the starting materials and products are unaffected by a catalyst.

How do catalysts affect the rate of a reaction without affecting the equilibrium? The answer lies in the existence of a high-energy intermediate or transition state which must be formed before the starting material can be converted to the product. The difference in energy between the transition state and the starting material is the activation energy, and it is the size of this activation energy which determines the rate of a reaction rather than the difference in energy between the starting material and the product (Fig. 4.2).

A catalyst acts to reduce the activation energy by helping to stabilize the transition state. The energy of the starting material and products are unaffected, and therefore the equilibrium ratio of starting material to product is unaffected.

We can relate energy, and the rate and equilibrium constants with the following equations:

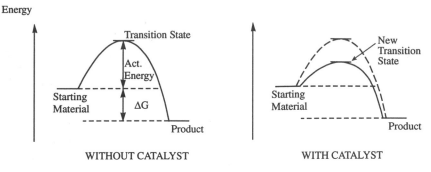

Fig. 4.2 Activation energy.

Energy difference = $\Delta G = -RT \ln K$ where K = equilibrium constant
$$= [\text{products}]/[\text{reactants}]$$
$$R = 8.314 \text{ J mol}^{-1} \text{ K}^{-1}$$
$$T = \text{temperature}$$
Rate constant = $k = Ae^{-E/RT}$ where E = activation energy
$$A = \text{frequency factor}$$

Note that the rate constant k has no dependence on the equilibrium constant K.

We have stated that catalysts (and enzymes) speed up reaction rates by lowering the activation energy, but we have still to explain how.

4.2 *How do catalysts lower activation energies?*

There are several factors at work.

• Catalysts provide a reaction surface or environment.
• Catalysts bring reactants together.
• Catalysts position reactants correctly so that they easily attain their transition state configurations.
• Catalysts weaken bonds.
• Catalysts may participate in the mechanism.

We can see these factors at work in our example of hydrogenation with a palladium–charcoal catalyst (Fig. 4.3).
In this reaction, the catalyst surface interacts with the hydrogen molecule and in doing so weakens the H–H bond. The bond is then broken and the hydrogen atoms are bound to the catalyst. The catalyst can then interact with the alkene molecule, so weakening the pi bond of the double bond. The hydrogen atoms and the alkene molecule are positioned on the catalyst conveniently close to each other to allow easy transfer of the hydrogens from catalyst to alkene. The alkane product then departs, leaving the catalyst as it was before the reaction.

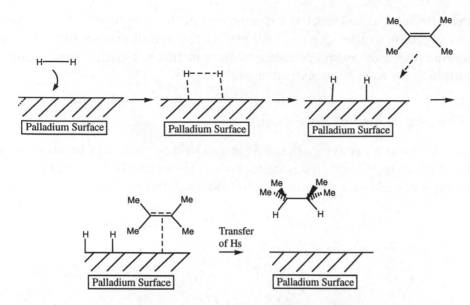

Fig. 4.3 Action of a palladium–charcoal catalyst.

Therefore, the catalyst helps the reaction by providing a surface to bring the two substrates together. It participates in the reaction by binding the substrates and breaking high-energy bonds, and then it holds the reagents close together to increase the chance of them reacting with each other.

Enzymes may have more complicated structures than, say, a palladium surface, but they catalyse reactions in the same way. They act as a surface or focus for the reaction, bringing the substrate or substrates together and holding them in the best position for reaction. The reaction takes place, aided by the enzyme, to give products which are then released (Fig. 4.4). Note again that it is a reversible process. Enzymes can catalyse both forward and backward reactions. The final equilibrium mixture will, however, be the same, regardless of whether we supply the enzyme with substrate or product.

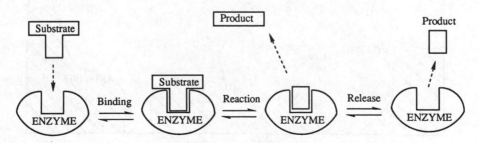

Fig. 4.4 Enzyme catalyst.

Substrates bind to and react at a specific area of the enzyme called the active site. The active site is usually quite a small part of the overall protein structure, but in considering the mechanism of enzymes we can make a useful simplification by concentrating on what happens at that site.

4.3 *The active site of an enzyme*

The active site of an enzyme (Fig. 4.5) is a 3D shape. It has to be on or near the surface of the enzyme if substrates are to reach it. However, the site could be a groove, hollow, or gully allowing the substrate to 'sink into' the enzyme.

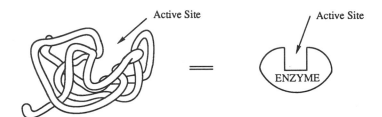

Fig. 4.5 The active site of an enzyme.

Because of the overall folding of the enzyme, the amino acid residues which are close together in the active site may be extremely far apart in the primary structure. For example, the important amino acids at the active site of lactate dehydrogenase are shown in Fig. 4.6. The numbers refer to their positions in the primary structure of the enzyme.

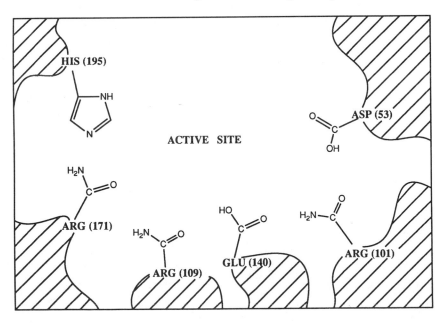

Fig. 4.6 The active site of lactate dehydrogenase.

The amino acids present in the active site play an important role in enzyme function and this can be demonstrated by comparing the primary structures of the same enzyme from different organisms. In such a study, we would find that the primary structure would differ from species to species as a result of mutations lasting over millions of years. The variability would be proportional to how far apart the organisms are on the evolutionary ladder and this is one method of determining such a relationship.

However, that does not concern us here. What does, is the fact that there are certain amino acids which remain constant, no matter the source of the enzyme. These are amino acids which are crucial to the enzyme's function and, as such, are often the amino acids which make up the active site. If one of these amino acids should be lost through mutation, the enzyme would become useless and an animal bearing this mutation would have a poor chance of survival. Thus, the mutation would not be preserved. (The only exception to this would be if the mutation either introduced an amino acid which could perform the same task as the original amino acid, or improved substrate binding.)

This consistency of active site amino acids can often help scientists determine which amino acids are present in an active site if this is not known already.

Amino acids present in the active site can have one of two roles.

1. Binding—the amino acid residue is involved in binding the substrate to the active site.
2. Catalytic—the amino acid is involved in the mechanism of the reaction.

We shall study these in turn.

4.4 *Substrate binding at an active site*

4.4.1 The binding forces

The forces which bind substrates to the active sites of enzymes are the same as those controlling the tertiary structure of enzymes—ionic, hydrogen, and van der Waals bonds. However, whereas ionic bonding plays a relatively minor role in protein tertiary structure compared to hydrogen bonding or van der Waals bonding, it can play a crucial role in the binding of a substrate to an active site—not too surprising since active sites are located on or near the surface of the enzyme.

Since we know the three bonding forces involved in substrate binding, it is possible to look at the structure of a substrate and postulate the probable interactions which it will have with its active site.

As an example, let us consider the substrate for lactate dehydrogenase—an enzyme which catalyses the reduction of pyruvic acid to lactic acid (Fig. 4.7).

If we look at the structure of pyruvic acid, we can propose three possible inter-

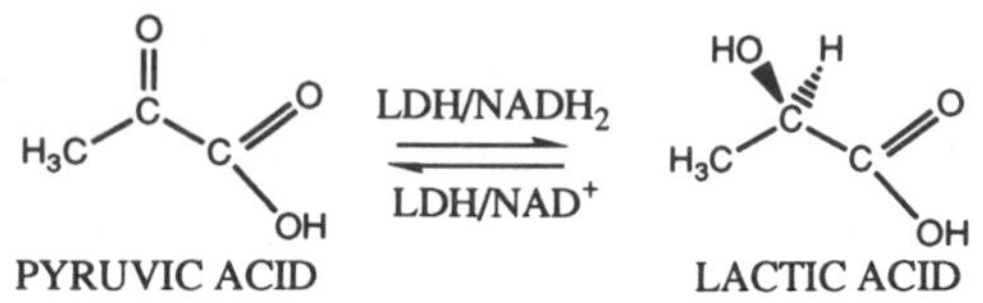

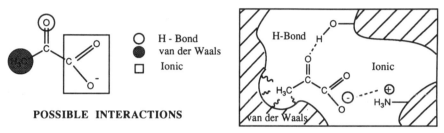

Fig. 4.7 Reduction of pyruvic acid to lactic acid.

PYRUVIC ACID   LACTIC ACID

actions with which it might bind to its active site—an ionic interaction involving the ionized carboxylate group, a hydrogen bond involving the ketonic oxygen, and a van der Waals interaction involving the methyl group (Fig. 4.8). If these postulates are correct, then it means that there must be suitable amino acids at the active site to take part in these bonds. A lysine residue, serine residue, and phenylalanine residue would fit the bill respectively.

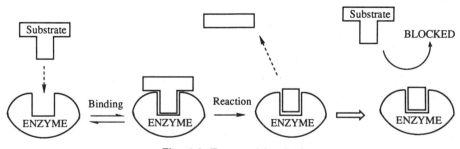

POSSIBLE INTERACTIONS

○   H - Bond
●   van der Waals
□   Ionic

Fig. 4.8 Interactions between pyruvic acid and lactate dehydrogenase.

4.4.2 Competitive (reversible) inhibitors

Binding interactions between substrate and enzyme are clearly important. If there were no interactions holding the substrate to the active site, then the substrate would drift in and drift out again before there was a chance for it to react.

Therefore, the more binding interactions there are, the better the substrate will be bound, and the better the chance of reaction. But there is a catch! What would happen if a substrate bound so strongly to the active site that it was not released again (Fig. 4.9)?

The answer, of course, is that the enzyme would become 'clogged up' and would be

Substrate   Substrate

BLOCKED

Binding   Reaction

ENZYME   ENZYME   ENZYME   ENZYME

Fig. 4.9 Enzyme 'clogging'.

unable to accept any more substrate. Therefore, the bonding interactions between substrate and enzyme have to be properly balanced such that they are strong enough to keep the substrate(s) at the active site to allow reaction, but weak enough to allow the product(s) to depart. This bonding balancing act can be turned to great advantage by the medicinal chemist wishing to inhibit a particular enzyme or to switch it off altogether. A molecule can be designed which is similar to the natural substrate and can fit the active site, but which binds more strongly. It may not undergo any reaction when it is in the active site, but as long as it is there, it blocks access to the natural substrate and the enzymatic reaction stops (Fig 4.10). This is known as competitive inhibition since the drug is competing with the natural substrate for the active site.

The longer the inhibitor is present in the active site, the greater the inhibition. Therefore, if the medicinal chemist has a good idea which binding groups are present in an active site and where they are, a range of molecules can be designed with different inhibitory strengths.

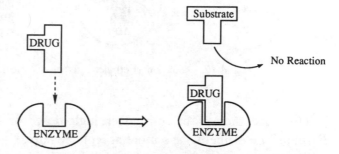

Fig. 4.10 Competitive inhibition.

Competitive inhibitors can generally be displaced by increasing the level of natural substrate. This feature has been useful in the treatment of accidental poisoning by antifreeze. The main constituent of antifreeze is ethylene glycol which is oxidized in a series of enzymatic reactions to oxalic acid (Fig. 4.11). It is the oxalic acid which is responsible for toxicity and if its synthesis can be blocked, recovery is possible.

The first step in this enzymatic process is the oxidation of ethylene glycol by alcohol dehydrogenase. Ethylene glycol is acting here as a substrate, but we can view it as a competitive inhibitor since it is competing with the natural substrate for the enzyme. If the levels of natural substrate are increased, it will compete far better with ethylene glycol and prevent it from reacting. Toxic oxalic acid would no longer be formed and the unreacted ethylene glycol would eventually be excreted from the body (Fig. 4.12).

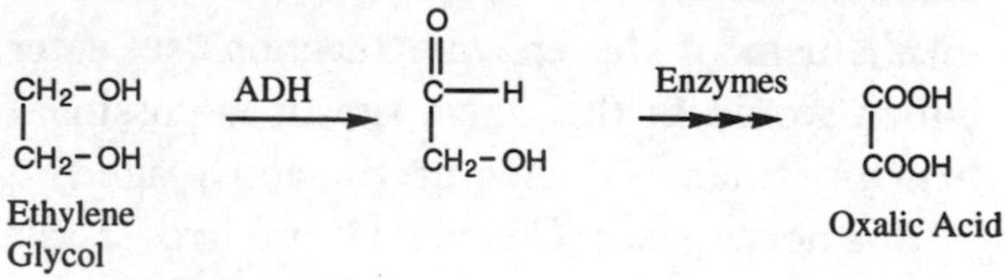

Fig. 4.11 Formation of oxalic acid from ethylene glycol.

OXIDATION OF ETHYLENE GLYCOL

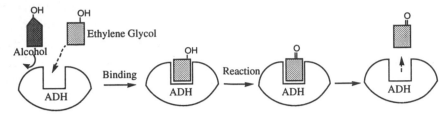

BLOCKING WITH EXCESS ALCOHOL

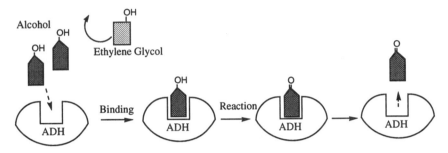

Fig. 4.12 Oxidation of ethylene glycol and blocking with excess alcohol.

The cure then is to administer high doses of the natural substrate—alcohol! Perhaps one of medicine's more acceptable cures?

There are many examples of useful drugs which act as competitive inhibitors. For example, the sulfonamides inhibit bacterial enzymes (Chapter 10), while anticholinesterases inhibit the mammalian enzyme acetylcholinesterase (Chapter 11). Many diuretics used to control blood pressure are competitive inhibitors.

4.4.3 Non-competitive (irreversible) inhibitors

To stop an enzyme altogether, the chemist can design a drug which binds irreversibly to the active site and blocks it permanently. This would be a non-competitive form of inhibition since increased levels of natural substrate would not be able to budge the unwanted squatter. The most effective irreversible inhibitors are those which can react with an amino acid at the active site to form a covalent bond. Amino acids such as serine and cysteine which bear nucleophilic residues (OH and SH respectively) are common inhabitants of enzyme active sites since they are frequently involved in the mechanism of the enzyme reaction (see later). By designing an electrophilic drug which would fit the active site, it is possible to alkylate these particular groups and hence permanently clog up the active site (Fig. 4.13).

The nerve gases (Chapter 11) are irreversible inhibitors of mammalian enzymes and

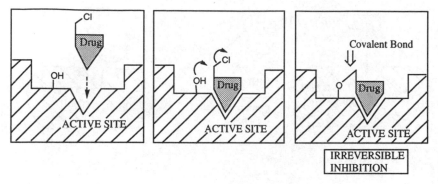

Fig. 4.13 Irreversible inhibition.

are therefore highly toxic. In the same way, penicillin (Chapter 10) is highly toxic to bacteria.

4.4.4 Non-competitive, reversible (allosteric) inhibitors

So far we have discussed inhibitors which bind to the active site and prevent the natural substrate from binding. We would therefore expect these inhibitors to have some sort of structural similarity to the natural substrate. We would also expect reversible inhibitors to be displaced by increased levels of natural substrate.

However, there are many enzyme inhibitors which appear to have no structural similarity to the natural substrate. Furthermore, increasing the amount of natural substrate has no effect on the inhibition. Such inhibitors are therefore non-competitive inhibitors, but unlike the non-competitive inhibitors mentioned above, the inhibition is reversible.

Non-competitive or allosteric inhibitors bind to a different region of the enzyme and are therefore not competing with the substrate for the active site. However, since binding is taking place, a moulding process takes place (Fig. 4.14) which causes the enzyme to change shape. If that change in shape hides the active site, then the substrate can no longer react.

Adding more substrate will not reverse the situation, but that does not mean that the inhibition is irreversible. Since the inhibitor uses non-covalent bonds to bind to the allosteric binding site, it will eventually depart in its own good time.

But why should there be this other binding site?

The answer is that allosteric binding sites are important in the control of enzymes. A biosynthetic pathway to a particular product involves a series of enzymes, all working efficiently to convert raw materials into final product. Eventually, the cell will have enough of the required material and will want to stop production. Therefore, there has to be some sort of control which says enough is enough. The most common control mechanism is one known as feedback control, where the final

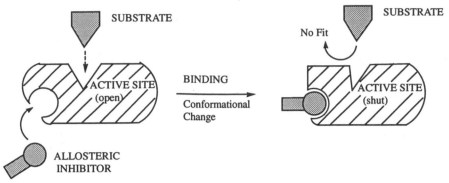

Fig. 4.14 Non-competitive, reversible (allosteric) inhibition.

product controls its own synthesis. It can do this by inhibiting the first enzyme in the biochemical pathway (Fig. 4.15). Therefore, when there are low levels of final product in the cell, the first enzyme in the pathway is not inhibited and works normally. As the levels of final product increase, more and more of the enzyme is blocked and the rate of synthesis drops off in a graded fashion.

We might wonder why the final product inhibits the enzyme at an allosteric site and not the active site itself. There are two explanations for this.

First of all, the final product has undergone many transformations since the original starting material and is no longer 'recognized' by the active site. It must therefore bind elsewhere on the enzyme.

Secondly, binding to the active site itself would not be a very efficient method of feedback control, since it would then have to compete with the starting material. If levels of the latter increased, then the inhibitor would be displaced and feedback control would fail.

An enzyme under feedback control offers the medicinal chemist an extra option in designing an inhibitor. The chemist can not only design drugs which are based on the structure of the substrate and which bind directly to the active site, but can also design drugs based on the structure of the final overall product and which bind to the allosteric binding site.

The drug 6-mercaptopurine, used in the treatment of leukaemia (Fig. 4.16), is

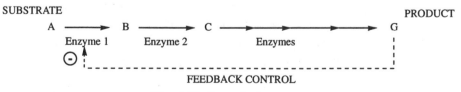

Fig. 4.15 Control of enzymes.

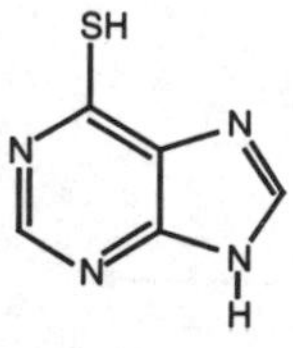

Fig. 4.16 6-Mercaptopurine.

an example of an allosteric inhibitor. It inhibits the first enzyme involved in the synthesis of purines and therefore blocks purine synthesis. This in turn blocks DNA synthesis.

4.5 *The catalytic role of enzymes*

We now move on to consider the mechanism of enzymes, and how they catalyse reactions.

In general, enzymes catalyse reactions by providing the following:

(1) binding interactions
(2) acid/base catalysis
(3) nucleophilic groups

4.5.1 Binding interactions

As mentioned previously, the rate of a reaction is increased if the energy of the transition state is lowered. This results from the bonding interactions between substrate and enzyme.

In the past, it was thought that a substrate fitted its active site in a similar way to a key fitting a lock. Both the enzyme and the substrate were seen as rigid structures with the substrate (the key) fitting perfectly into the active site (the lock) (Fig. 4.17). However, such a scenario does not explain how some enzymes can catalyse a reaction on a range of different substrates. It implies instead that an enzyme has an optimum substrate which fits it perfectly and which can be catalysed very efficiently, whereas all other substrates are catalysed less efficiently. Since this is not the case, the lock and key analogy is invalid.

It is now believed that the substrate is nearly a good fit for the active site but not a perfect fit. It is thought instead that the substrate enters the active site and forces it to change shape—a kind of moulding process. This theory is known as Koshland's Theory of Induced Fit since the substrate induces the active site to take up the ideal shape to accommodate it (Fig. 4.17).

For example, a substrate such as pyruvic acid might interact with its active site via one hydrogen bond, one ionic bond, and one van der Waals interaction. However, the fit might not be quite right and the three bonding interactions might be a bit too long

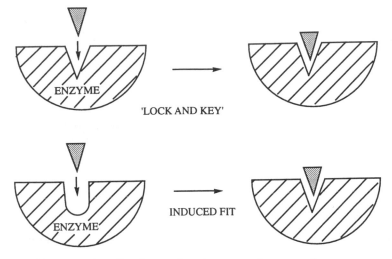

Fig. 4.17 Bonding interactions between substrate and enzyme.

to be perfect. In order to maximize the strength of these bonds, the enzyme would have to change shape so that the amino acid residues involved in the binding move closer to the substrate (Fig. 4.18).

This theory of induced fit helps to explain why enzymes can catalyse a wide range of substrates. Each substrate induces the active site into a shape which is ideal for *it*, and as long as the moulding process does not distort the active site too much, such that the reaction mechanism proves impossible, then reaction can proceed.

But note this. The substrate is not just a passive spectator to the moulding process going on around it. As the enzyme changes shape to maximize bonding interactions, the same thing is going to happen to the substrate. It too will alter shape. Bond rotation may occur to fix the substrate in a particular conformation—not necessarily the most stable conformation. Bonds may even be stretched and weakened. Consequently, this moulding process designed to maximize binding interactions may force the substrate into the ideal conformation for the reaction to follow (i.e. the transition state) and may also weaken the very bonds which have to be broken.

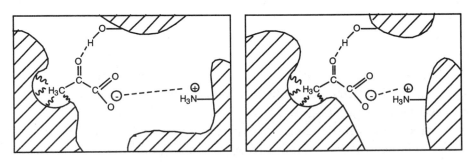

Fig. 4.18 Example of induced fit.

Once bound to an active site, the substrate is now held ready for the reaction to follow. Binding has fixed the 'victim' so that it cannot evade attack and this same binding has weakened its defences (bonds) so that reaction is easier (lower activation energy).

4.5.2 Acid/base catalysis

Usually acid/base catalysis is provided by the amino acid histidine. Histidine is a weak base and can easily equilibrate between its protonated form and its free base form (Fig. 4.19). In doing so, it can act as a proton 'bank'; that is, it has the capability to accept and donate protons in the reaction mechanism. This is important since active sites are frequently hydrophobic and will therefore have a low concentration of water and an even lower concentration of protons.

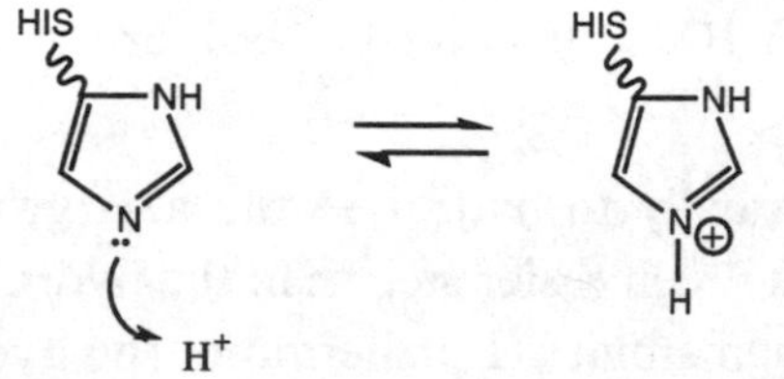

Fig. 4.19 Histidine.

4.5.3 Nucleophilic groups

The amino acids serine and cysteine are common inhabitants of active sites. These amino acids have nucleophilic residues (OH and SH respectively) which are able to participate in the reaction mechanism. They do this by reacting with the substrate to form intermediates which would not be formed in the uncatalysed reaction. These intermediates offer an alternative reaction pathway which may avoid a high-energy transition state and hence increase the rate of the reaction.

Normally, an alcoholic group such as that on serine is not a good nucleophile. However, there is usually a histidine residue close by to catalyse the reaction. For example, the mechanism by which chymotrypsin hydrolyses peptide bonds is shown in Fig. 4.20.

The presence of a nucleophilic serine residue means that water is not required in the initial stages of the mechanism. This is important since water is a poor nucleophile and may also find it difficult to penetrate a hydrophobic active site. Secondly, a water molecule would have to drift into the active site, and search out the carboxyl group before it could attack it. This would be something similar to a game of blind man's buff. The enzyme, on the other hand, can provide a serine OH group, positioned in exactly the right spot to react with the substrate. Therefore, the nucleophile has no need to search for its substrate. The substrate has been delivered to it.

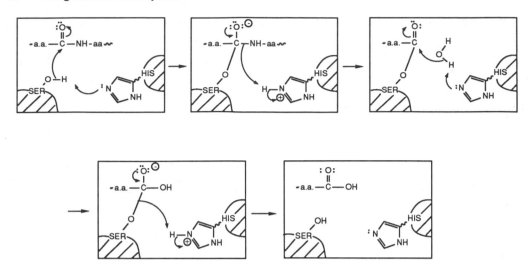

Fig. 4.20 Hydrolysis of peptide bonds by chymotrypsin.

Water is required eventually to hydrolyse the acyl group attached to the serine residue. However, this is a much easier step than the hydrolysis of a peptide link since esters are more reactive than amides. Furthermore, the hydrolysis of the peptide link means that one half of the peptide can drift away from the active site and leave room for a water molecule to enter.

As far as the medicinal chemist is concerned, an understanding of the mechanism can help in the design of more powerful inhibitors.

First of all, if the mechanism is known, it is possible to design antagonists which bind so strongly to the active site by non-covalent forces that they are effectively irreversible inhibitors—a bit like inviting the mother-in-law for dinner and finding her moving in on a permanent basis.

The logic is as follows. We have already seen that the enzyme alters the shape and the bond lengths of the substrate such that it effectively converts it to the transition state for the reaction. Therefore, if the enzyme was given a compound which was already the same shape as the transition state, then that compound should be an ideal binding group for the active site, and should bind very efficiently. If the compound was unable to react further, then the enzyme would be strongly inhibited.

Such compounds are known as transition-state analogues.

The beauty of the tactic is that it can be used effectively against enzyme reactions involving two substrates. With such enzymes, inhibitors based on one substrate or the other could be designed, but neither will be as good as an inhibitor based on a transition-state analogue where the two are linked together. The latter is bound to have more bonding interactions.

One interesting example of a transition-state inhibitor is the drug 5-fluorouracil

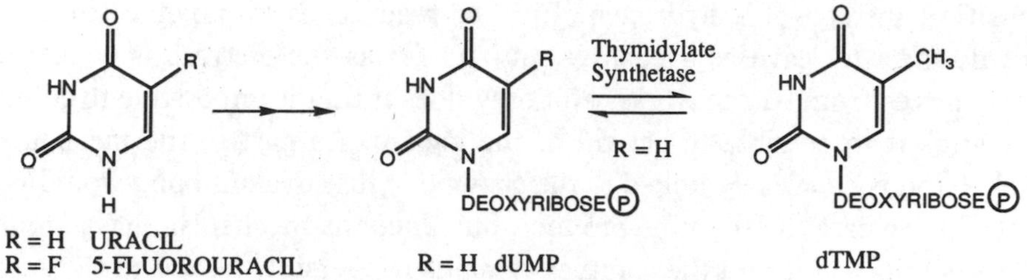

Fig. 4.21 Biosynthesis of dTMP.

(Fig. 4.21)—used to treat breast cancer and skin cancer. The target enzyme is thymidylate synthetase which catalyses the conversion of 2'-deoxyuridylic acid to dTMP (Figs 4.21 and 4.22).

5-Fluorouracil is not the transition-state analogue itself. It is converted in the body to the fluorinated analogue of 2'-deoxyuridylic acid which then combines with a second substrate (tetrahydrofolate) to form a transition-state analogue *in situ*. Up until this point, nothing unusual has happened and the reaction mechanism has been proceeding normally. The tetrahydrofolate has formed a covalent bond to the uracil skeleton via the methylene unit which is to be transferred. At this stage, the loss of a proton from the 5-position is required. However, 5-fluorouracil has a fluorine atom at

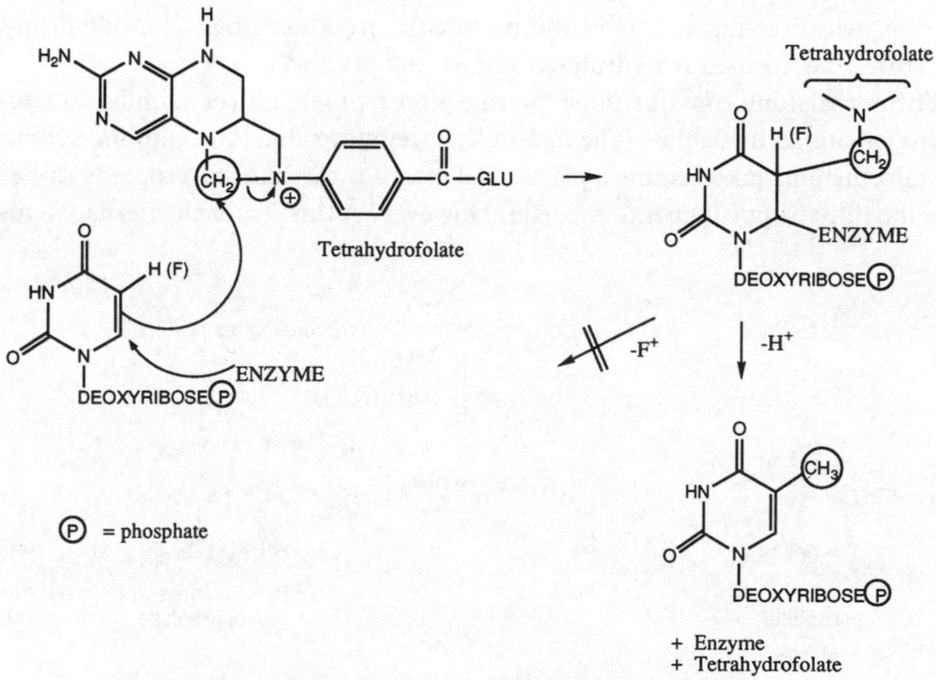

Fig. 4.22 Use of 5-fluorouracil as a transition-state inhibitor.

that position instead of a hydrogen. Further reaction is impossible since it would require fluorine to leave as a positive ion. As far as the enzyme is concerned, the situation moves from bad to worse. Not only does it find it impossible to complete its task, it finds it impossible to get rid of the logjam. As part of the mechanism, the uracil skeleton is covalently linked to the enzyme. This covalent bond would normally be cleaved to release the thymine product, but since the mechanism has jammed, this now proves impossible and the complex remains irreversibly bound to the active site. Synthesis of thymidine is terminated, which is turn stops the synthesis of DNA. Result: replication and cell division are blocked.

5-Fluorouracil is a particularly useful drug for the treatment of skin cancer since it shows a high level of selectivity for cancer cells over normal skin cells.

In the above example, the enzyme accepted the drug as a bona fide visitor, only to find that it gained an awkward squatter impossible to move. Other apparently harmless visitors can turn into lethal assassins which actively attack the enzyme. Once again, it is the enzyme mechanism itself which causes the transformation. One example of this is provided by the irreversible inhibition of the enzyme alanine transaminase by trifluoroalanine (Fig. 4.23).

The normal mechanism for the transamination reaction is shown in Fig. 4.24 (R=H) and involves the condensation of alanine and pyridoxal phosphate to give an imine. A proton is lost from the imine to give a dihydropyridine intermediate. This reaction is catalysed by a basic amino acid provided by the enzyme as well as the electron withdrawing effects of the protonated pyridine ring. The dihydropyridine structure now formed is hydrolysed to give the products.

Trifluoroalanine contains three fluorine atoms which are very similar in size to the hydrogen atoms in alanine. The molecule is therefore able to fit into the active site of the enzyme and take alanine's place. The reaction mechanism proceeds as before to give the dihydropyridine intermediate. However, at this stage, an alternative mechan-

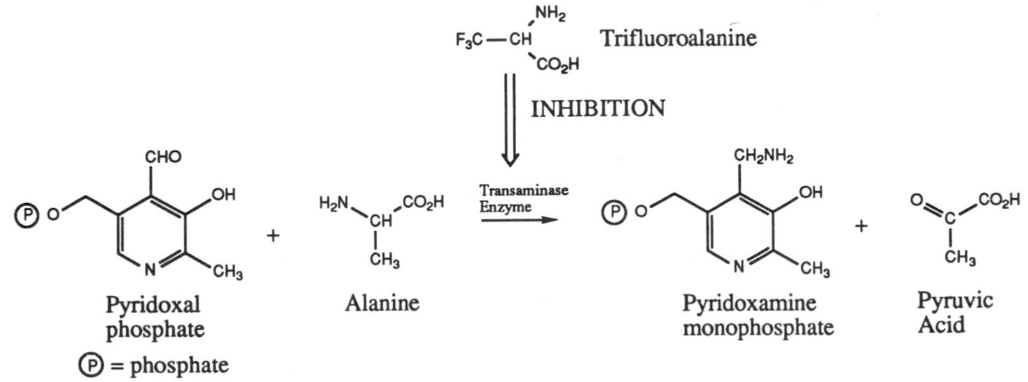

Fig. 4.23 Irreversible inhibition of the enzyme alanine transaminase.

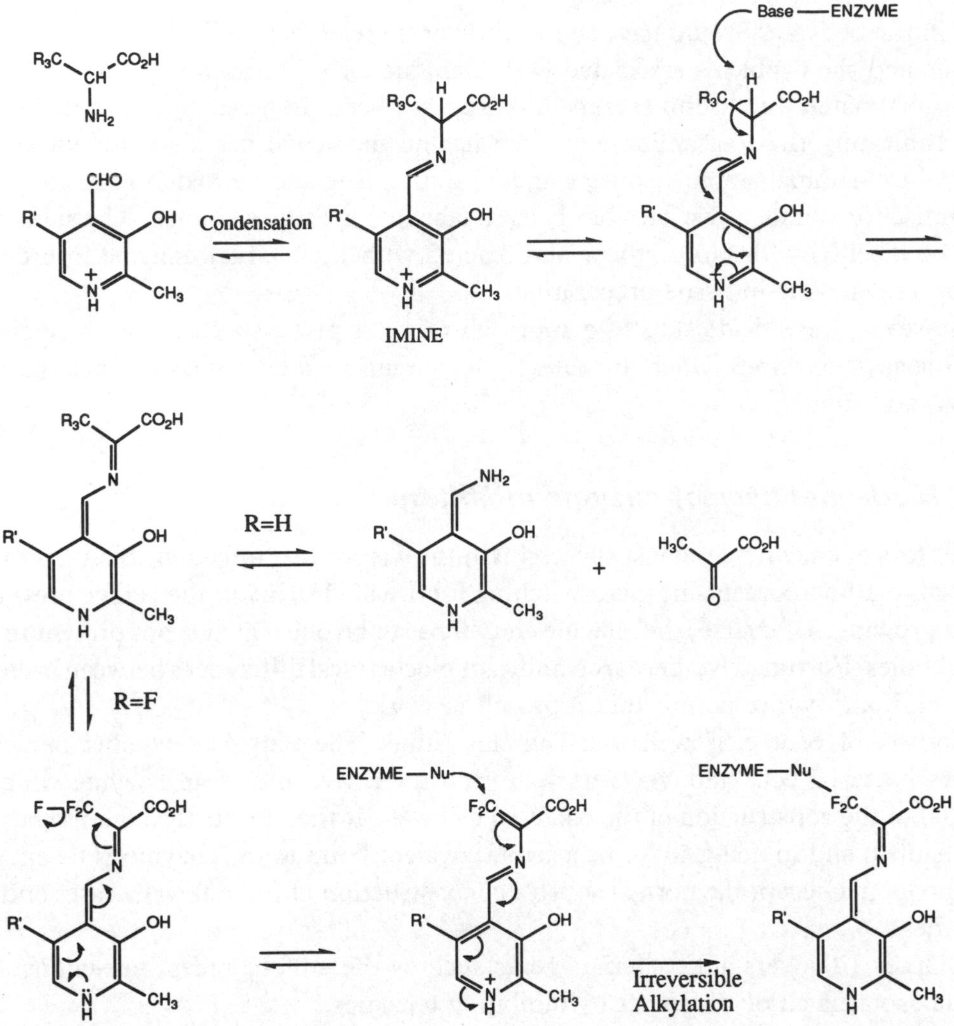

Fig. 4.24 Mechanism for the transamination reaction and its inhibition.

ism now becomes possible (R=F). A fluoride atom is electronegative and can therefore act as a leaving group. When this happens, a powerful alkylating agent is formed
which can irreversibly alkylate any nucleophilic group present in the enzyme's active
site. A covalent bond is now formed and the active site is unable to accept further
substrate. As a result, the enzyme is irreversibly inhibited.

Drugs which operate in this way are often called suicide substrates since the enzyme
is committing suicide by reacting with them. The great advantage of this approach is
that the alkylating agent is generated at the site where it is meant to act and is
therefore highly selective for the target enzyme. If the alkylating group had not been
disguised in this way, the drug would have alkylated the first nucleophilic group it

met in the body and would have shown little or no selectivity. (The uses of alkylating agents and the problems associated with them are also discussed in Chapter 6.)

Unfortunately, no useful therapeutic drug has been designed by this approach so far. Inhibiting the transaminase enzyme has no medicinal use since the enzyme is crucial to mammalian biochemistry and inhibiting it would be toxic to the host. The main use for suicide substrates has been in labelling specific enzymes. The substrates can be labelled with radioactivity and reacted with their target enzyme in order to locate the enzyme in tissue preparations.

However, the suicide substrate approach may yet prove successful in medicine if used against enzymes which are unique to 'foreign invaders' such as bacteria, protozoa, and fungi.

4.6 *Medicinal uses of enzyme inhibitors*

Inhibitors of enzymes are most successful in the war against infection. If an enzyme is crucial to a microorganism, then switching it off will clearly kill the cell or prevent it from growing. Of course, the enzyme chosen has to be one which is not present in our own bodies. Fortunately, there are significant biochemical differences between bacterial cells and our own to permit this approach to work.

Nature, of course, is well ahead in this game. The fungal metabolite penicillin enters bacterial cells and 'fools its way' into the active site of an enzyme which is crucial to the construction of the bacterial cell wall. It then reacts through the normal mechanism and in doing so forms a stable covalent bond to the enzyme. The enzyme can no longer accept the normal substrate, construction of the cell wall ceases and the cell dies.

Chapter 10 covers antibacterial agents such as the sulfonamides, penicillins, and cephalosporins all of which act by inhibiting enzymes.

Chapter 11 considers agents known as anticholinesterases which inhibit the enzyme responsible for the hydrolysis of the neurotransmitter acetylcholine.

5 ▪ Drug action at receptors

5.1 *The receptor role*

Enzymes are one major target for drugs. Receptors are another. Drugs which interact with receptors are amongst the most important in medicine and provide treatment for ailments such as pain, depression, Parkinson's disease, psychosis, heart failure, asthma, and many other problems.

What are these receptors and what do they do?

Cells are all individual, but in a complex organism such as ourselves, they have to 'get along' with their neighbours. There has to be some sort of communication system. After all, it would be pointless if individual heart cells were to contract at different times. The heart would then be a wobbly jelly and totally useless in its function as a pump. Communication is essential to ensure that all heart muscle cells contract at the same time. The same holds true for all the body's organs and functions. Communication is essential if the body is to operate in a coordinated and controlled fashion.

Control and communication come primarily from the brain and spinal column (the central nervous system—CNS) which receives and sends messages via a vast network of nerves (Fig. 5.1). The detailed mechanism by which nerves transmit messages along their length need not concern us here (see Appendix 2). It is sufficient for our purposes to think of the message as being an electrical 'pulse' which travels down the nerve cell towards the target, whether that be a muscle cell or another nerve. If that was all there was to it, it would be difficult to imagine how drugs could affect this communication system. However, there is one important feature of this system which is crucial to our understanding of drug action. The nerves do not connect directly to their target cells. They stop just short of the cell surface. The distance is minute, about 100 Å, but it is a space which the electrical impulse is unable to 'jump'.

Therefore, there has to be a way of carrying the message across the gap between the nerve ending and the cell. The problem is solved by the release of a chemical messenger (neurotransmitter) from the nerve cell (Fig. 5.2). Once released, this

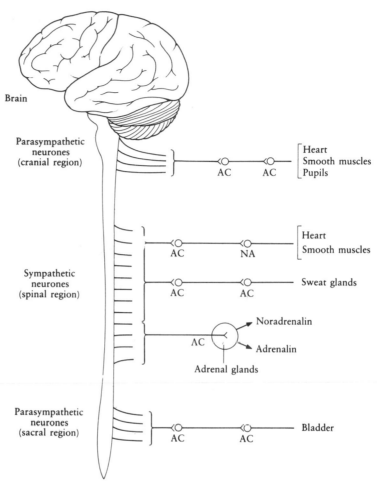

Fig. 5.1 The central nervous system. Taken from J. Mann, *Murder, magic, and medicine*, Oxford University Press (1992), with permission.

chemical messenger can diffuse across the gap to the target cell, where it can bind and interact with a specific protein (receptor) embedded in the cell membrane. This process of binding leads to a series or cascade of secondary effects which result either in a flow of ions across the cell membrane or in the switching on (or off) of enzymes inside the target cell. A biological response then results, such as the contraction of a muscle cell or the activation of fatty acid metabolism in a fat cell.

We shall consider these secondary effects and how they result in a biological action at a later stage, but for the moment the important thing to note is that the communication system depends crucially on a chemical messenger. Since a chemical process is

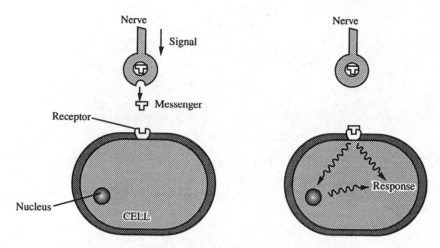

Fig. 5.2 Neurotransmitter action.

involved, it should be possible for other chemicals (drugs) to interfere or to take part in the process.

5.2 *Neurotransmitters*

Let us now look a bit closer at neurotransmitters and receptors, and consider first the messengers. What are they?

There are a large variety of messengers, many of them quite simple molecules. Neurotransmitters include such compounds as acetylcholine, noradrenaline, dopamine, γ-aminobutanoic acid (GABA), serotonin, 5-hydroxytryptophan, and even glycine (Fig. 5.3).

In general, a nerve releases only one type of neurotransmitter[1] and the receptor which awaits it on the target cell will be specific for that messenger. However, that does not mean that the target cell has only one type of receptor protein. Each target cell has a large number of nerves communicating with it and they do not all use the same neurotransmitter (Fig. 5.4). Therefore, the target cell will have other types of receptors specific for those other neurotransmitters. It may also have receptors waiting to receive messages from chemical messengers which have longer distances to

[1] In the past, it has been assumed that only one type of neurotransmitter is released from any one type of nerve cell. This is now known not to be true. Certainly as far as the amine neurotransmitters (i.e. acetylcholine, noradrenaline, glycine, serotonin, GABA, and dopamine) are concerned, it is generally true that only one of these messengers is released by any one nerve cell.

However, there is now a growing list of peptide cotransmitter substances which appear to be released from nerve cells along with the above neurotransmitters. For example, somatostatin, cholecystokinin, vasointestinal peptide, substance P, and neurotensin have all been identified as cotransmitters of acetylcholine in a variety of situations.

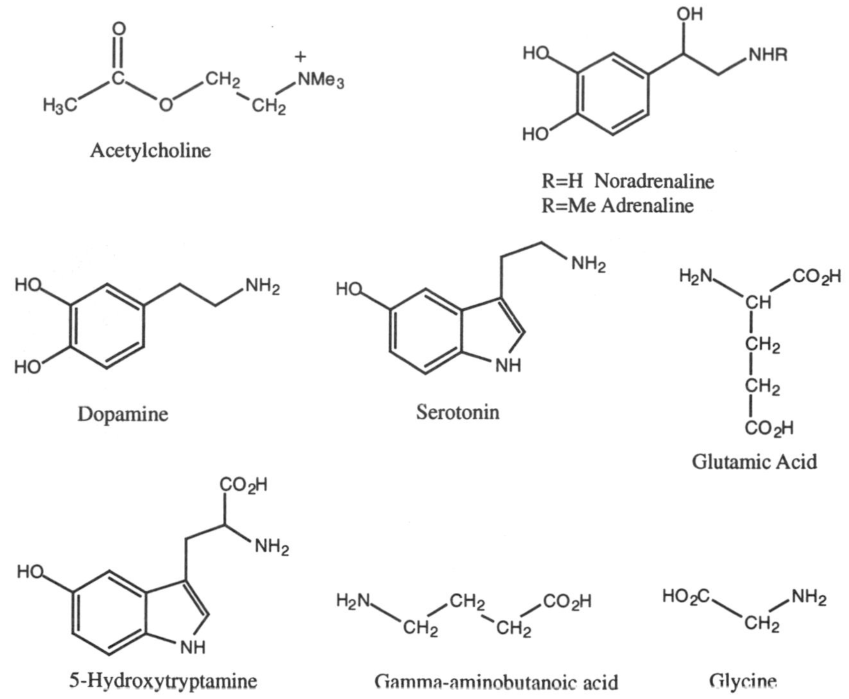

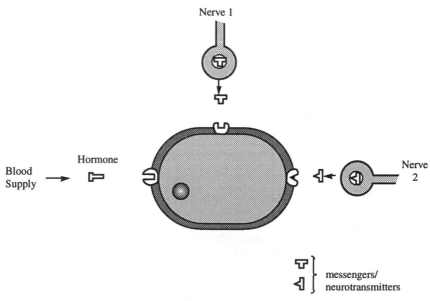

Acetylcholine

R=H Noradrenaline
R=Me Adrenaline

Dopamine

Serotonin

Glutamic Acid

5-Hydroxytryptamine

Gamma-aminobutanoic acid

Glycine

Fig. 5.3 Examples of neurotransmitters.

Nerve 1

Blood
Supply

Hormone

Nerve
2

messengers/
neurotransmitters

Fig. 5.4 Target cell containing receptors specific to each type of messenger.

travel. These are the hormones released into the circulatory system by various glands in the body. The best known example of a hormone is adrenaline. When danger or exercise is anticipated, the adrenal medulla gland releases adrenaline into the bloodstream where it is carried round the body, preparing it for violent exercise.

Hormones and neurotransmitters can be distinguished by the route they travel and by the way they are released, but their action when they reach the target cell is the same. They both interact with a receptor and a message is received. The cell responds to that message, adjusts its internal chemistry accordingly and a biological response results.

Communication is clearly essential for the normal working of the human body, and if the communication should become faulty then it could lead to such ailments as depression, heart problems, schizophrenia, muscle fatigue, and many other problems. What sort of things *could* go wrong?

One problem would be if too many messengers were released. The target cell could start to 'overheat'. Alternatively, if too few messengers were sent out, the cell could become 'sluggish'. It is at this point that drugs can play a role by either acting as replacement messengers (if there is a lack of the body's own messengers), or by blocking the receptors for the natural messengers (if there are too many host messengers). Drugs of the former type are known as agonists. Those of the latter type are known as antagonists.

What determines whether a drug is an agonist or an antagonist and is it possible to predict whether a new drug will act as one or the other?

In order to answer that, we have to move down once more to the molecular level and understand what happens when a small molecule such as a drug or a neurotransmitter interacts with a receptor protein.

Let us first look at what happens when one of the body's own messengers (neurotransmitters or hormones) interacts with its receptor.

5.3 *Receptors*

A receptor is a protein molecule embedded within the cell membrane with part of its structure facing the outside of the cell. The protein surface will be a complicated shape containing hollows, ravines, and ridges, and somewhere amidst this complicated geography, there will be an area which has the correct shape to accept the incoming messenger. This area is known as the binding site and is analogous to the active site of an enzyme. When the chemical messenger fits into this site, it 'switches on' the receptor molecule and a message is received (Fig. 5.5).

However, there is an important difference between enzymes and receptors in that the chemical messenger does not undergo a chemical reaction. It fits into the binding site of the receptor protein, passes on its message and then leaves unchanged. If no

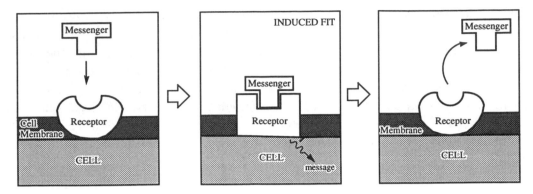

Fig. 5.5 Binding of a messenger to a receptor.

reaction takes place, what has happened? How does the chemical messenger tell the receptor its message and how is this message conveyed to the cell?

5.4 *How does the message get received?*

It all has to do with shape. Put simply, the messenger binds to the receptor and induces it to change shape. This change subsequently affects other components of the cell membrane and leads to a biological effect.

There are two main components involved:

(1) ion channels
(2) membrane-bound enzymes

5.4.1 Ion channels and their control

Some neurotransmitters operate by controlling ion channels. What are these ion channels and why are they necessary? Let us look again at the structure of the cell membrane.

As described in Chapter 2, the membrane is made up of a bilayer of fatty molecules, and so the middle of the cell membrane is 'fatty' and hydrophobic. Such a barrier makes it difficult for polar molecules or ions to move in or out of the cell. Yet it is important that these species should cross. For example, the movement of sodium and potassium ions across the membrane is crucial to the function of nerves (Appendix 2), while polar compounds such as amino acids are needed by the cell to build essential macromolecules such as proteins. It seems an intractable problem, but once again the ubiquitous proteins come up with the answer.

There are proteins present in the cell membrane which can smuggle polar molecules such as amino acids across the unfriendly medium of the cell membrane. These so-

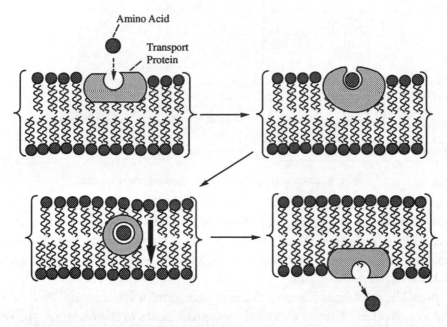

Fig. 5.6 Transport proteins.

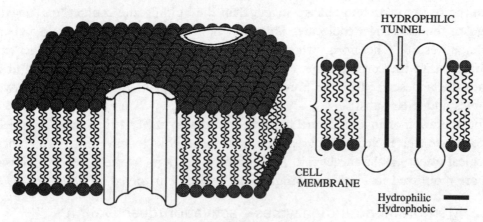

Fig. 5.7 Ion channel protein structure.

called 'transport proteins' bind the polar molecule at the outside of the cell, 'wrap it up' and ferry it across the membrane to release it on the other side (Fig. 5.6).

The ions, meanwhile, are assisted by a protein structure called an ion channel (Fig. 5.7). This structure traverses the cell membrane and consists of a protein complex made up of several subunits. The centre of the complex is hollow and is lined with polar amino acids to give a hydrophilic pore.

Ions can now cross the fatty barrier of the cell membrane by moving through these

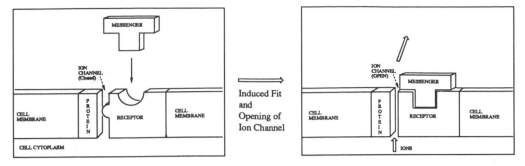

Fig. 5.8 Lock-gate mechanism for opening ion channels.

hydrophilic channels or tunnels. But there has to be some control. In other words, there has to be a 'lock-gate' which can be opened or closed as required. It makes sense that the control of this lock-gate should be associated with the message received from a neurotransmitter or hormone.

One possible mechanism is that the receptor protein itself is the lock-gate which seals the ion channel. When a chemical messenger binds to the receptor, the resulting change of shape could pull the lock-gate open and allow ions to pass (Fig. 5.8).

It is worth emphasizing one important point at this stage. If the messenger is to make the receptor protein change shape, then the binding site cannot be a 'negative image' of the messenger molecule. This must be true, otherwise how could a change of shape result? Therefore, when the binding site receives its messenger it is 'moulded' into the correct shape for the ideal fit. (This theory of an induced fit has already been described in the previous chapter to explain the interaction between enzymes and their substrates.)

The operation of an ion channel helps to explain why the relatively small number of neurotransmitter molecules released by a nerve is able to have such a significant biological effect on the target cell. By opening a few ion channels, several thousand ions are mobilized for each neurotransmitter molecule involved.

5.4.2 Membrane-bound enzymes—activation/deactivation

This is the second possible mechanism by which neurotransmitters can pass on their message to a cell. The receptor protein is situated on the outer surface of the cell membrane as before. This time, however, it is associated with a protein or enzyme situated at the inner surface of the membrane. When the receptor protein binds to its neurotransmitter, it changes shape and this forces the enzyme to change shape as well. Such a change might then reveal an active site in the enzyme which had previously been concealed, and thus start a new reaction within the cell (Fig. 5.9).

Alternatively, the membrane-bound enzyme may be working normally and the

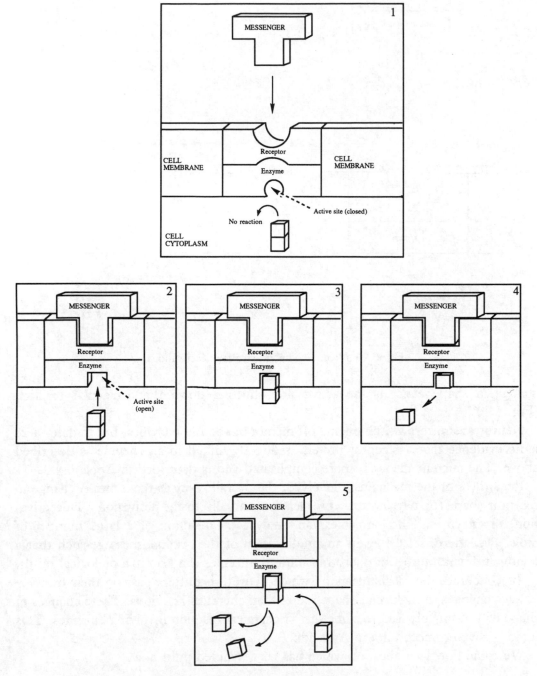

Fig. 5.9 Membrane-bound enzyme activation.

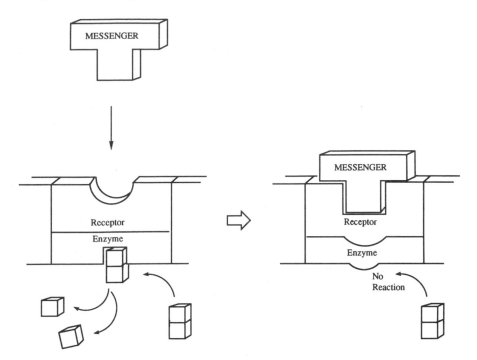

Fig. 5.10 Membrane-bound enzyme deactivation.

change in shape conceals the active site, shutting down that particular reaction (Fig. 5.10).

Neurotransmitters switch on and off membrane-bound enzymes, but to date there is no evidence that a receptor protein is directly linked to an enzyme as described above. The current theory is more complicated and is described in Appendix 3.

Regardless of the mechanism involved, the overall result is the same. A change in receptor shape (or tertiary structure) leads eventually to the activation (or deactivation) of enzymes. Since enzymes can catalyse the reaction of a large number of molecules, there is once again an amplification of the original message such that a relatively small number of neurotransmitter molecules can lead to a biological result.

To conclude, the mechanisms by which neurotransmitters pass on their message involve changes in molecular shape rather than chemical reactions. These changes in shape will ultimately lead to some sort of chemical reaction involving enzymes. This topic is covered more fully in Appendix 3.

We come now to a question which has been avoided until now.

5.5 *How does a receptor change shape?*

We have seen already that it is the messenger molecule which induces the receptor to change shape, but how does it do it? It is not simply, as one might think, a moulding process whereby the receptor wraps itself around the shape of the messenger molecule. The answer lies rather in specific binding interactions between messenger and receptor. These are the same interactions already described in Chapter 4 for enzyme/substrate binding, i.e. ionic bonding, hydrogen-bonding, and van der Waals bonding. The messenger and the receptor protein both take up conformations or shapes to maximize these bonding forces. As with enzyme substrate binding, there is a fine balance involved in receptor/messenger bonding. The bonding forces must be large enough to change the shape of the receptor in the first place but not so strong that the messenger is unable to leave again. Most neurotransmitters bind quickly to their receptors, then 'shake themselves loose again' as soon as the message has been received.

As an example of the various binding forces involved, let us consider a scenario involving a hypothetical neurotransmitter and receptor as shown in Fig. 5.11. The neurotransmitter has an aromatic ring which can interact with a hydrophobic binding site by van der Waals forces, an alcoholic group which can interact by hydrogen bonding, and a charged nitrogen centre which can interact by ionic forces.

The hypothetical receptor protein is positioned in the cell membrane such that it is sealing an ion channel and contains three binding areas (Fig. 5.12).

If the binding site has complementary binding groups for the groups described above, then the drug can fit into the binding site and bind strongly. This is all very well, but now that it has docked, how does it make the receptor change shape? As before, we have to propose that the fit is not quite exact or else there would be no reason for the receptor to change shape. In this example, we can envisage our messenger molecule fitting into the binding site and binding well to two of the three possible binding sites. The third binding site, however, (the ionic one), is not quite in the right position (Fig. 5.13). It is close enough to have a weak interaction, but not close enough for the optimum interaction. The receptor protein is therefore forced to alter shape in order to get the best binding interaction. The carboxylate group is pulled closer to the positively charged nitrogen on the messenger molecule and, as a result, the lock-gate is opened and will remain open until the messenger molecule detaches from the binding site, and allows the receptor to return to its original shape.

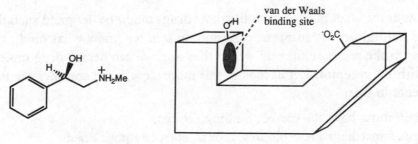

Fig. 5.11 A hypothetical neurotransmitter and receptor.

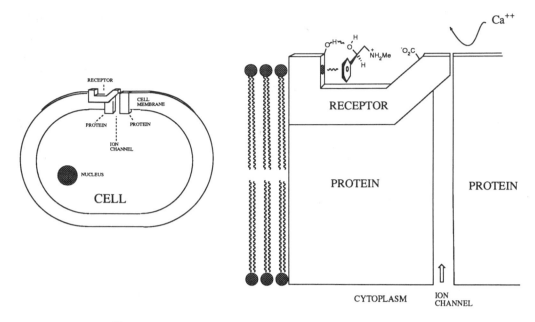

Fig. 5.12 Receptor protein positioned in the cell membrane.

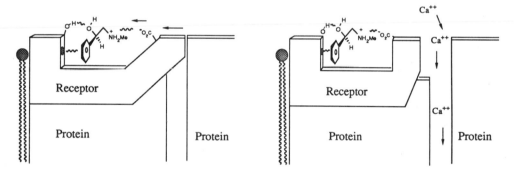

Fig. 5.13 Opening of the 'lock gate'.

5.6 *The design of agonists*

We are now at the stage of understanding how drugs might be designed such that they mimic the natural neurotransmitters. Assuming that we know what binding groups are present in the receptor site and where they are, we can design drug molecules to interact with the receptor. Let us look at this more closely and consider the following requirements in turn.

1. The drug must have the correct binding groups.
2. The drug must have these binding groups correctly positioned.
3. The drug must be the right size for the binding site.

5.6.1 Binding groups

If we consider our hypothetical receptor and its natural neurotransmitter, then we might reasonably predict which of a series of molecules would interact with the receptor and which would not.

For example, consider the structures in Fig. 5.14. They all look different, but they all contain the necessary binding groups to interact with the receptor. Therefore, they may well be potential agonists or alternatives for the natural neurotransmitter.

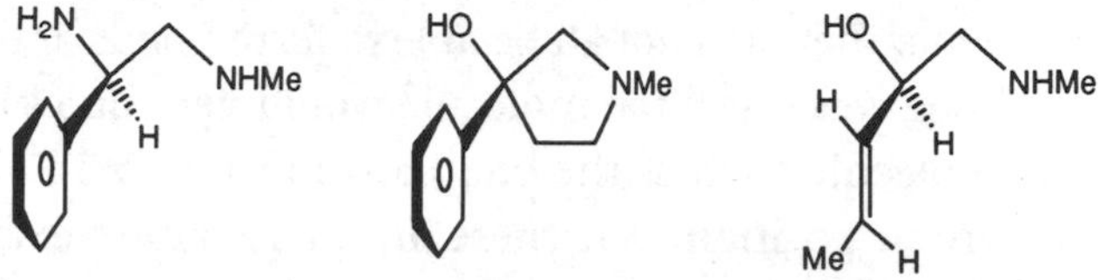

Fig. 5.14 Possible agonists.

However, the structures in Fig. 5.15 lack one or more of the required binding groups and might therefore be expected to have poor activity. We would expect them to drift into the receptor site and then drift back out again, binding only weakly if at all.

Of course, we are making an assumption here; that all three binding groups are essential. It might be argued that a compound such as structure II in Fig. 5.15 might be effective even though it lacks a suitable hydrogen bonding group. Why, for example, could it not bind initially by van der Waals forces alone and then alter the shape of the receptor protein via ionic bonding?

In fact, this seems unlikely when we consider that neurotransmitters appear to bind, pass on their message and then leave the binding site very quickly. In order to do that, there must be a fine balance in the binding forces between receptor and

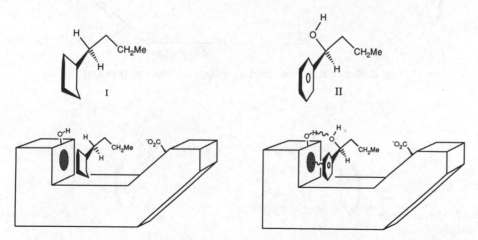

Fig. 5.15 Structures possessing fewer than the required number of binding sites.

neurotransmitter. The binding forces must be strong enough to bind the neurotransmitter effectively such that the receptor changes shape. However, the binding forces cannot be too strong, or else the neurotransmitter would not be able to leave and the receptor would not be able to return to its original shape. Therefore, it is reasonable to assume that a neurotransmitter needs all of its binding interactions to be effective. The lack of even one of these interactions would lead to a significant loss in activity.

5.6.2 Position of binding groups

The molecule may have the correct binding groups, but if they are in the wrong positions, they will not all be able to form bonds at the same time. As a result, bonding would be weak and the molecule would very quickly drift away. Result—no activity.

A molecule such as the one shown in Fig. 5.16 obviously has its binding groups in the wrong position, but there are more subtle examples of molecules which do not have the correct arrangement of binding groups. For example, the mirror image of our hypothetical neurotransmitter would not fit (Fig. 5.17). The structure has the same formula and the same constitutional structure as our original structure. It will have the same physical properties and undergo the same chemical reactions, but it is not the same shape. It is a non-superimposable mirror image and it cannot fit the receptor site (Fig. 5.18).

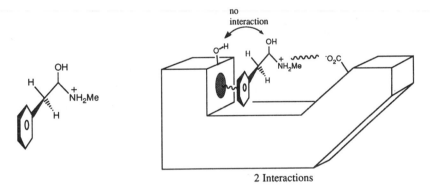

Fig. 5.16 Molecule with binding groups in incorrect positions.

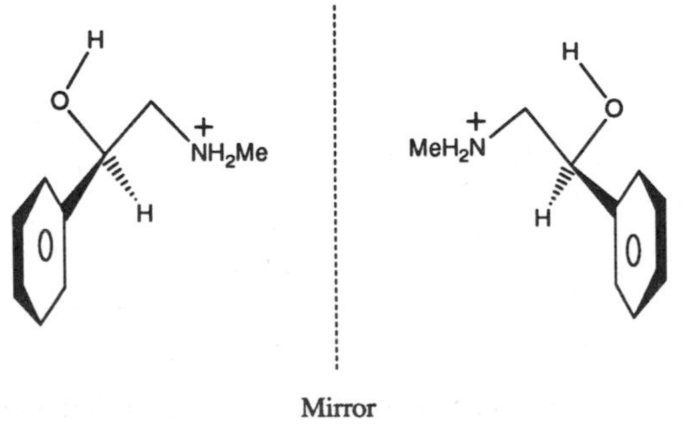

Fig. 5.17 Mirror image of hypothetical neurotransmitter.

Mirror

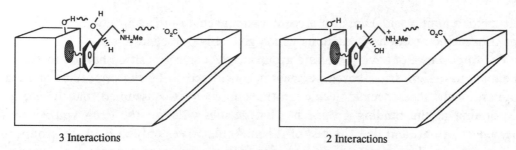

3 Interactions   2 Interactions

Fig. 5.18 Interactions between the hypothetical neurotransmitter and its mirror image with the receptor site.

Compounds which have non-superimposable mirror images are termed chiral or asymmetric. There are only two detectable differences between the two mirror images (or enantiomers) of a chiral compound. They rotate plane polarized light in opposite directions and they interact differently with other chiral systems such as enzymes. This has very important consequences for the pharmaceutical industry.

Pharmaceutical agents are usually synthesized from simple starting materials using simple achiral (symmetrical) chemical reagents. These reagents are incapable of distinguishing between the two mirror images of a chiral compound. As a result, most chiral drugs are synthesized as a mixture of both mirror images (a racemate). However, we have seen from our own simple example that only one of these enantiomers is going to interact with a receptor. What happens to the other enantiomer?

At best, it floats about in the body doing nothing. At worst, it interacts with a totally different receptor and results in an undesired side-effect. Herein lies the explanation for the thalidomide tragedy. One of the enantiomers was an excellent sedative. The other reacted elsewhere in the body as a poison and was teratogenic (induced abnormalities in human embryos). If the two enantiomers had been separated, then the tragedy would not have occurred.

Even if the 'wrong' enantiomer does not do any harm, it seems to be a great waste of time, money and effort to synthesize drugs which are only 50 per cent efficient. That is why one of the biggest areas of chemical research in recent years has been in the field of asymmetric synthesis—the synthesis in the laboratory of a single enantiomer of a chiral compound.

Of course, nature has been at it for millions of years. Since nature has chosen to work with only the 'left-handed' enantiomer of amino acids,[2] enzymes (made up of

[2] Naturally occurring amino acids exist as the one enantiomer, termed the L-enantiomer. This terminology is historical and defines the absolute configuration of the chiral carbon present at the 'head-group' of the amino acid. The current terminology for chiral centres is to define them as R or S according to a set of rules known as the Cahn–Ingold–Prelog rules. Naturally occurring amino acids exist as the (S)-configuration, but the older terminology still dominates in the case of amino acids.

Experimentally, the L-amino acids were found to rotate plane polarized light anticlockwise or to the left. Hence the expression left-handed amino acids.

left-handed amino acids) are also present as single enantiomers and therefore catalyse enantiospecific reactions—reactions which give only one enantiomer.

The importance of having binding groups in the correct position has led medicinal chemists to design drugs based on what is considered to be the important pharma-cophore of the messenger molecule. In this approach, it is assumed that the correct positioning of the binding groups is what decides whether the drug will act as a messenger or not, and that the rest of the molecule serves only to hold the groups in these positions. Therefore, the activity of apparently disparate structures at a receptor can be explained if they all contain the correct binding forces at the correct positions. The design of totally novel structures or molecular frameworks to hold these binding groups in the correct positions could then be proposed, leading to a new series of drugs. There is, however, a limiting factor to this which will now be discussed.

5.6.3 Size and shape

It is possible for a compound to have the correct binding groups in the correct positions and yet fail to interact effectively if it has the wrong size or shape.

As an example, let us consider the structure shown in Fig. 5.19 as a possible candidate for our hypothetical receptor system.

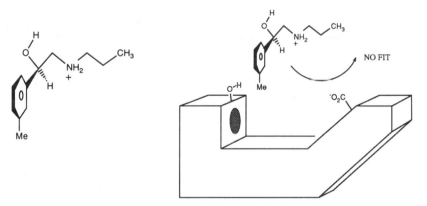

Fig. 5.19 Structure with a *para*-methyl group.

The structure has a *meta*-methyl group on the aromatic ring and a long alkyl chain containing the nitrogen atom. By considering size factors alone, we could conclude that both these features would prevent this molecule from binding effectively to the receptor.

The *meta*-methyl group would act as a buffer and prevent the structure from 'sinking' deep enough into the binding site for effective binding. Furthermore, the long alkyl chain on the nitrogen atom would make that part of the molecule too long for the space available to it.

A thorough understanding of the space available in the binding site is therefore necessary when designing analogues which will fit it.

5.7 *The design of antagonists*

5.7.1 Antagonists acting at the binding site

We have seen how it might be possible to design drugs to mimic the natural neurotransmitters (agonists) and how these would be useful in treating a shortage of the natural neurotransmitter. However, suppose that we have too much neurotransmitter operating in the body. How could a drug counteract that?

There are several strategies, but in theory we could design a drug (an antagonist) which would be the right shape to bind to the receptor site, but which would either fail to change the shape of the receptor protein or would distort it too much. Consider the following scenario.

The compound shown in Fig. 5.20 fits the binding site perfectly and as a result does not cause any change of shape. Therefore, there is no biological effect and the binding site is blocked to the natural neurotransmitter.

In such a situation, the antagonist has to compete with the agonist for the receptor, but usually the antagonist will get the better of this contest since it often binds more strongly.

To sum up, if we know the shape and make-up of receptor binding sites, then we should be able to design drugs to act as agonists or antagonists. Unfortunately, it is not as straightforward as it sounds. Finding the receptor and determining the layout of its binding site is no easy task. In reality, the theoretical shape of many receptor sites have been worked out by synthesizing a large number of compounds and considering those molecules which fit and those which do not—a bit like a 3D jigsaw.

However, the recent advent of computer-based molecular graphics and the availability of X-ray crystallographic data now allow a more accurate representation of proteins and their binding sites (Chapter 7) and promise to spark off a new phase of drug development.

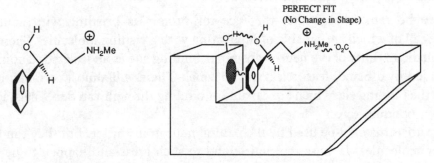

Fig. 5.20 Compound acting as an antagonist at the binding site.

5.7.2 Antagonists acting outwith the binding site

Even if the 'layout' of a binding site is known, it may not help in the design of antagonists. There are many example of antagonists which bear no apparent structural similarity to the native neurotransmitter and could not possibly fit the geometrical requirements of the binding site. Such antagonists frequently contain one or more aromatic rings, suggesting van der Waals interactions are important in their binding, yet there may not be any corresponding area in the binding site. How then do such antagonists work? There are two possible explanations.

Allosteric antagonists

The antagonist may bind to a totally different part of the receptor. The process of binding could alter the shape of the receptor protein such that the neurotransmitter binding site is distorted and is unable to recognize the natural neurotransmitter (Fig. 5.21). Therefore, binding between neurotransmitter and receptor would be prevented and the message lost. This form of antagonism is a non-competitive form of antagonism since the antagonist is not competing with the neurotransmitter for the same binding site (compare allosteric inhibitors of enzymes—Chapter 4).

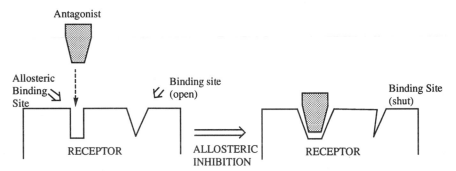

Fig. 5.21 Allosteric antagonists.

Antagonism by the 'umbrella' effect

It has to be remembered that the receptor protein is bristling with amino acid residues, all of which are capable of interacting with a visiting molecule. Therefore, it is unrealistic to think of the neurotransmitter binding site as an isolated 'landing pad', surrounded by a bland, featureless 'no go zone'. There will almost certainly be areas close to the binding site which are capable of binding through van der Waals, ionic, or hydrogen bonding forces.

These areas may not be used by the natural neurotransmitter, but they can be used by other molecules. If these molecules bind to such areas and happen to lie over or partially lie over the neurotransmitter binding site, then they will act as antagonists

Fig. 5.22 Antagonism by the 'umbrella effect'.

and prevent the neurotransmitter reaching its binding site (Fig. 5.22). This form of antagonism has also been dubbed the 'umbrella effect' and is a form of competitive antagonism since the normal binding site is directly affected.

Many antagonists are capable of binding to both the normal binding site and the neighbouring sites. Such antagonists will clearly bind far more strongly than agonists. Because of this stronger binding, antagonists have been useful in the isolation and identification of specific receptors present in tissues. A further tactic in this respect is to incorporate a highly reactive chemical group—usually an electrophilic group—into such a powerful antagonist. The electrophilic group will then react with any convenient nucleophilic group on the receptor surface and alkylate it to form a strong covalent bond. The antagonist will then be irreversibly tied to the receptor and can act as a molecular label. One example is tritium-labelled propylbenzilylcholine mustard —used to label the muscarinic acetylcholine receptor (Fig. 5.23) (see also Chapter 10).

5.8 *Partial agonists*

Frequently drugs are discovered which cannot be defined as pure antagonists or pure agonists.

Such compounds bind to the receptor site and block access to the natural neurotransmitter, and so in this sense they are antagonists. However, they also activate the receptor very weakly such that a slight signal is received. In our hypothetical situation (Fig. 5.24), we could imagine a partial agonist being a molecule which is almost a perfect fit for the binding site, such that binding results in only a very slight distortion of the receptor. This would then only partly open the ion channel.

An alternative explanation for partial agonism is that the molecule in question

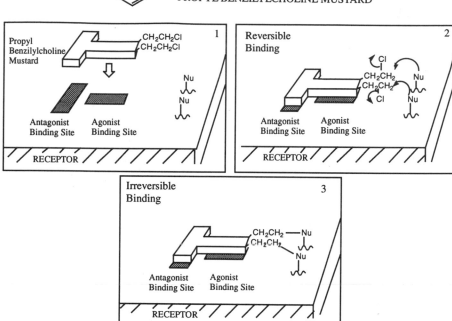

Fig. 5.23 Antagonist used as a molecular label.

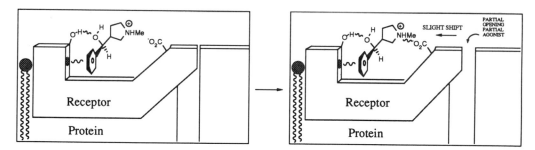

Fig. 5.24 Partial agonism.

might be capable of binding to a receptor in two different ways by using different binding groups. One method of binding would activate the receptor, while the other would not. The balance of agonism versus antagonism would then depend on the relative proportions of molecules binding by either method. Examples of partial agonists are discussed in Chapters 11 and 12.

5.9 *Desensitization*

Some drugs bind relatively strongly to a receptor, switch it on, but then subsequently block the receptor. Thus, they are acting as agonists, then antagonists. The mechanism of how this takes place is not clear. One theory is that receptors can only remain activated for a certain period of time. Once that period is up, another change in the tertiary structure takes place which switches the receptor off, despite the binding site being occupied (Fig. 5.25). This altered tertiary structure is then maintained as long as the binding site is occupied. When the drug eventually leaves, the receptor returns to its original resting shape.

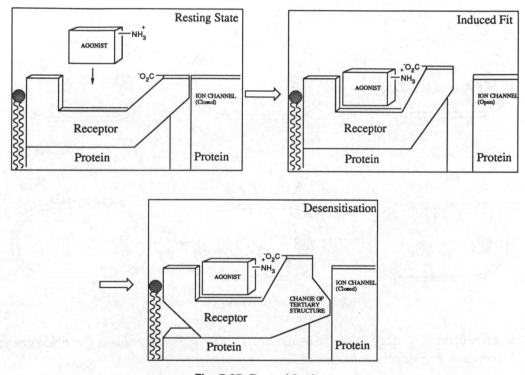

Fig. 5.25 Desensitization.

In conclusion, it is thought that the best agonists bind swiftly to the receptor, pass on their message and then leave quickly. Antagonists, in contrast, tend to be slow to add and slow to leave.

5.10 *Tolerance and dependence*

It has been discovered that 'starving' a target cell of a certain neurotransmitter induces that cell to synthesize more receptors. By doing so, the cell gains a greater sensitivity

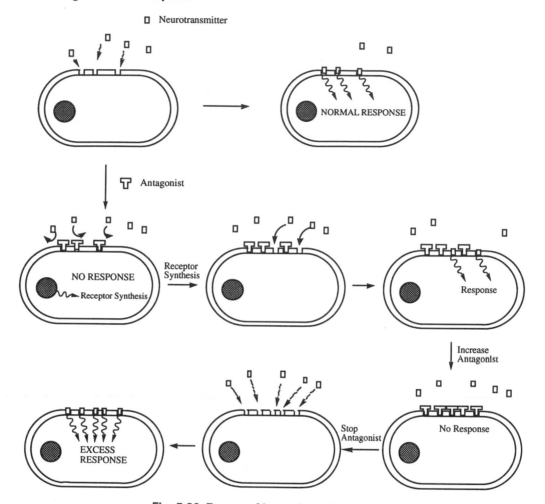

Fig. 5.26 Process of increasing cell sensitivity.

for what little neurotransmitter is available. This process can explain the phenomena of tolerance and dependence (Fig. 5.26).

Tolerance is a situation where higher levels of a drug are required to get the same biological response. If a drug is acting to suppress the binding of a neurotransmitter, then the cell may respond by increasing the number of receptors. This would require increasing the dose to regain the same level of antagonism.

If the drug was suddenly stopped, then all the receptors would suddenly become available. There would now be an excess of receptors which would make the cell supersensitive to normal levels of neurotransmitter. This would be equivalent to receiving a drug overdose. The resulting biological effects would explain the distressing withdrawal symptoms resulting from stopping certain drugs. These withdrawal symptoms would continue until the number of receptors returned to their original level. During this period, the patient may be tempted to take the drug again in order to 'return to normal' and will have then gained a dependence on the drug.

The problem of tolerance is also discussed in Appendix 3.

It is only in recent years that medicinal chemists have begun to understand receptors and drug receptor interactions. This increased understanding should revolutionize the subject in the years to come.

6 ▪ Nucleic acids

Although the majority of drugs act on protein structures, there are several examples of important drugs which act directly on nucleic acids to disrupt replication, transcription, and translation.

There are two types of nucleic acid—DNA (deoxyribonucleic acid) and RNA (ribonucleic acid). We shall first consider the structure of DNA and the drugs which act on it.

6.1 *Structure of DNA*

As with proteins, DNA has a primary, secondary, and tertiary structure.

6.1.1 The primary structure of DNA

The primary structure of DNA is the way in which the DNA building blocks are linked together. Whereas proteins have over twenty building blocks to choose from, DNA has only four—the nucleosides deoxyadenosine, deoxyguanosine, deoxycytidine, and deoxythymidine (Fig. 6.1).

Each nucleoside is constructed from two components—a deoxyribose sugar and a base. The sugar is the same in all four nucleosides and only the base is different. The four possible bases are two bicyclic purines (adenine and guanine), and two smaller pyrimidine structures (cytosine and thymine) (Fig. 6.2).

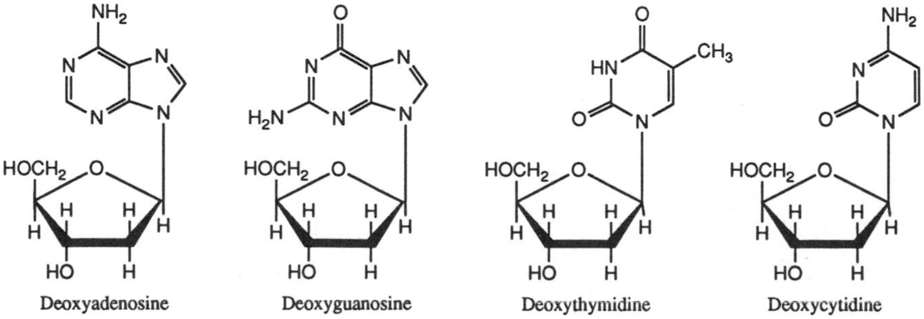

Fig. 6.1 The building blocks of DNA—nucleosides.

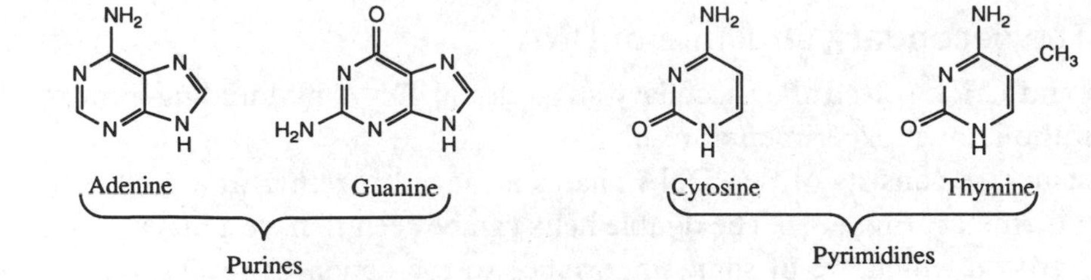

Fig. 6.2 The four bases of nucleosides.

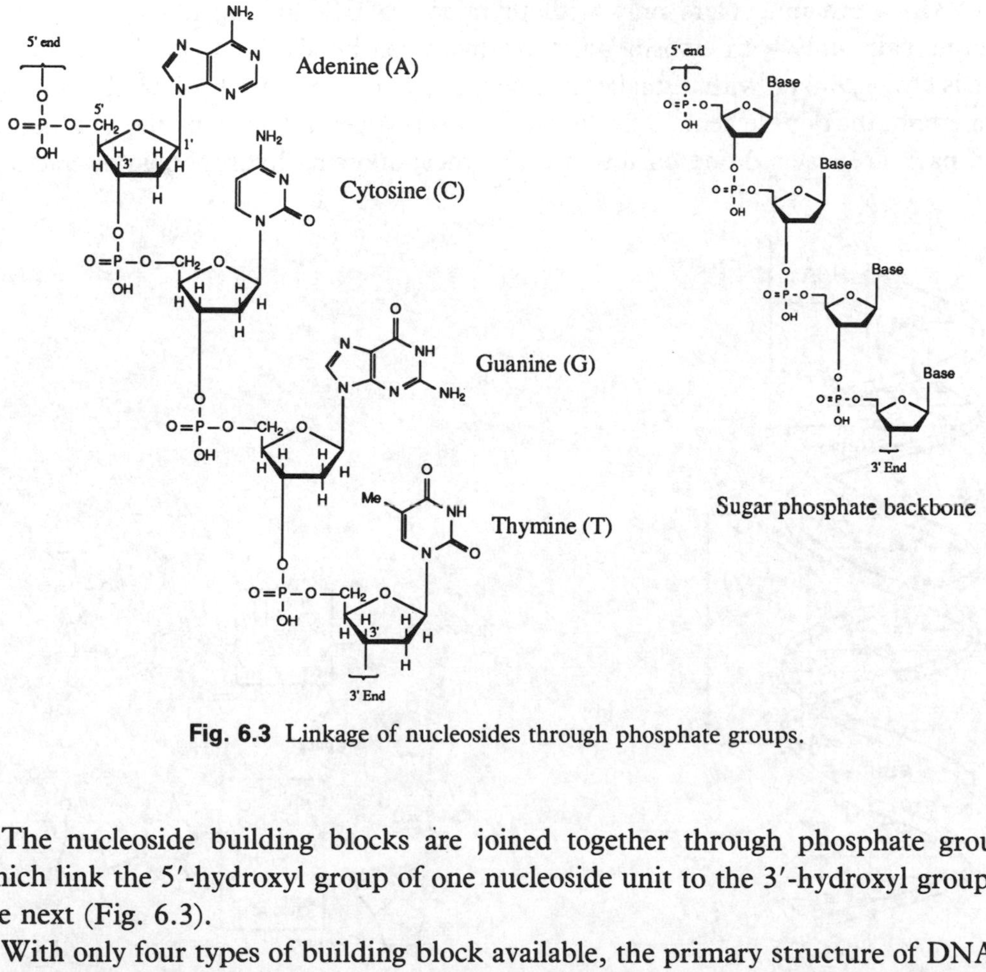

Fig. 6.3 Linkage of nucleosides through phosphate groups.

The nucleoside building blocks are joined together through phosphate groups which link the 5'-hydroxyl group of one nucleoside unit to the 3'-hydroxyl group of the next (Fig. 6.3).

With only four types of building block available, the primary structure of DNA is far less varied than the primary structure of proteins. As a result, it was long thought that DNA only had a minor role to play in cell biochemistry, since it was hard to see how such an apparently simple molecule could have anything to do with the mysteries of the genetic code.

The solution to this mystery lies in the secondary structure of DNA.

6.1.2 The secondary structure of DNA

Watson and Crick solved the secondary structure of DNA by building a model that fitted all the known experimental results.

The structure consists of two DNA chains arranged together in a double helix of constant diameter (Fig. 6.4). The double helix can be seen to have a major groove and a minor groove which are of some importance to the action of several antibacterial agents (see later).

The structure relies crucially on the pairing up of nucleic acid bases between the two chains. Adenine pairs only with thymine via two hydrogen bonds, whereas guanine pairs only with cytosine via three hydrogen bonds. Thus, a bicyclic purine base is always linked with a smaller monocyclic pyrimidine base to allow the constant diameter of the double helix. The double helix is further stabilized by the fact that the base pairs are stacked one on top of each other, allowing hydrophobic interactions

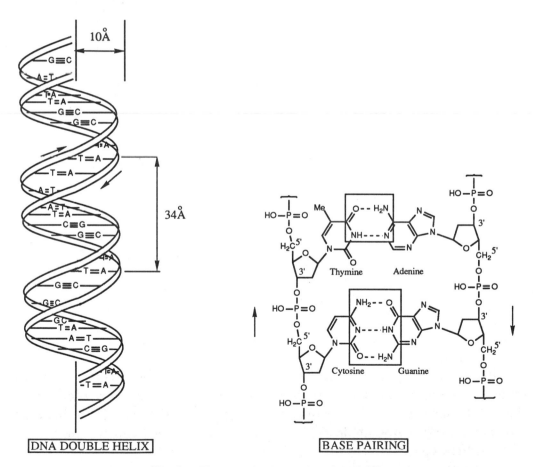

Fig. 6.4 The secondary structure of DNA.

between the heterocyclic rings. The polar sugar phosphate backbone is placed to the outside of the structure and therefore can form favourable polar interactions with water.

The fact that adenine always binds to thymine, and cytosine always binds to guanine means that the chains are complementary to each other. In other words, one chain can be visualized as a negative image of its partner. It is now possible to see how replication (the copying of the genetic information) is feasible. If the double helix unravels, then a new chain can be constructed on each of the original chains (Fig. 6.5). In other words, each of the original chains can act as a template for the construction of a new and identical double helix.

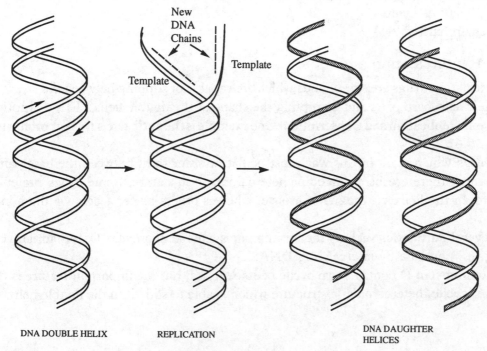

Fig. 6.5 Replication of DNA chains.

It is less obvious how DNA can code for proteins. How can four nucleotides code for over twenty amino acids?

The answer lies in the triplet code. In other words, an amino acid is not coded by one nucleotide but by a set of three. There are sixty-four ways in which four nucleotides can be arranged in sets of three—more than enough for the task required.

6.1.3 The tertiary structure of DNA

The tertiary structure of DNA is often neglected or ignored, but it is important to the action of the quinolone group of antibacterial agents (Chapter 10). The double helix is able to coil into a 3D shape and this is known as supercoiling. During replication, the double strand of DNA must unravel, but due to the tertiary supercoiling this leads to a high level of strain which has to be relieved by the temporary cutting, then repair, of the DNA chain. These procedures are enzyme-catalysed and these are the enzymes which are inhibited by the quinolone antibacterial agents.

6.2 Drugs acting on DNA

In general, we can classify the drugs which act on DNA into three groups:

(1) intercalating cytostatic agents
(2) alkylating agents
(3) chain 'cutters'

6.2.1 Intercalating agents

Intercalating drugs are compounds which are capable of slipping between the layers of nucleic acid base pairs and disrupting the shape of the double helix. This disruption prevents replication and transcription. One example is the antibacterial agent proflavine (Fig. 6.6).

Drugs which work in this way must be flat in order to fit between the base pairs, and must therefore be aromatic or heteroaromatic in nature. Some drugs prefer to approach the helix via the major groove, whereas others prefer access via the minor groove.

Several antibiotics such as the antitumour agents actinomycin D and adriamycin (Fig. 6.7) operate by intercalating DNA.

Actinomycin D contains two cyclic pentapeptides, but the important feature is the flat, tricyclic, heteroaromatic structure which is able to slide into the double helix. It

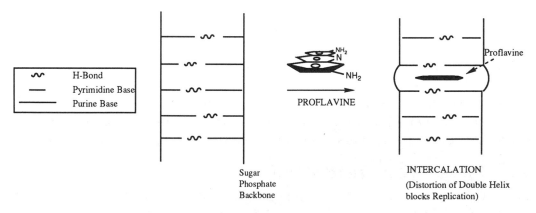

Fig. 6.6 Action of intercalating drugs.

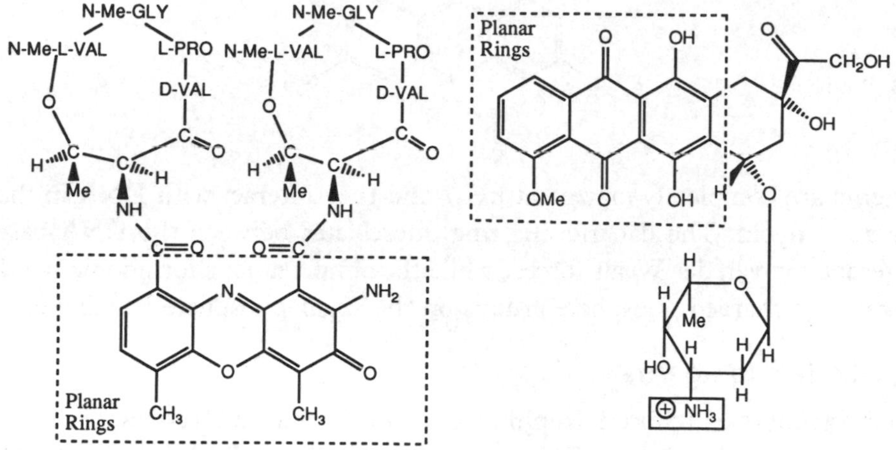

ACTINOMYCIN D ADRIAMYCIN

Fig. 6.7 Antibiotics which operate by intercalating DNA.

appears to favour interactions with guanine–cytosine base pairs and, in particular, between two adjacent guanine bases on alternate strands of the helix. Actinomycin D is further held in position by hydrogen bond interactions between the nucleic acid bases of DNA and the cyclic pentapeptides positioned on the outside of the helix.

Adriamycin has a tetracyclic system where three of the rings are planar and are able to fit into the double helix. The drug approaches DNA via the major groove of the double helix. The amino group attached to the sugar is important in helping to lock the antibiotic into place since it can ionize and form an ionic bond with the negatively charged phosphate groups on the DNA backbone.

The highly effective antimalarial agent chloroquine—a drug developed from quinine—can attack the malarial parasite by blocking DNA transcription as part of its action. Once again a flat heteroaromatic structure is present which can intercalate DNA (Fig. 6.8).

Aminoacridine agents such as proflavine (Fig. 6.9) are topical antibacterial agents which were used particularly in the Second World War to treat surface wounds. The

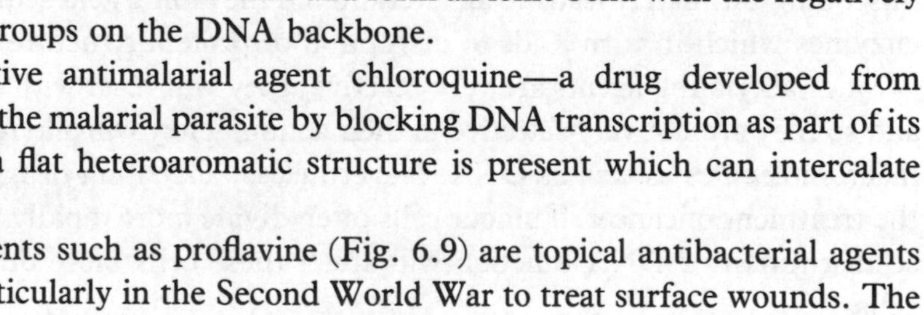

QUININE CHLOROQUINE

Fig. 6.8 Intercalating antimalarial drugs.

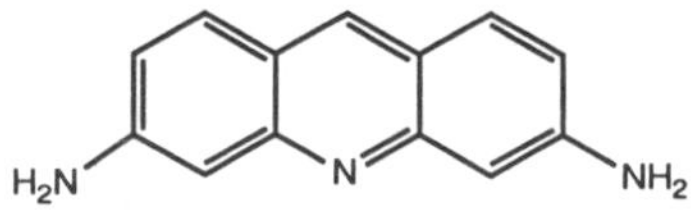

Fig. 6.9 Proflavine.

best agents are completely ionized at pH 7 and they interact with DNA in the same way as adriamycin. The flat tricyclic ring intercalates between the DNA base pairs and interacts by van der Waals forces, while the amino cations form ionic bonds with the negatively charged phosphate groups on the sugar phosphate backbone.

6.2.2 Alkylating agents

Alkylating agents are highly electrophilic compounds which will react with nucleophiles to form strong covalent bonds. There are several nucleophilic groups in DNA and in particular the 7-nitrogen of guanine. Drugs with two such alkylating groups could therefore react with a guanine on each chain and cross-link the strands such that they cannot unravel during replication or transcription.

Alternatively, the drug could link two guanine groups on the same chain such that the drug is attached like a limpet to the side of the DNA helix. Such an attachment would mask that portion of DNA and block access to the necessary enzymes required for DNA function.

Miscoding due to alkylated guanine units is also possible. The guanine base usually exists as the keto tautomer and base pairs with cytosine. Once alkylated however, guanine prefers the enol tautomer and is more likely to base pair with thymine. Such miscoding ultimately leads to an alteration in the amino acid sequence of proteins and enzymes which in turn leads to disruption of protein structure and function.

Since alkylating agents are very reactive, they will react with any good nucleophile and so they are not very selective in their action. They will alkylate proteins and other macromolecules as well as DNA. Nevertheless, alkylating drugs have been useful in the treatment of cancer. Tumour cells often divide more rapidly than normal cells and so disruption of DNA function will affect these cells more drastically than normal cells.

The nitrogen mustard compound mechlorethamine (Fig 6.10) was the first alkylating agent to be used (1942). The nitrogen atom is able to displace a chloride ion intra-molecularly to form the highly electrophilic aziridine ion. Alkylation of DNA can then take place. Since the process can be repeated, cross-linking between chains will occur.

The side-reactions mentioned above can be reduced by reducing the reactivity of the alkylating agent. For example, putting an aromatic ring on the nitrogen atom instead of a methyl group (Fig. 6.11) has such an effect. The lone pair of the nitrogen is 'pulled into' the ring and is less available to displace the chloride ion. As a result,

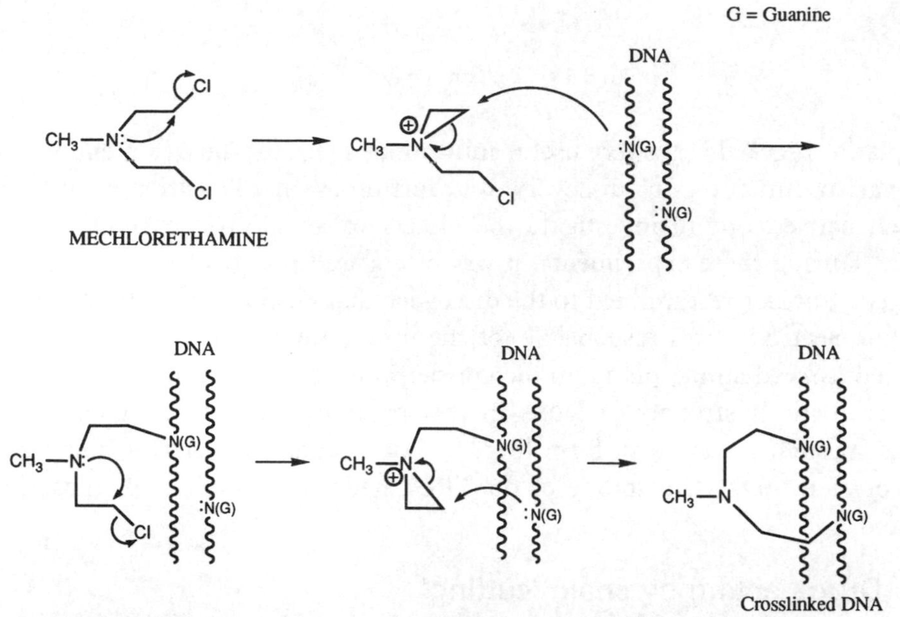

G = Guanine

MECHLORETHAMINE

Crosslinked DNA

Fig. 6.10 Alkylation of DNA by the nitrogen mustard compound, mechlorethamine.

the intermediate aziridine ion is less easily formed and only strong nucleophiles such as guanine will now react with it.

Another approach which has been used to direct these alkylating agents more specifically to DNA has been to attach a nucleic acid building block onto the molecule. For example, uracil mustard (Fig. 6.12) contains one of the nucleic acid bases. This drug has been used successfully in the treatment of chronic lymphatic leukaemia and has a certain amount of selectivity for tumour cells over normal cells. This is because tumour cells generally divide faster than normal cells. As a result, nucleic acid synthesis is faster and tumour cells are 'hungrier' for the nucleic acid building blocks. The tumour cells therefore take more than their share of the building blocks and of any cytotoxic drug which mimics the building blocks. Unfortunately, this approach has not so far succeeded in achieving the high levels of selectivity desired for effective eradication of tumour cells.

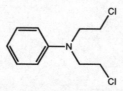

Fig. 6.11 Method of reducing reactivity of the alkylating agent.

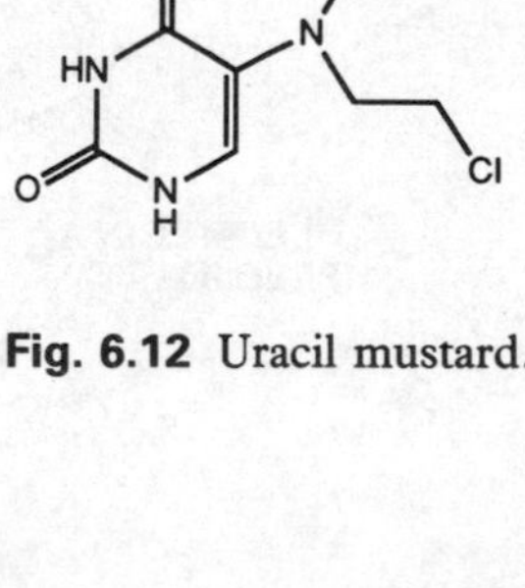

Fig. 6.12 Uracil mustard.

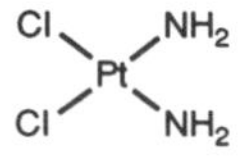

Fig. 6.13 Cisplatin.

Cisplatin (Fig. 6.13) is a very useful antitumour agent for the treatment of testicular and ovarian tumours. Its discovery was fortuitous in the extreme, arising from research carried out to investigate the effects of an electric current on bacterial growth. During these experiments, it was discovered that bacterial cell division was inhibited. Further research led to the discovery that an electrolysis product from the platinum electrodes was responsible for the inhibition and the agent was eventually identified as *cis*-diamino platinum dichloride, known as cisplatin.

Cisplatin binds strongly to DNA in regions containing several guanidine units, binding in such a way as to form links within strands (intrastrand binding) rather than between them. Unwinding of the DNA helix takes place and transcription is inhibited.

6.2.3 Drugs acting by chain 'cutting'

Bleomycin (Fig. 6.14) is a large glycoprotein which appears to be able to cut the strands of DNA and then prevent the enzyme DNA ligase from repairing the damage. It appears to act by abstracting hydrogen atoms from DNA. The resultant radicals

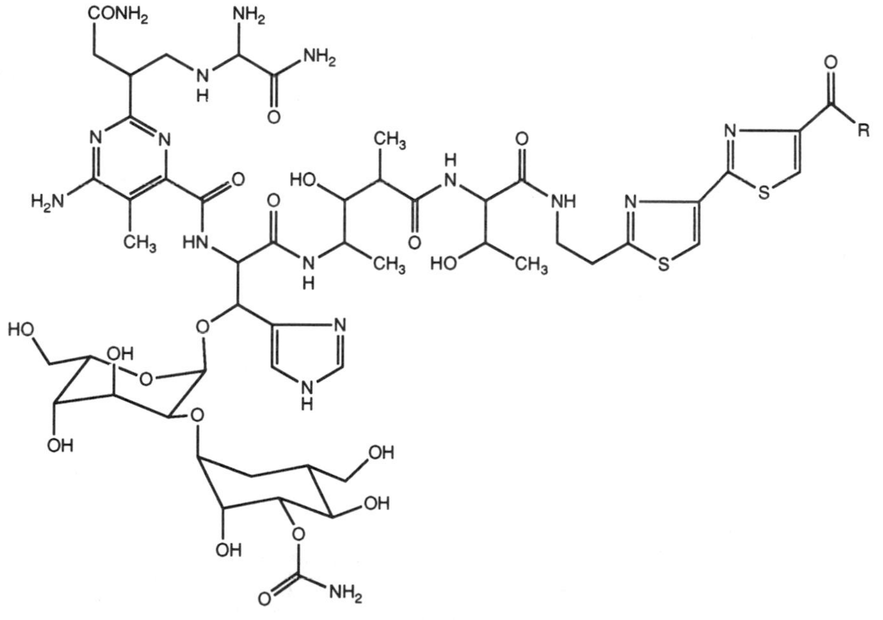

BLEOMYCIN A$_2$ R = NHCH$_2$CH$_2$CH$_2$SMe$_2$
BLEOMYCIN B$_2$ R = NHCH$_2$CH$_2$CH$_2$CH$_2$NHC(NH$_2$)=NH

Fig. 6.14 Bleomycin.

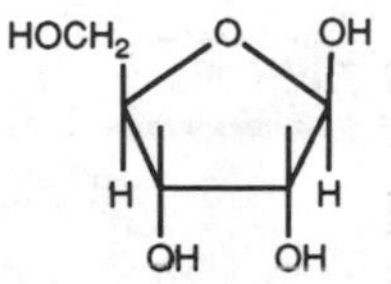

HOCH2
O
OH
H
H
OH
OH

Fig. 6.15 Ribose.

O
HN
O
N
H

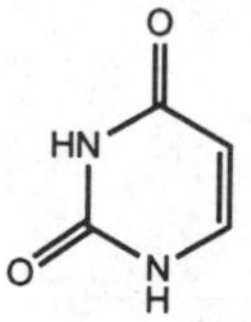

Fig. 6.16 Uracil.

react with oxygen to form peroxy species which then fragment. The drug is useful against certain types of skin cancer.

6.3 *Ribonucleic acid*

The primary structure of RNA is the same as DNA, with two exceptions. Ribose (Fig. 6.15) is the sugar component rather than deoxyribose, while uracil (Fig. 6.16) replaces thymine as one of the bases.

Base pairing between nucleic acid bases can occur in RNA with adenine pairing with uracil, and cytosine pairing with guanine. However, the pairing is between bases within the same chain, and it does not occur for the whole length of the molecule (e.g. Fig. 6.18). Therefore, RNA is not a double helix, but it does have regions of helical secondary structure.

Since the secondary structure is not uniform along the length of the RNA chain, more variety is allowed in RNA tertiary structure. Three types of RNA molecules have been identified with different cell functions. The three are messenger RNA (mRNA), transfer RNA (tRNA), and ribosomal RNA (rRNA).

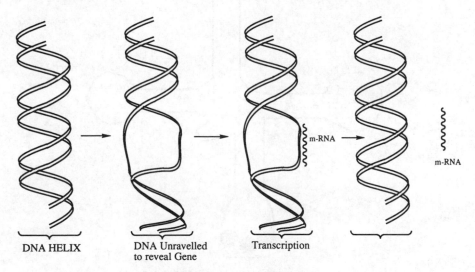

m-RNA
m-RNA
DNA HELIX
DNA Unravelled
to reveal Gene
Transcription

Fig. 6.17 Formation of m-RNA.

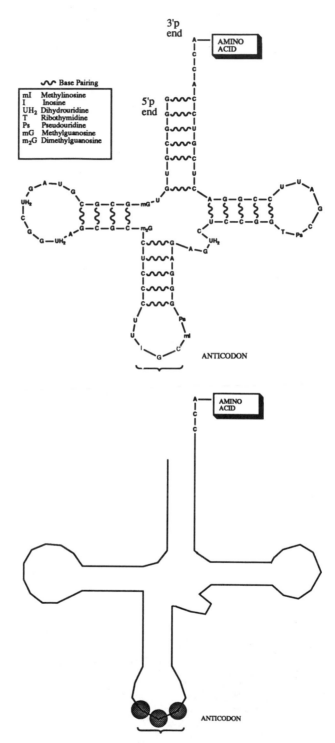

Fig. 6.18 Yeast alanine tRNA.

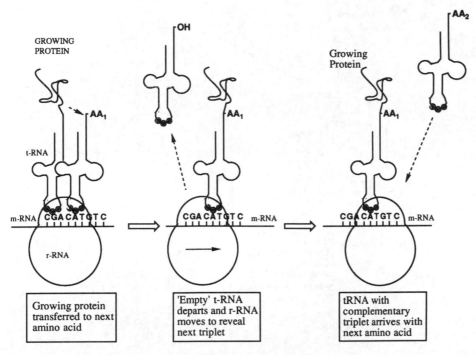

Fig. 6.19 Protein synthesis.

Messenger RNA is responsible for relaying the code for one particular protein from the DNA genetic bank to the protein 'production site'. The segment of DNA required is copied by a process called transcription. The DNA double helix unravels and the stretch which is exposed acts as a template on which the mRNA can be built (Fig. 6.17). Once complete, the mRNA leaves to seek out rRNA, while the DNA reforms the double helix.

Ribosomal RNA can be looked upon as the production site for protein synthesis. It binds to one end of the mRNA molecule, then travels along it to the other end, reading the code and constructing the protein molecule one amino acid at a time as it moves along (Fig. 6.19). There are two segments to the rRNA known as the 50S and 30S subunits.

Transfer RNA is the crucial adaptor unit which links the triplet code on mRNA to a specific amino acid. Therefore, there has to be a different tRNA for each amino acid. All the tRNAs are clover-leaf in shape with two different binding sites at opposite ends of the molecule (Fig. 6.18). One binding site is for a specific amino acid where the amino acid is covalently linked to the terminal adenosyl residue. The other is a set of three nucleic acid bases (anticodon) which will base pair with a complementary triplet on the mRNA molecule.

As rRNA travels along mRNA it reveals the triplet codes on mRNA (Fig. 6.19). As

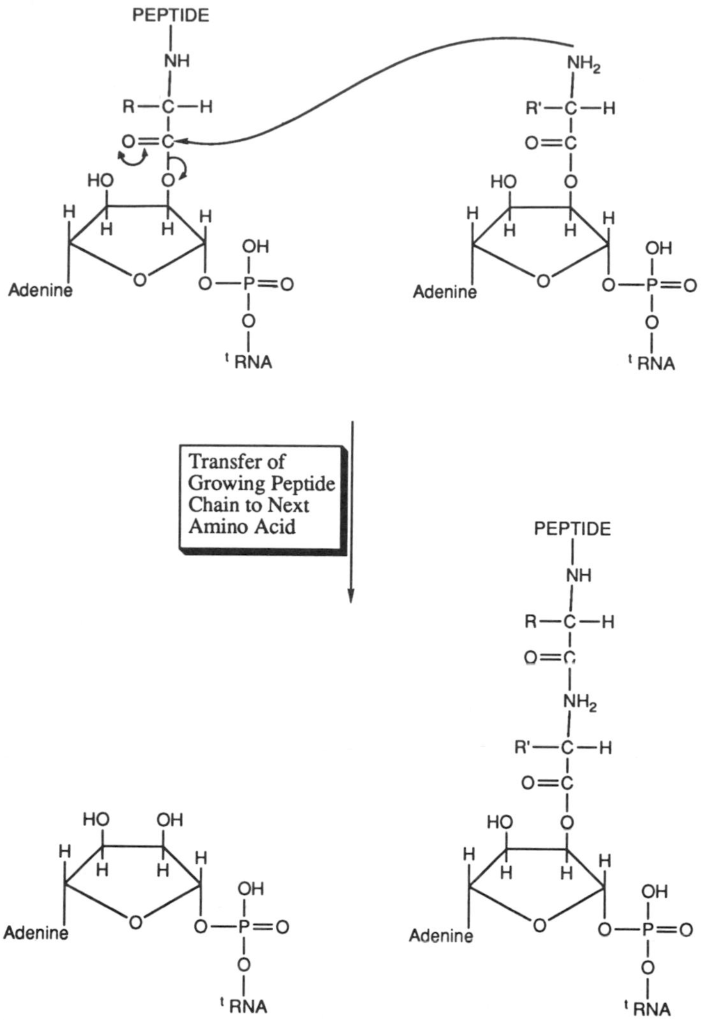

Fig. 6.20 Mechanism of transfer.

a triplet code is revealed (e.g. CAT), a tRNA with the complementary GTA triplet will bind to it and bring the specific amino acid coded by that triplet. The growing peptide chain will then be grafted on to that amino acid (Fig. 6.20). The rRNA will shift along the chain to reveal the next triplet and so the process continues until the whole strand is read. The new protein is then released from rRNA which is now available to start the process (translation) again.

6.4 *Drugs acting on RNA*

Several antibiotic agents are capable of acting on RNA molecules and interfering with transcription and translation. These are discussed in Chapter 10.

6.5 *Summary*

In this chapter we have considered drugs which act on transcription, translation, and replication by acting *directly* on DNA and RNA. There are other drugs (e.g. nalidixic acid) which affect these processes, but since these drugs work by inhibiting enzymes rather than by a direct interaction with DNA or RNA, they have not been mentioned here.

7 ▪ Drug development

For several thousand years, man has used herbs and potions as medicines, but it is only since the mid-nineteenth century that serious efforts were made to isolate and purify the active principles of these remedies. Since then, a large variety of biologically active compounds have been obtained and their structures determined (e.g. morphine from opium, cocaine from coca leaves, quinine from the bark of the cinchona tree).

These natural products became the lead compounds for a major synthetic effort where chemists made literally thousands of analogues in an attempt to improve on what Nature had provided. The vast majority of this work was carried out with no real design or reason, but out of the results came an appreciation of certain tactics which generally worked. A pattern for drug development evolved. This chapter attempts to show what that pattern is and the useful tactics which can be employed for developing drugs.

Nowadays, the development of a novel drug from natural sources might follow the following pattern.

- Screening of natural compounds for biological activity.
- Isolation and purification of the active principle.
- Determination of structure.
- Structure–activity relationships (SARs).
- Synthesis of analogues.
- Receptor theories.
- Design and synthesis of novel drug structures.

7.1 *Screening of natural products*

The screening of natural products became highly popular following the discovery of penicillin from a mould. Plants, fungi, and bacterial strains were collected from all round the world in an effort to find other metabolites with useful biological activities. This led in particular to an impressive arsenal of antibacterial agents (Chapter 9).

Screening of natural products from plant and microbial sources continues today in the never-ending quest to find new lead compounds. In recent years, organisms from marine sources have given novel compounds with interesting biological activity and this is a field likely to expand.

7.2 *Isolation and purification*

The ease with which the active principle can be isolated and purified depends very much on the structure, stability, and quantity of the compound.

Penicillin proved a difficult compound to isolate and purify. Although Fleming recognized the antibiotic qualities of penicillin and its remarkable non-toxic nature to man, he disregarded it as a useful drug since it appeared too unstable. He could isolate it in solution, but whenever he tried to remove the solvent, the drug was destroyed. Now that we know the structure of penicillin (Chapter 9), its instability under the purification procedures of the day is understandable and it was not until the development of a new procedure called freeze-drying that a successful isolation of penicillin was achieved.

Other advances in isolation techniques have occurred since those days and in particular in the field of chromatography. There are now a variety of chromatographic techniques available to help in the isolation and purification of a natural product.

7.3 *Structure determination*

In the past, determining the structure of a new compound was a major hurdle to overcome. It is sometimes hard for present-day chemists to appreciate how difficult structure determinations were before the days of NMR and IR spectroscopy. A novel structure which may now take a week's work to determine would have provided two or three decades of work in the past. For example, the microanalysis of cholesterol was carried out in 1888 to get its molecular formula, but its chemical structure was not fully established until an X-ray crystallographic study was carried out in 1932.

Structures had to be degraded to simpler compounds, which were further degraded to recognizable fragments. From these scraps of evidence, possible structures were proposed, but the only sure way of proving the theory was to synthesize these structures and to compare their chemical and physical properties with those of the natural compound or its degradation products.

Today, structure determination is a relatively straightforward process and it is only when the natural product is obtained in minute quantities that a full synthesis is required to establish its structure.

In cases where there is not enough sample for an IR or NMR analysis, mass spectroscopy can be helpful. The fragmentation pattern can give useful clues about the structure, but it does not, however, prove the structure. A full synthesis is still required as final proof.

Vinblastine (Fig. 7.1), an alkaloid used against advanced teratomas and lymphomas, is an example of how complex the structures of natural products can be. However, analytical skills and instruments have advanced to such an extent that even this structure is rel-

Fig. 7.1 Vinblastine.

atively simple compared to some of the natural product structures being studied today.

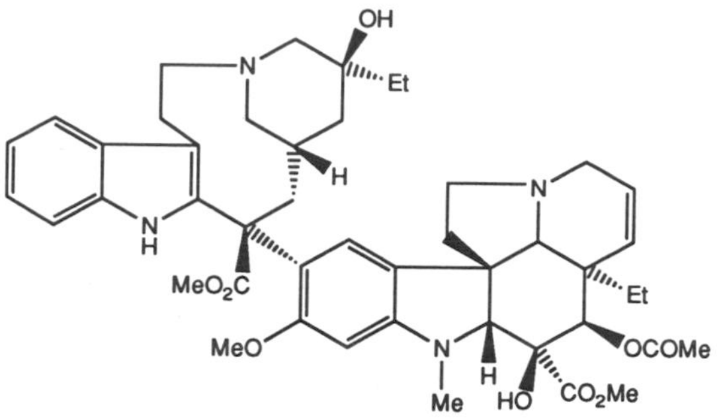

7.4 *Structure–activity relationships*

Once the structure of a biologically active compound is known, the medicinal chemist is ready to move on to study the structure–activity relationships of the compound.

The aim of such a study is to discover which parts of the molecule are important to biological activity and which are not. The chemist makes a selected number of compounds, which vary slightly from the original molecule, and studies what effect that has on the biological activity.

One could imagine the drug as a chemical knight entering the depths of a forest (the body) in order to make battle with an unseen dragon (the body's affliction) (Fig. 7.2). The knight (Sir Drugalot) is armed with a large variety of weapons and armour, but since his battle with the dragon goes unseen, it is impossible to tell which weapon he uses or whether his armour is essential to his survival. We only know of his success if he returns unscathed with the dragon slain. If the knight declines to reveal how he slew the dragon, then the only way to find out how he did it would be to remove some of his weapons and armour and to send him in against other dragons to see if he can still succeed.

As far as a drug is concerned, the weapons and armour are the various chemical functional groups present in the structure, which can bind to the receptor or enzyme. We have to be able to recognize these functional groups and determine which ones are important.

Let us imagine that we have isolated a natural product with the structure shown in Fig. 7.3. We shall name it Glipine. There are a variety of groups present in the structure and the diagram shows the potential bonding interactions which are possible with a receptor.

It is unlikely that all of these interactions take place, so we have to identify those which do. By synthesizing compounds (such as the examples shown in Fig. 7.4) where

Fig. 7.2

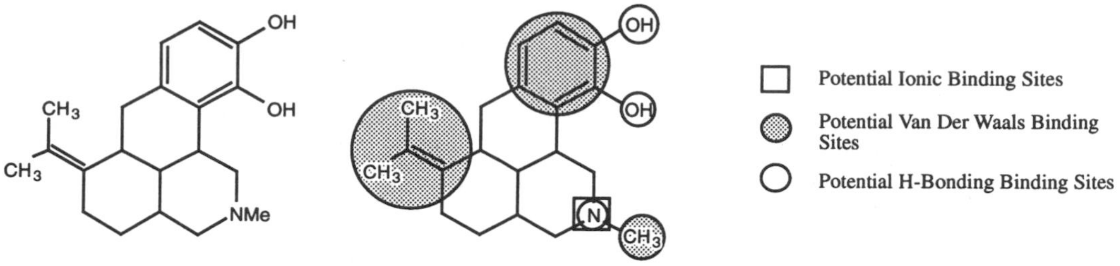

Fig. 7.3 Glipine.

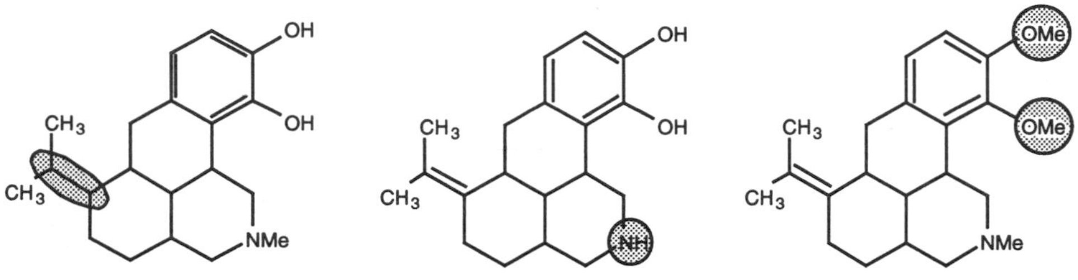

Fig. 7.4 Modifications of glipine.

one particular group of the molecule is removed or altered, it is possible to find out which groups are essential and which are not.

The ease with which this task can be carried out depends on how easily we can carry out the necessary chemical transformations to remove or alter the relevant group. For example, the importance or otherwise of an amine group is relatively easy to establish, whereas the importance of an aromatic ring might be more difficult. Hydroxyl groups, amino groups, and aromatic rings are particularly common binding groups in medicinal chemistry, so let us consider what analogues could be synthesized to establish whether they are involved or not.

7.4.1 The binding role of hydroxyl groups

Hydroxyl groups are commonly involved in hydrogen bonding. Converting such a group to a methyl ether or an ester is straightforward (Fig. 7.5) and will usually destroy or weaken such a bond.

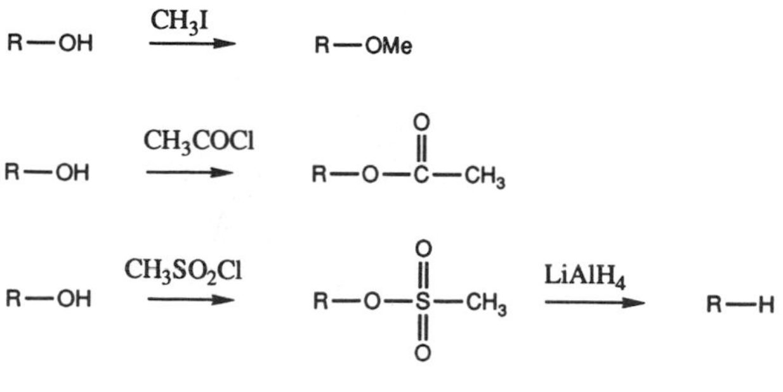

Fig. 7.5 Conversions of hydroxyl groups.

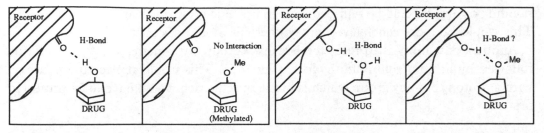

Fig. 7.6 Possible hydrogen bond interactions.

There are several possible explanations for this. The obvious explanation is that the proton of the hydroxyl group is involved in the hydrogen bond to the receptor and if it is removed, the hydrogen bond is lost (Fig. 7.6). However, suppose it is the oxygen atom which is hydrogen bonding to a suitable amino acid residue?

The oxygen is still present in the ether or the ester analogue, so could we really expect there to be any effect on hydrogen bonding? Well, yes we could. The hydrogen bonding may not be completely destroyed, but we could reasonably expect it to be weakened, especially in the case of an ester.

The reason is straightforward. When we consider the electronic properties of an ester compared to an alcohol, then we observe an important difference. The carboxyl group can 'pull' electrons from the neighbouring oxygen to give the resonance structure shown in Fig. 7.7 Since the lone pair is involved in such an interaction, it cannot take part so effectively in a hydrogen bond.

Steric factors also count against the hydrogen bond. The extra bulk of the acyl group will hinder the close approach which was previously attainable.

This steric hindrance also explains how a methyl ether could disrupt hydrogen bonding.

If there is still some doubt over whether a hydroxyl group is involved in hydrogen

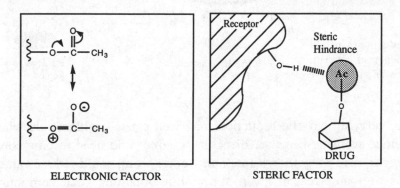

ELECTRONIC FACTOR   STERIC FACTOR

Fig. 7.7 Factors by which an ester group can disrupt hydrogen bonding.

bonding or not, it could be replaced with an isosteric group such as methyl (see later). This would be more conclusive, but synthesis is more difficult.

Another possibility is to react the hydroxyl group with methanesulfonyl chloride followed by lithium aluminum hydride (Fig. 7.5). This would replace the hydroxyl with a proton, but any group which is prone to reduction would have to be protected first.

7.4.2 The binding role of amino groups

Amines may be involved in hydrogen bonding or ionic bonding, but the latter is more common. The same strategy used for hydroxyl groups works here too. Converting the amine to an amide will prevent the nitrogen's lone pair taking part in hydrogen bonding or taking up a proton to form an ion.

Tertiary amines have to be dealkylated first, before the amide can be made. Dealkylation is normally carried out with cyanogen bromide or a chloroformate such as vinyloxycarbonyl chloride (Fig. 7.8).

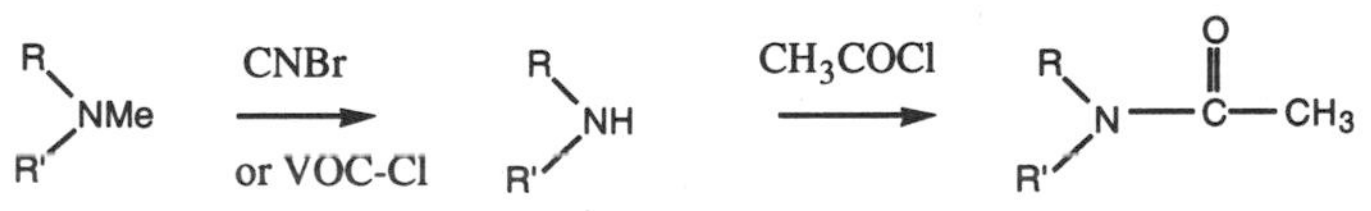

Fig. 7.8 Dealkylation of tertiary amines.

7.4.3 The binding role of aromatic rings

Aromatic rings are commonly involved in van der Waals interactions with flat hydrophobic regions of the binding site. If the ring is hydrogenated to a cyclohexane ring, the structure is no longer flat and interacts far less efficiently with the binding site (Fig. 7.9).

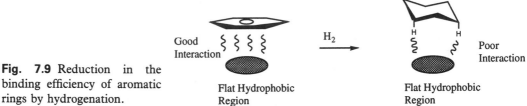

Fig. 7.9 Reduction in the binding efficiency of aromatic rings by hydrogenation.

However, carrying out the reduction may well cause problems elsewhere in the structure, since aromatic rings are difficult to reduce and need forcing conditions.

Replacing the ring altogether with a bulky alkyl group could reduce van der Waals bonding for the same reason given above, but obtaining such compounds could involve a major synthetic effort.

7.4.4 The binding role of double bonds

Unlike aromatic rings, double bonds are easy to reduce and this has a significant effect on the shape of that part of the molecule. The planar double bond is converted into a bulky alkyl group.

If the original alkene was involved in van der Waals bonding with a flat surface on the receptor, reduction should weaken that interaction, since the bulkier product is less able to approach the receptor surface (Fig. 7.10).

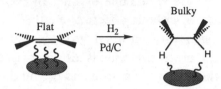

Fig. 7.10 The binding role of double bonds.

Once it is established which groups are important for a drug's activity, the medicinal chemist can move on to the next stage—the synthesis of analogues which still contain these essential features.

7.5 *Synthetic analogues*

Why is this stage necessary? If a natural compound such as our hypothetical Glipine has useful biological activity, why bother making synthetic analogues? The answer is that very few drugs are ideal. Many have serious side-effects and there is a great advantage in finding analogues which lack them. In general, the medicinal chemist is developing drugs with three objectives in mind:

- to increase activity
- to reduce side-effects
- to provide easy and efficient administration to the patient

Drug development in the past has mostly been a hit or a miss affair with a large number of compounds being synthesized at random. Luck has played a great part, but we can now recognize strategies which have evolved over the years:

- variation of substituents
- extension of the structure
- chain extensions/contractions
- ring expansions/contractions
- ring variations
- isosteres
- simplification of the structure
- rigidification of the structure

7.5.1 Variation of substituents

Once the essential groups for biological activity have been identified, substituents are varied since this is usually quite easy to do synthetically. The aim here is to fine tune the molecule and to optimize its activity. Biological activity may depend not only on how well the compound interacts with its receptor, but also on a whole range of physical features such as basicity, lipophilicity, electronic distribution, and size (see Chapter 8). The idea of varying substituents is to attach a series of substituents such that these physical features are varied one by one. In reality, it is rarely possible to change one physical feature without affecting another. For example, replacing a methyl group on a nitrogen with an ethyl group could affect the basicity of the nitrogen atom, but the size of the molecule is also increased. Either of these changes might have an effect on the activity of a drug and it would be difficult to know which was more important without more results.

The following are routine variations which can be carried out.

Alkyl substituents

If the molecule has an easily accessible functional group such as an alcohol, phenol, or amino group, then alkyl chains of various lengths and bulks such as methyl, ethyl, propyl, butyl, isopropyl, isobutyl or *tert*-butyl can be attached.

Different alkyl groups on a nitrogen atom may alter the basicity and/or lipophilicity of the drug and thus affect how strongly the drug binds to its binding site or how easily the compound crosses membrane barriers (see Chapter 8).

Larger alkyl groups, however, increase the bulk of the compound and this may confer selectivity on the drug. For example, in the case of a compound which interacts with two different receptors, a bulkier alkyl substituent may prevent the drug from binding to one of those receptors and so cut down side-effects (Fig. 7.11).

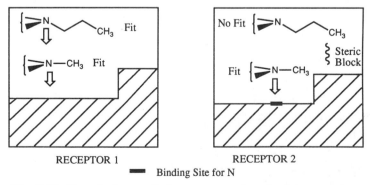

Fig. 7.11 Use of a larger alkyl group to confer selectivity on a drug.

Aromatic substitutions

A favourite approach for aromatic compounds is to vary the substitution pattern. This may give increased activity if the relevant binding groups are not already in the ideal positions for bonding (Fig. 7.12).

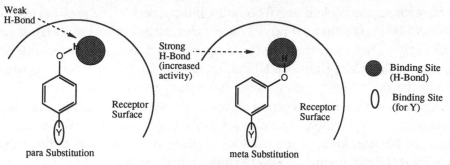

Fig. 7.12 Aromatic substitutions.

Electronic effects may also be involved. For example, an electron withdrawing nitro group will affect the basicity of an aromatic amine more significantly if it is in the *para* position rather than the *meta* position (Fig. 7.13). We have noted already that varying the basicity of a nitrogen atom may have a biological effect.

If the substitution pattern is ideal, then we can try varying the substituents themselves. Substituents of different sizes and electronic properties are usually tried to see if steric and electronic factors have any effect on activity. It may be that activity is improved by having a more electron withdrawing substituent, in which case a nitro substituent might be tried in place of a chloro substituent.

The chemistry involved in these procedures is usually straightforward and so these analogues are made as a matter of course whenever a novel drug structure is dis-

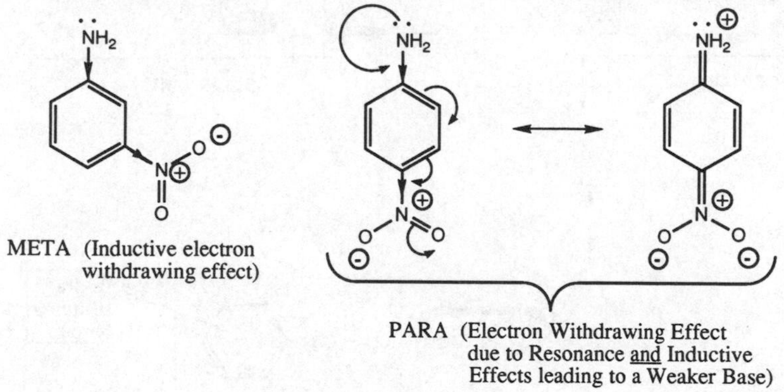

META (Inductive electron withdrawing effect)

PARA (Electron Withdrawing Effect due to Resonance and Inductive Effects leading to a Weaker Base)

Fig. 7.13 Electronic effects of aromatic substitutions.

covered or developed. Furthermore, the variation of aromatic or aliphatic substituents is open to quantitative structure–activity studies (QSARs) as described in Chapter 9.

7.5.2 Extension of the structure

This strategy has been used successfully on natural products such as morphine. It might seem strange that a natural product which is produced in a plant or a fungus should have important biological effects in the human body. One possible explanation for this could be that the natural product is present in the body as well. However, this seems unlikely. Therefore, we have to conclude that it is a happy coincidence that morphine has the right shape and binding groups to interact with a painkilling receptor in the body. This leads to some interesting conclusions.

Since there is a painkilling receptor in the body we have to accept that there is a neurotransmitter (or hormone) which switches on that receptor. We already know that it cannot be morphine, so the painkilling molecule has to have a different shape and possibly different binding groups. Assuming that the body's own painkiller is the ideal molecule for its receptor, then we must also conclude that morphine is not the ideal molecule. For example, it is perfectly possible that the natural painkiller has four important binding interactions with its receptor, whereas morphine has only three (Fig. 7.14). Therefore, why not add binding groups to the morphine skeleton to search for that fourth binding site? This tactic has been employed successfully to produce compounds such as the phenethyl analogue of morphine which has 14 times greater activity. Such a result suggests that the extra binding site on the receptor is hydrophobic, interacting with the aromatic ring by van der Waals interactions.

Frequently, this extension tactic has resulted in a compound which acts as an antagonist rather than as an agonist. In this case, the extra binding site is not one used

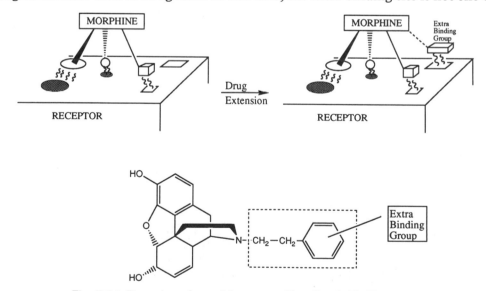

Fig. 7.14 Extension of morphine to provide a fourth binding group.

by the natural agonist or substrate. The binding interaction is different and no biological response results.

7.5.3 Chain extensions/contractions

Some drugs have two important binding groups linked together by a chain. Many of the natural neurotransmitters are like this. It is possible that the chain length is not ideal for the best interaction. Therefore, shortening or lengthening the chain length is a useful tactic to try (Fig. 7.15).

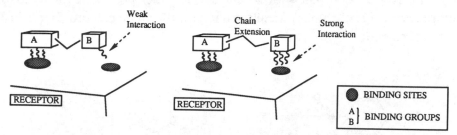

Fig. 7.15 Chain extension.

7.5.4 Ring expansions/contractions

If a drug has a ring, it is generally worth synthesizing analogues where one of these rings is expanded or contracted by one unit. The principle behind this approach is much the same as varying the substitution pattern of an aromatic ring. Expanding or contracting the ring puts the binding groups in different positions relative to each other and, with luck, may lead to better interactions with the binding site (Fig. 7.16).

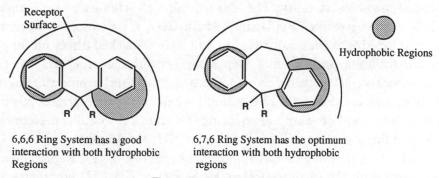

6,6,6 Ring System has a good interaction with both hydrophobic Regions

6,7,6 Ring System has the optimum interaction with both hydrophobic regions

Fig. 7.16 Ring expansion.

7.5.5 Ring variations

A further popular approach for compounds containing an aromatic ring is to try replacing the aromatic ring with a range of heteroaromatic rings of different ring size and heteroatom positions. Admittedly, a lot of these changes are merely ways of avoiding patent restrictions and do not result in significant improvements, but sometimes there are significant advantages in changing a ring system.

One of the major advances in the development of the selective beta blockers was the replacement of the aromatic ring in adrenaline with a naphthalene ring system (pronethalol) (Fig. 7.17). This resulted in a compound which was able to distinguish between two very similar receptors, the alpha and beta receptors for adrenaline. One possible explanation for this could be that the beta receptor has a larger van der Waals binding area for the aromatic system than the alpha receptor and can interact more strongly with pronethalol than with adrenaline. Another possible explanation is that the naphthalene ring system is sterically too big for the alpha receptor but is just right for the beta receptor.

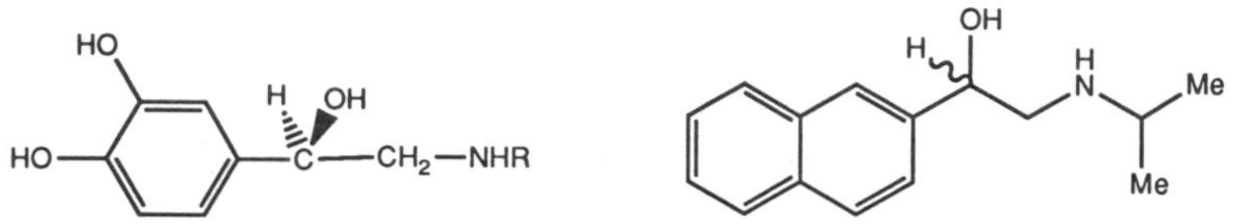

R = Me   ADRENALINE
R = H   NORADRENALINE

PRONETHALOL

Fig. 7.17 Ring variation of adrenaline.

7.5.6 Isosteres

Isosteres are atoms or groups of atoms which have the same valency (or number of outer shell electrons). For example, SH, NH_2, and CH_3 are isosteres of OH, while S, NH, and CH_2 are isosteres of O. Isosteres have often been used to design an inhibitor or to increase metabolic stability. The idea is to alter the character of the molecule in as subtle a way as possible. Replacing O with CH_2, for example, will make little difference to the size of the analogue, but will have a marked effect on its polarity, electronic distribution, and bonding. Replacing OH with the larger SH may not have such an influence on the electronic character, but steric factors become more significant.

Isosteric groups could be used to determine whether a particular group is involved in hydrogen bonding. For example, replacing OH and CH_3 would completely destroy hydrogen bonding, whereas replacing OH with NH_2 would not.

The beta blocker propranolol has an ether linkage (Fig. 7.18). Replacement of the OCH_2 segment with the isosteres CH=CH, SCH_2, or CH_2CH_2 eliminates activity, whereas replacement with $NHCH_2$ retains activity (though reduced). These results show that the ether oxygen is important to the activity of the drug and suggests that it is involved in hydrogen bonding with the receptor.

Replacing the methyl of a methyl ester group with NH_2 has been a useful tactic in stabilizing esters which are susceptible to enzymatic hydrolysis (Fig. 7.19). The NH_2

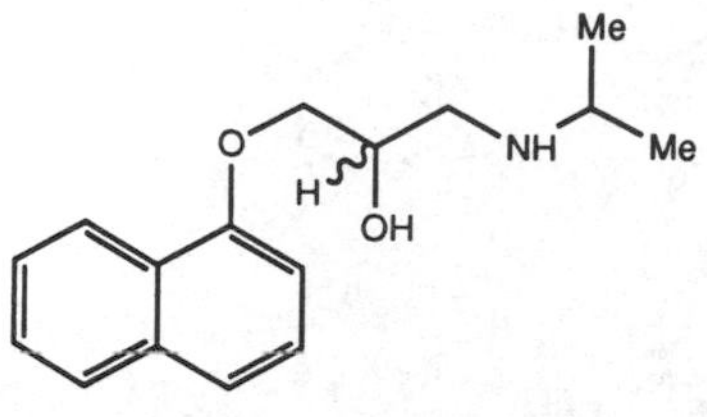

Fig. 7.18 Propranolol.

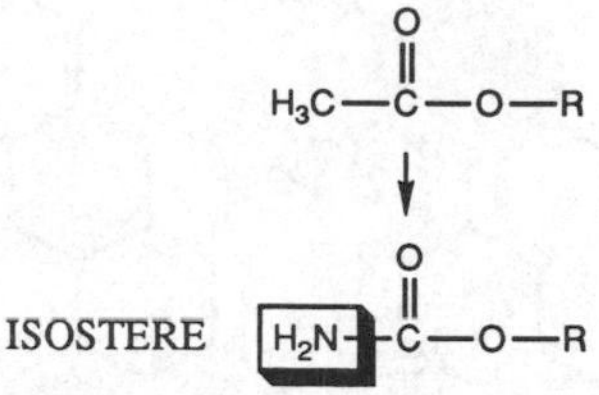

Fig. 7.19 Isosteric replacement of a methyl with an amino group.

group is the same size as the methyl and therefore has no steric effect. However, it has totally different electronic properties, and as such can feed electrons into the carboxyl group and stabilize it from hydrolysis (see Chapter 11).

Although fluorine does not have the same valency as hydrogen, it is often considered an isostere of that atom since it is virtually the same size. Replacement of a hydrogen atom with a fluorine atom will therefore have little steric effect, but since the fluorine is strongly electronegative, the electronic effect may be dramatic.

The use of fluorine as an isostere for hydrogen has been highly successful in recent years. One example is the antitumour drug 5-fluorouracil described in Section 4.5.3. The drug is accepted by the target enzyme since it appears little different from the normal substrate (uracil). However, the mechanism of the enzyme-catalysed reaction is totally disrupted. Fluorine has replaced a hydrogen atom which must be lost as a proton during the mechanism. There is no chance of fluorine departing as a positively charged species.

7.5.7 Simplification of the structure

If the essential groups of a drug have been identified, then by implication, it might be possible to discard non-essential parts of the structure without losing activity. The advantage would be in gaining a far simpler compound which would be much easier and cheaper to synthesize in the laboratory. For example, let us consider our hypothetical natural product Glipine (Fig. 7.20). The essential groups have been marked and so we might aim to synthesize compounds such as those shown in Fig. 7.20. These have simpler structures, but still retain the groups which we have identified as being essential.

This tactic was used successfully with the alkaloid cocaine (Fig. 7.21). It was well known that cocaine had local anaesthetic properties and it was hoped to develop a local anaesthetic based on a simplified structure of cocaine which could be easily synthesized in the laboratory. Success resulted with the discovery of procaine (or Novocaine) in 1909.

However, there is a trade-off involved when simplifying molecules. The advantage in obtaining simpler compounds may be outweighed by the disadvantage of increased

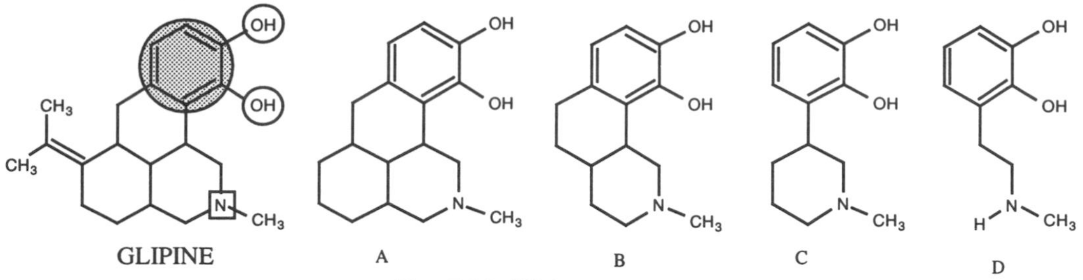

Fig. 7.20 Glipine analogues.

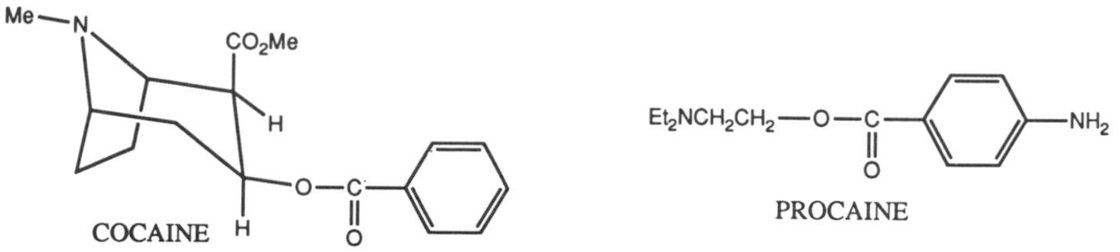

Fig. 7.21 Cocaine and procaine.

side-effects and reduced selectivity. We shall see below how these undesirable properties can creep in with simpler molecules and why the opposite tactic of rigidification can be just as useful as that of simplification.

7.5.8 Rigidification of the structure

Rigidification has been a popular tactic used to increase the activity of a drug or to reduce its side-effects. In order to understand why, let us consider again our hypothetical neurotransmitter from Chapter 5 (Fig. 7.22). This is quite a simple molecule and highly flexible. Bond rotation can lead to a large number of conformations or shapes. However, as seen from the receptor/messenger interaction, conformation I is the conformation accepted by the receptor. Other conformations such as II have the ionized amino group too far away from the anionic centre to interact efficiently and so this is an inactive conformation for our model receptor site. However, it is quite possible that there exists a different receptor which is capable of binding conformation II. If this is the case, then our model neurotransmitter could switch on two different receptors and give two different biological responses.

The body's own neurotransmitters are highly flexible molecules (Chapter 5), but fortunately the body is quite efficient at releasing them close to their target receptors, then quickly inactivating them so that they do not make the journey to other receptors. However, this is not the case for drugs. They have to be sturdy enough to travel through the body and consequently will interact with all the receptors which are prepared to accept them. The more flexible a drug molecule is, the more likely it will

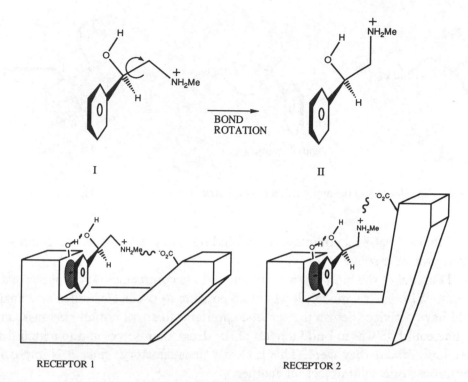

Fig. 7.22 Two conformations of a neurotransmitter which are capable of binding with different receptors.

interact with more than one receptor and produce other biological responses (side-effects).

The strategy of rigidification is to 'lock' the drug molecule into a more rigid conformation such that it cannot take up these other shapes or conformations. Consequently, other receptor interactions and side-effects are eliminated. This same strategy should also increase activity since, by locking the drug into the active conformation, the drug is ready to fit its target receptor site more readily and does not need to 'find' the correct conformation. Incorporating the skeleton of a flexible drug into a ring is the usual way of 'locking' a conformation and so for our model compound the analogue shown in Fig. 7.23 would be suitably rigid.

The sedative etorphine (Fig. 7.24) was designed by this approach (Chapter 12).

7.6 *Receptor theories*

The synthesis of a large number of analogues not only gives compounds with improved activity and reduced side-effects, but can also give information about the protein with which the drugs interact. Clearly, if a drug has an important binding

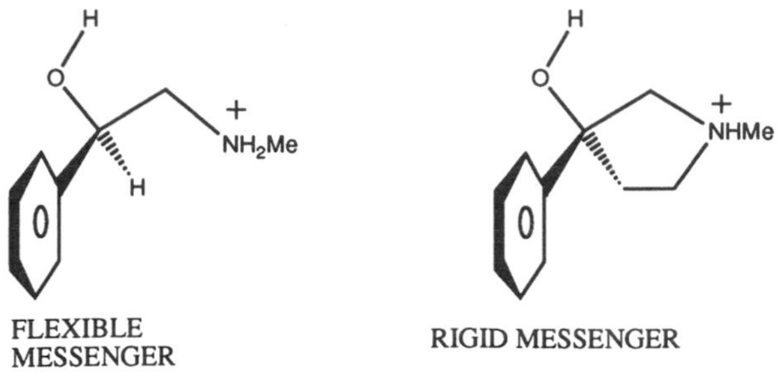

Fig. 7.23 Rigidification of the hypothetical neurotransmitter.

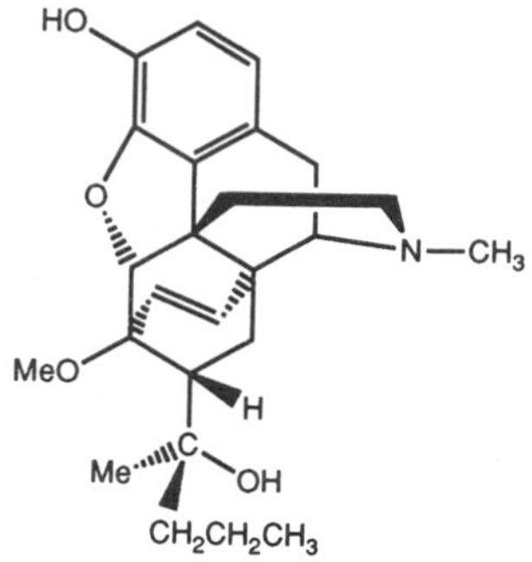

Fig. 7.24 Etorphine.

group, there must be a complementary binding group present in the binding site of the receptor or enzyme.

A 3D model of the binding site containing these complementary groups could then be built. With such a model, it would be possible to predict whether new analogues would have activity. Before the age of computers, this was not an easy task, and the best one could do was to build models of the drugs themselves and to match them up to see how similar they were. This is clearly unsatisfactory, since it is impossible to superimpose one solid object on another.

The introduction of computer graphics changed all that and revolutionized the field of medicinal chemistry such that the goal of rational, scientific drug design is now feasible.

At the simplest level, the computer can be used to compare drugs and to see how similar they are. The structures are built on the screen and displayed in a form suited to the operator. They can be viewed as ball-and-stick models or as space-filled models. Each model can be colour coded, then superimposed on each other to see how well they match up. The operator can then rotate the superimposed models to study them from different angles and then accurately measure important distances and angles. By doing this, the 'pharmacophore' can be determined for a particular group of drugs. The pharmacophore is the relative position of the important groups and ignores the molecular skeleton which holds them in position. The skeleton is considered irrelevant unless it sterically prevents the molecule from fitting the receptor site. In theory, all drugs acting in the same way and on the same receptor/enzyme should have the same pharmacophore.

However, this approach is not very satisfactory especially when dealing with flexible molecules which can take up various conformations. The computer is clever, but it is not a miracle worker. It has no way of knowing which conformation the drugs will adopt to fit the binding site, and so the structures which are built and energy-minimized on the screen will be constructed such they have the most stable conformation, not necessarily the most active conformation.

One way round this is to compare drugs against a rigid molecule (usually an antagonist) which binds to the receptor site and contains the pharmacophore within its skeleton. It is then assumed that more flexible drugs take up the conformation which most closely matches this pharmacophore.

However, the computer is capable of more powerful things.

An X-ray structure of a receptor or protein is of enormous benefit to the whole modelling process, since the X-ray data can be fed into the computer to build a 3D space-filled model of the macromolecule.

Unfortunately, this does not reveal where the active site or binding site is and so it is better to get an X-ray structure of the macromolecule with an antagonist irreversibly bound to the binding site. The antagonist then acts as a label for the binding site.

Once the protein is on the computer screen, the computer operator can peel away layers until a cross-sectional view of the binding site is revealed.

Drugs both known and planned can then be introduced to see whether they fit. Substituents and conformations can be changed at will to see whether or not the fit is improved. By doing each studies, it is possible to rule out a number of proposed new structures and so cut down the amount of synthetic work to be done in the laboratory. For it is important to realize that computers, powerful though they may be, are unlikely to replace the bench chemist. Compounds still have to be made to test the theories.

The computer can be used to study molecular features which cannot be studied in any other way.

For example, electron densities can be calculated throughout the molecule and displayed as numbers or by colour coding. This will show areas of high and low electron density in the molecule and can be compared with similar areas in the binding site. This is a far more scientific approach to the study of ionic interactions than just comparing a positively charged nitrogen atom bound to a negatively charged carboxylate anion, where both species are fixed at one point in space.

These quantum mechanical calculations have led to rather startling results which challenge the chemist's traditional viewpoint of charged molecules. For example, the calculated electronic distribution for ionized histamine is shown in Fig. 7.25. The numbers in each sphere represent the overall charge of that sphere (i.e. electronic charge minus nuclear charge). In this calculation, the electron density outwith these spheres is not included, which is why the numbers do not add up to $+1$. This is not important. What is important are the relative values of the charge throughout the molecule.

Conventionally, the chemist would consider the positive charge to be localized on the nitrogen atom. However, the calculations above show that the charge has been spread over the molecule and that the nitrogen atom is no more positively charged than neighbouring carbon atoms. This has important consequences to the way we

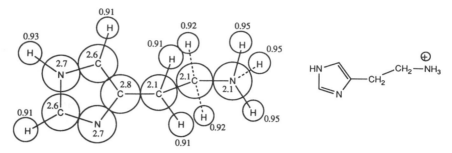

Fig. 7.25 Histamine.

think of ionic interactions between drugs and receptors. It implies that charged areas both on the receptor and the drug are more diffuse than originally thought. This suggests that we have wider scope in designing novel drugs. For example, in the classical viewpoint of charge distribution, a certain molecule might be considered to have its charged centre too far away from the corresponding 'centre' in the receptor binding site. If these charged areas are actually more diffuse, then this assumption is false (Fig. 7.26).

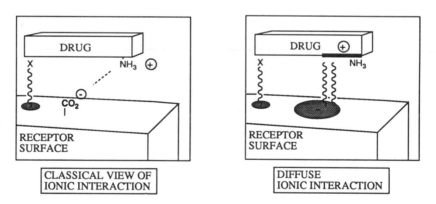

Fig. 7.26 Ionic interactions.

It is worth pointing out, however, that the above calculations were carried out on a histamine structure in isolation from its environment. In the body, histamine is in an aqueous environment and would be surrounded by water molecules which would solvate the charge and consequently have an effect on charge distribution.

If an X-ray structure of a protein is not available, we are thrown back to designing a theoretical binding site based on the structures of the compounds which interact with it. This, of course, introduces the uncertainty of which conformations are adopted by flexible molecules. One cannot assume that the most stable conformation is adopted, since the energy gained from bonding interactions between protein and drug may be sufficient to force the drug into a less favoured conformation. This complicates the mapping of receptor sites, since it means that all reasonable conformations of the drug have to be considered.

Despite this, programmes have been designed to try and solve these problems.

For example, by considering the various conformations of a range of active compounds, it is possible to find the common volume or space into which the active compounds can be placed in order to interact with the receptor. These 'spaces' can then be compared with 'no go areas'—areas which must not be occupied. These areas are determined by considering similar structures which do not fit the binding site. By subtracting one from the other, a more accurate shape or volume is obtained, which, in turn, can reveal which conformations are not permitted for active drugs. Clearly, the more drugs studied, the more accurate the picture.

These achievements have now resulted in a new phase of drug development whereby the chemist can design completely new structures totally unrelated to the structure of the original compounds. Put simply, the chemist can ensure that the important groupings are still present, but design a totally new framework to hold them in the correct positions. Rather than research being directed by the original drug, research is now directed by the receptor which the scientist wants to influence.

A warning! It is important to appreciate that the computer graphic studies described above are tackling only one part of a much bigger problem—the design of an *effective* drug. True, one could design the perfect compound for a particular enzyme or receptor, but if the compound never reaches the target protein in the body, it is a useless drug. The various hurdles and obstacles which a drug has to overcome are described in Chapter 8, but for the moment, we should appreciate that these hurdles are many and varied, and that there is no way of predicting whether a particular compound will win through to its target site.

Therefore, finding the ideal drug for a protein target is still only part of the overall battle. A great deal of fine-tuning still requires to be done, and the compound with the most effective interactions at the protein target will not necessarily be the most effective drug.

7.7 *The elements of luck and inspiration*

It is true to say that drug design is becoming more rational, but it has not yet eliminated the role of chance or the need for hard-working, mentally alert bench chemists.

The vast majority of drugs still on the market were developed by a mixture of rational design, trial and error, hard graft, and pure luck. The drugs which were achieved by purely rational design are severely limited and are exemplified by cimetidine (Chapter 13) and pralidoxime (Chapter 11).

Frequently, the development of drugs has been based on 'ringing the changes', watching the literature to see what works on related compounds and what doesn't, then trying out similar alterations to one's own work. It is very much a case of groping in the dark, with the chemist asking whether adding a group at a certain position will have a steric effect, an electronic effect, or a bonding effect.

Even when drug design is followed on rational lines, good fortune often has a role to play.

The development of the beta blocker propranolol (Fig. 7.27) was aided by such a slice of good fortune. Chemists at ICI were trying to improve on a drug called pronethalol (Fig. 7.17), the first beta blocker to reach the market. It was known that the naphthalene ring and the ethanolamine segment were important both to activity and selectivity, so these groups had to be retained. Therefore, it was decided to study what would happen if the distance between these two groups was extended. Perhaps by doing so, the two groups would interact with their respective binding sites more efficiently.

Various segments were to be inserted, one of which was the OCH_2 moiety. The analogue which would have been obtained is shown as structure I in Fig. 7.27. β-Naphthol was the starting material, but was not immediately available, and rather than waste the day, α-naphthol was used instead. The result was propranolol which has proved a successful drug in the treatment of angina for many years.

When the original target was eventually made, it showed little improvement over the original compound—pronethalol.

A further interesting point concerning this work is that the propranolol skeleton had been synthesized some years earlier. However, the workers involved had not been searching for beta blocking activity and had not recognized the potential of the compound.

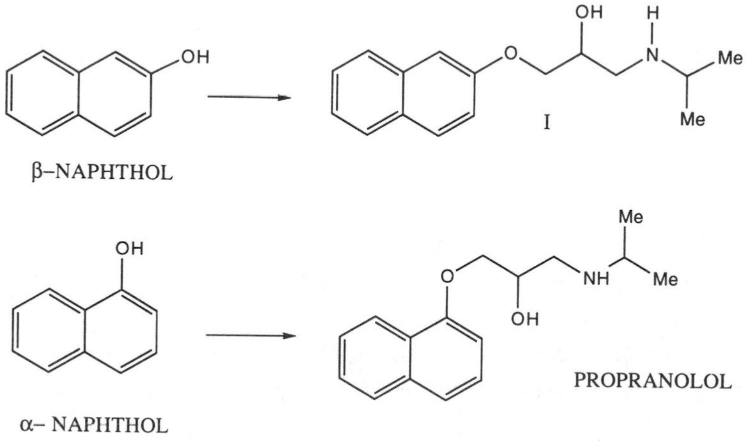

Fig. 7.27 The development of the beta-blocker propranolol.

This point emphasizes the importance of keeping an open mind, especially when testing for biological activity. Frequently, an analogue made in one field of medicinal chemistry is found to have an unexpected application in another field altogether.

Alternatively, work carried out in order to solve a specific problem in a certain field may produce a compound which solves a completely different problem in the same field. Once again, the beta blockers provide an illustration of this. We have seen how propranolol and its analogues are effective beta blocking drugs. However, they are also lipophilic, which means that they can enter the CNS (central nervous system) and cause side-effects. In an attempt to cut down this entry into the CNS, it was decided to add a hydrophilic amide group to the molecule and so inhibit passage through the blood brain barrier. One of the compounds made was practolol (Fig. 7.28). As expected, this compound had less CNS side-effects, but more importantly, it was found to be a selective antagonist for the beta receptors of the heart over beta receptors in other organs—a result that was highly desirable, but not the one that was being looked for at the time.

7.8 *Lead compounds*

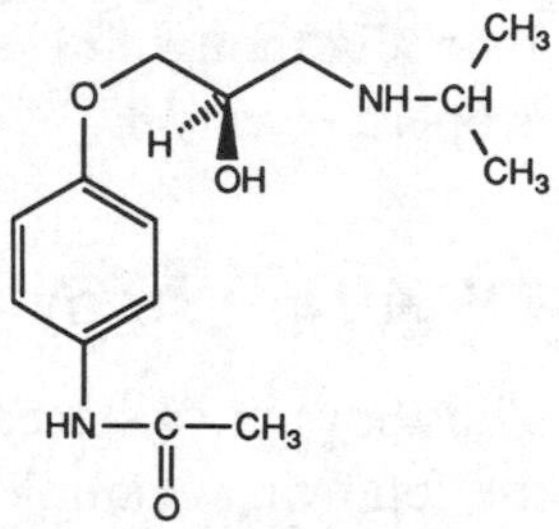

In order to design a drug with a particular biological activity, the medicinal chemist requires a lead compound—a compound which shows a useful pharmaceutical activity. The level of activity may not be very great and there may be undesirable side-effects, but the lead compound provides a start. By altering the structure using the strategies already mentioned, a useful drug may be developed with improved

Fig. 7.28 Practolol.

activity and reduced side-effects. Lead compounds are often found from natural sources or herbs used in traditional medicine, but these are not the only sources. Pharmaceutical companies routinely screen a large variety of novel compounds synthesized in industrial and academic laboratories. These compounds may be intermediates in a purely synthetic research study, but there is always the chance that they may have useful biological activity.

On occasions, the useful biological activity may only be a minor property or a side-effect of a compound. The aim then would be to enhance the side-effect and eliminate the major biological activity. The story of the antiulcer agent cimetidine (Fig. 7.29) is a case in point. The desired biological property was selective *antagonism* of histamine receptors in the stomach. The lead compound was a histamine *agonist* with a very weak antagonism for the receptors of interest.

In such situations, the medicinal chemist wants to alter the molecule such that the major biological activity is eliminated and the side-effect is boosted until it becomes the dominant effect. Once this has been achieved, the drug can be 'fine-tuned' as

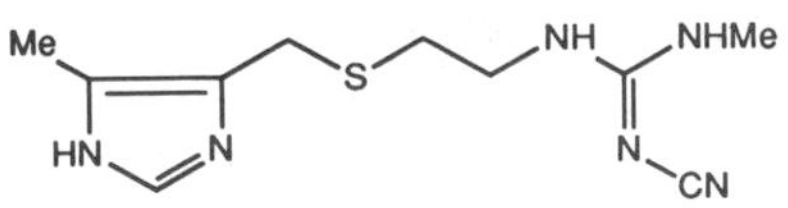

Fig. 7.29 Cimetidine.

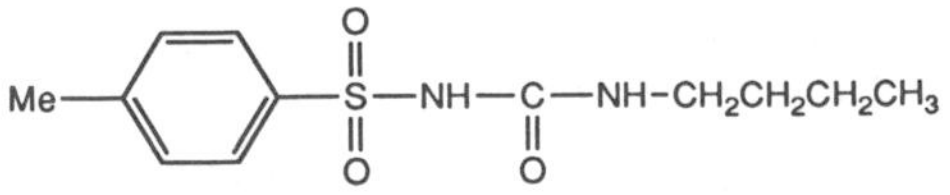

Fig. 7.30 Tolbutamide.

described above. Clearly, this is a more demanding objective, but the cimetidine story proves that it is attainable (Chapter 13).

A further example is provided by the antidiabetic agent tolbutamide (Fig. 7.30) which was developed from a sulfonamide structure. Most sulfonamides are used as antibacterial agents, but some proved unsatisfactory since they led to convulsions brought on by hypoglycaemia (low glucose levels in the blood). Structural alterations were made to eliminate the antibacterial activity and to enhance the hypoglycaemic activity and this led to tolbutamide.

The moral of the story is that an unsatisfactory drug in one field may provide the lead compound for another field. It is not a good idea to think of a structural group of compounds as having only one type of biological activity. The sulfonamides are generally thought of as antibacterial agents, but we have seen that they can have other properties as well.

7.9 *A case study—oxamniquine*

The above tactics need not necessarily be used in the order given. We shall see this in the following example—the development of oxamniquine (Fig. 7.31).

Oxamniquine is an important Third World drug used in the treatment of schistosomiasis (bilharzia). This disease affects an estimated two hundred million people and is contracted by swimming or wading in infected water. The disease is carried by a snail which can penetrate human skin and enter the blood supply. There, it produces eggs which become trapped in organs and tissues, and this in turn leads to the symptoms of the disease.

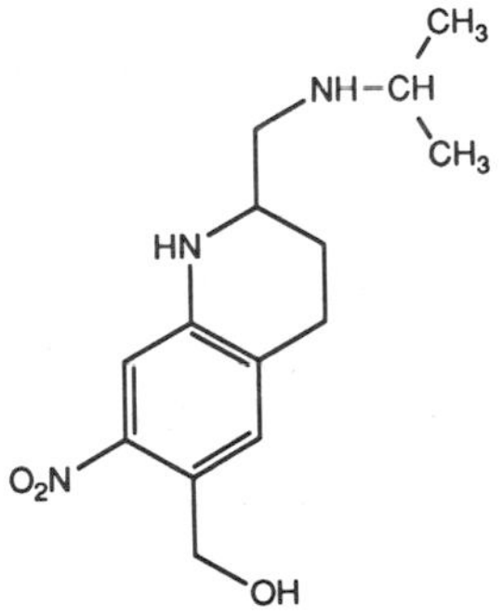

The first stage in the development of oxamniquine was to find a lead compound, and so a study was made of compounds which were active against the parasite. The tricyclic structure lucanthone (Fig. 7.32) was chosen. It was known to be effective against some forms of the disease, but it was

Fig. 7.31 Oxamniquine.

also toxic and had to be injected at regular intervals to remain effective. Therefore, the goal was to increase the activity of the drug, broaden its activity, reduce side-effects, and make it orally active.

Having found a lead compound, it was decided to try *simplifying* the structure to see

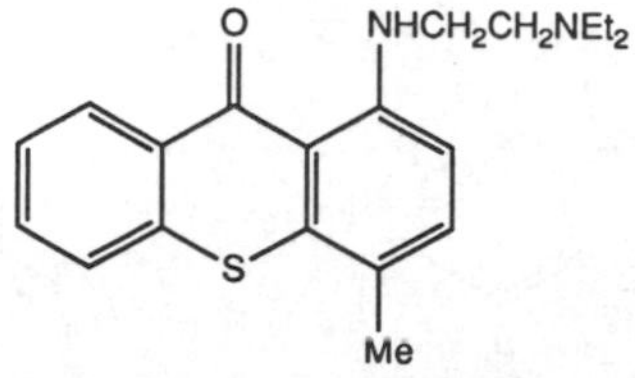

Fig. 7.32 Lucanthone.

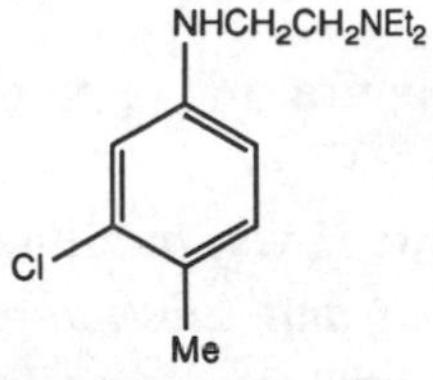

Fig. 7.33 Mirasan.

whether the tricyclic system was really necessary. Several compounds were made, and the most interesting structure was one where the two 'left-hand' rings had been removed.

This gave a compound called mirasan (Fig. 7.33) which retained the 'right-hand' aromatic ring containing the methyl and β-aminoethylamino side-chains *para* to each other. *Varying substituents* showed that an electronegative chloro substituent, positioned where the sulfur atom had been, was beneficial to activity. Mirasan was active against the bilharzia parasite in mice, but not in humans.

It was now reasoned that the β-aminoethylamino side-chain was important to receptor binding and would adopt a particular conformation in order to bind efficiently. This conformation would only be one of the many conformations which are available to a flexible molecule such as mirasan and so there would only be a limited chance of it being adopted at any one time.

Therefore, it was decided to try and restrict the number of possible conformations by incorporating the side-chain into a ring (*rigidification*). This would cut down the number of available conformations and increase the chance of the molecule having the correct conformation when it approached the receptor.

There was the risk, however, that the active conformation itself would be disallowed by this tactic. Therefore, rather than incorporate the whole side-chain into a ring structure, compounds were designed initially such that only portions of the chain were included.

The bicyclic structure (I) (Fig. 7.34) contains one of the side-chain bonds fixed in a ring to prevent rotation round that bond. It was found that this gave a dramatic improvement in activity. The compound was still not active in man, but unlike mirasan, it was active in monkeys. This gave hope that the chemists were on the right track. Further *rigidification* led to structure II (Fig. 7.34) where two of the side-chain bonds were constrained. This compound showed even more activity in mouse studies and it was decided to concentrate on this compound.

By now, it can be seen that the structure of the compound has been altered significantly from mirasan. In general, when a breakthrough has been achieved and a novel structure has been obtained, it is advisable to check whether past results still hold true. For example, does the chloro group still have to be ortho to the methyl? Can we change the chloro group for something else? Novel structures may fit the binding site slightly differently from the lead compound such that the binding groups are no longer in the optimum positions for binding.

Therefore, structure II was modified by *varying substituents and substitution patterns* on the aromatic ring, and by *varying alkyl substituents* on the amino groups. Chains were also *extended* to search for other possible binding sites.

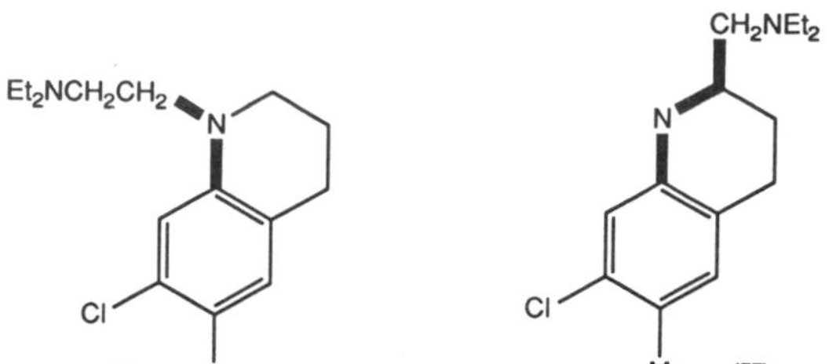

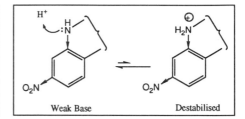

Fig. 7.34 Bicyclic structures I and II.

The results and possible conclusions were as follows.

- The substitution pattern on the aromatic ring could not be altered and was essential for activity. Altering the substitution pattern presumably places the essential binding groups out of position with respect to their binding sites.

- Replacing the chloro substituent with more electronegative substituents improved activity, with the nitro group being the best substituent. Therefore, an electron deficient aromatic ring is beneficial to activity. One possible explanation for this could be the effect of the neighbouring aromatic ring on the basicity of the nitrogen atom. A strongly electron deficient aromatic ring would 'pull' the cyclic nitrogen's lone pair of electrons into the ring, thus reducing its basicity (Fig. 7.35). This in turn might improve the pK_a of the drug such that it is less easily ionized and is able to pass through cell membranes more easily (see Chapter 8).

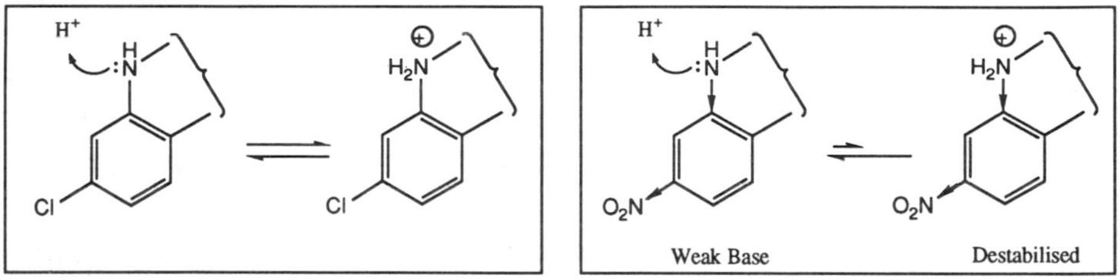

Fig. 7.35 Effect of aromatic substituents on pK_a.

- The best activities were found if the amino group on the side-chain was secondary rather than primary or tertiary (Fig. 7.36).

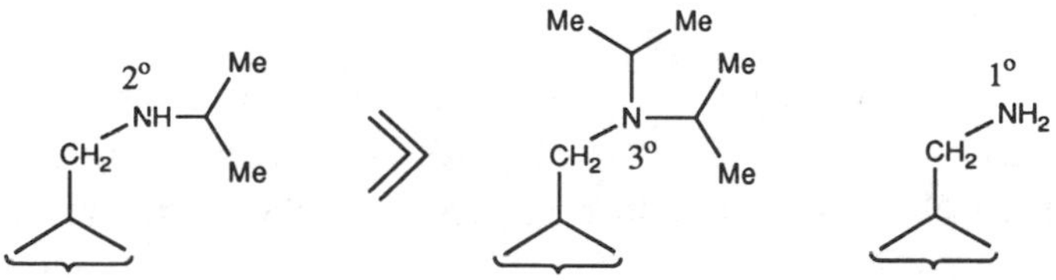

Fig. 7.36 Activity of secondary amino side chain.

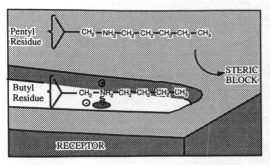 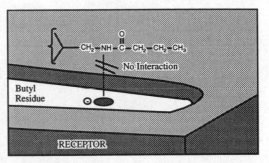

Fig. 7.37 Proposed ionic binding interaction.

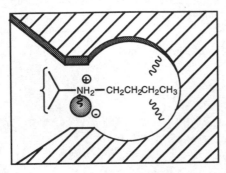 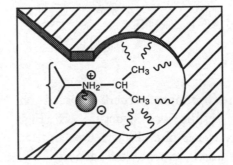

Fig. 7.38 Branching of the alkyl chain.

- The alkyl group on this nitrogen could be increased up to four carbon units with a corresponding increase in activity. Longer chains led to a reduction in activity. The latter result might imply that large substituents are too bulky and prevent the drug from binding to the binding site. Acyl groups eliminated activity altogether, emphasizing the importance of this nitrogen atom. Most likely, it is ionized and interacts with the receptor through an ionic bond (Fig. 7.37).

- Branching of the alkyl chain increased activity. A possible explanation could be that branching increases van der Waals interactions to a hydrophobic region of the binding site (Fig. 7.38). Alternatively, the lipophilicity of the drug might be increased, allowing easier passage through cell membranes.

- Putting a methyl group on the side-chain eliminated activity (Fig. 7.39). A methyl group is a bulky group compared with a proton and it is possible that it prevents the side-chain taking up the correct binding conformation.

- Extending the length of the side-chain by an extra methylene group eliminated activity (Fig. 7.40).

  This tactic was tried in case the binding groups were not far enough apart for optimum binding. This result suggests the opposite.

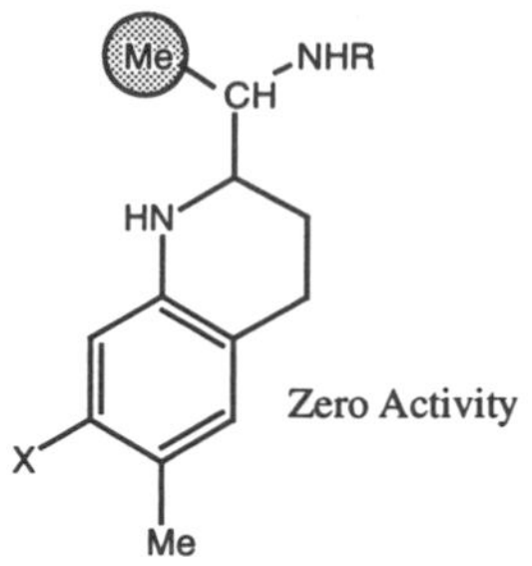

Fig. 7.39 Addition of a methyl group.

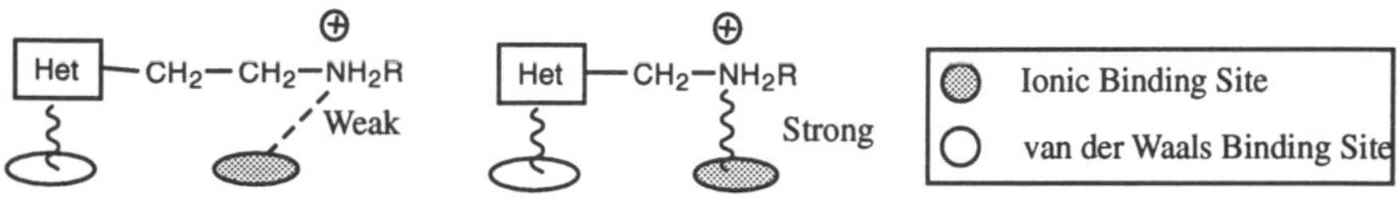

Fig. 7.40 Effect of extension of the side chain.

The optimum structure based on these results was structure III (Fig. 7.41). It has one chiral centre and, as one might expect, the activity was much greater in one enantiomer than it was in the other.

The tricyclic structure IV (Fig. 7.41) was also constructed. In this compound, the side-chain is fully incorporated into a ring structure, restricting the number of possible conformations drastically. As mentioned earlier, there was a risk that the active conformation

Fig. 7.41 The optimum structure (III) and the tricyclic structure (IV). *, Chiral centre.

would no longer be allowed, but in this case good activity was still obtained. The same variations as above were carried out to show that a secondary amine was essential and that an electronegative group on the aromatic ring was required. However, some conflicting results were obtained compared with the previous results for structure III. A chloro substituent on the aromatic ring was better than a nitro, and it could be in either of the two possible *ortho* positions relative to the methyl group.

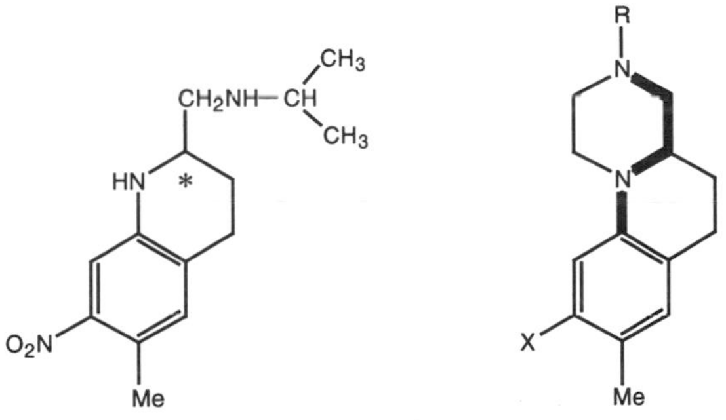

These results demonstrate that optimizing substituents in one structure does not necessarily mean that they will be optimum in a different skeleton.

One possible explanation for the chloro substituent being better than the nitro is that a less electronegative substituent is required to produce the optimum pK_a or basicity for membrane permeability.

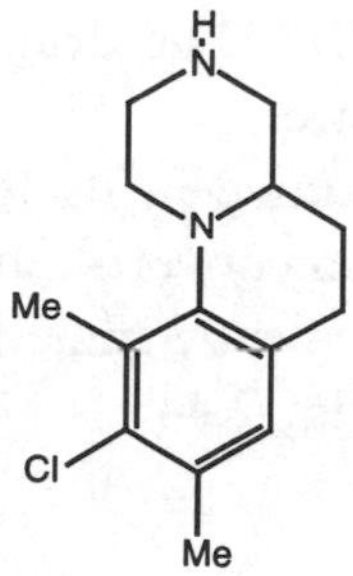

Fig. 7.42 Structure V.

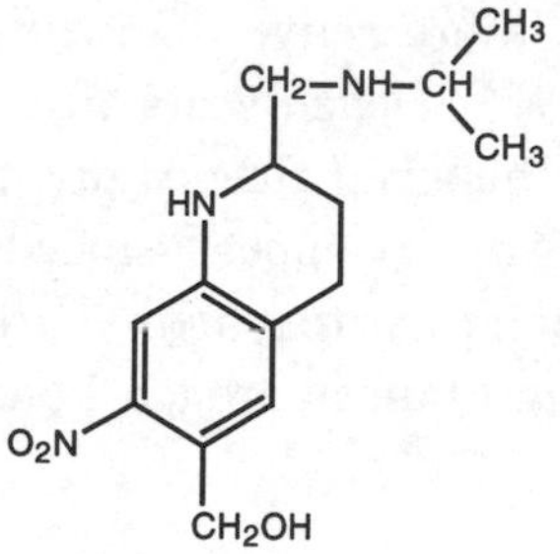

Fig. 7.43 Oxamniquine.

Adding a further methyl group on the aromatic ring to give the structure shown in Fig. 7.42 increased activity. It was proposed that the bulky methyl group was interacting with the piperazine ring and causing it to twist out of the plane of the other two rings. The increase in activity which resulted suggests that a better fitting conformation is obtained for the receptor.

This compound V (Fig. 7.42) was three times more active than structure III (Fig. 7.41). However, structure III was chosen for further development. The decision to choose III over V was based on preliminary toxicity results as well as the cost of producing the compounds. The cost of synthesizing III would be expected to be cheaper since it is a simpler molecule.

Further studies on the metabolism of related compounds then revealed that the methyl group on these compounds was oxidized in the body to a hydroxymethyl group and that this was in fact the active compound. The methyl group on III was replaced with a hydroxymethyl group to give oxamniquine (Fig. 7.43) which was

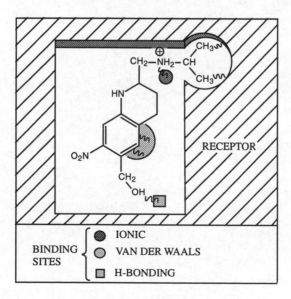

Fig. 7.44 Bonding interactions of oxamniquine.

BINDING SITES
- ● IONIC
- ◎ VAN DER WAALS
- ▦ H-BONDING

RECEPTOR

found to be more active than the methyl structure (III). The drug was put on the market in 1975, eleven years after the start of the project.

It is now believed that compound III is totally inactive in itself. This is not as surprising as it may appear, since the metabolic reaction converts a non-polar methyl group to a polar hydroxymethyl group. Presumably the newly gained hydroxyl group forms an important hydrogen bond to the receptor (Fig. 7.44).

8 · Pharmacodynamics

8.1 *Drug distribution and 'survival'*

In Chapter 7, we concentrated on the interaction of drugs with binding sites. However, the compound which has the best binding interaction with its receptor is not necessarily the best drug to use in medicine. There are other variables which have to be taken into account.

The drug has to be stable enough to survive a rather tortuous journey through the body's circulatory system. It also has to be capable of negotiating barriers put in its way and not be diverted from its target.

For example, consider a drug taken as a pill. It has to dissolve in aqueous solution. It has to survive the acid of the stomach, then be absorbed from the gastrointestinal tract into the bloodstream. To do that it has to negotiate barriers in the form of cell membranes. It has to survive the destructive tendencies of the liver and its enzymes. It has to survive the enzymes present in the blood. If it is a lipophilic drug, it may be taken up by fat tissue. If it is anionic, it may get bound by plasma protein and if it is cationic, it may be bound by nucleic acids. It has to avoid being excreted by the kidneys or the bile duct. If the drug is aimed at the brain, it has to cross another cell barrier known as the blood–brain barrier. If it is to react with an enzyme, it has to negotiate another cell membrane to reach that.

Only then will the drug interact with its receptor or enzyme. As far as the drug is concerned, it is a long, strenuous, and dangerous journey.

Many of these problems can be avoided by giving the drug as an intravenous or intramuscular injection, but clearly orally administered drugs are preferred by the patient and if at all possible, drug design aims at an orally active compound. Let us look again at the journey which has to be followed by an orally administered drug.

The success of the journey depends principally on the physical properties of the drug. First of all, it has to be chemically stable and not break down in the acid conditions of the stomach. Secondly, it has to be metabolically stable so that it survives the hydrolytic enzymes present in the digestive system, liver, and bloodstream. Thirdly, it has to have the correct balance of hydrophilic to hydrophobic character. Let us consider each of these factors in turn.

8.1.1 Chemical stability

There are several useful drugs with chemically labile functional groups. Penicillins have a chemically labile β-lactam ring which is susceptible to acid hydrolysis. Cholinergic agents have a susceptible ester group which is also susceptible to acid hydrolysis.

One way round the problem is to inject the drug in order to avoid the acid conditions of the stomach. However, there are strategies available which can be used to make the offending functional group less labile (see Section 8.3.2.).

8.1.2 Metabolic stability

Drugs are foreign substances[1] as far as the body is concerned and the body has its own method of getting rid of such chemical invaders. Non-specific enzymes (particularly in the liver) are able to add polar functional groups to a wide variety of drugs. Once the polar functional group has been added, the overall drug is more polar and water soluble, and is therefore more likely to be excreted when it passes through the kidneys.

An alternative set of non-specific enzymatic reactions can reveal 'masked' polar functional groups which might be present in a drug. For example, there are enzymes which can demethylate a methyl ether to reveal a more polar hydroxyl group. Once again, the more polar product (metabolite) is excreted more efficiently.

These reactions are classed as phase I reactions in the overall process of drug metabolism. They generally involve oxidation, reduction, and hydrolysis (Fig. 8.1).

The structures most prone to oxidation are N-methyl groups, aromatic rings, the terminal positions of alkyl chains, and the least hindered positions of alicyclic rings.

Nitro and carbonyl groups are prone to reduction by reductases, whilst amides and esters are prone to hydrolysis by esterases.

There is also a series of metabolic reactions classed as phase II reactions (Fig. 8.2). These are conjugation reactions whereby a polar molecule is attached to a suitable polar 'handle' which is either already present on the drug or has been placed there by a phase I reaction. The resulting conjugate has increased polarity, thus increasing its excretion rate in urine or bile even further.

Phenols, alcohols, and amines form O- or N-glucuronides by reaction with UDP-glucose such that the highly polar glucose molecule is attached to the drug.

Phenols, epoxides, and halides can react with the tripeptide glutathione to give mercapturic acids and some steroids can react with sulfates.

---

[1] Substances which are foreign to the particular biological system under study are known as xenobiotics. The word is derived from the Greek words 'xenos' meaning foreign and 'bios' meaning life.

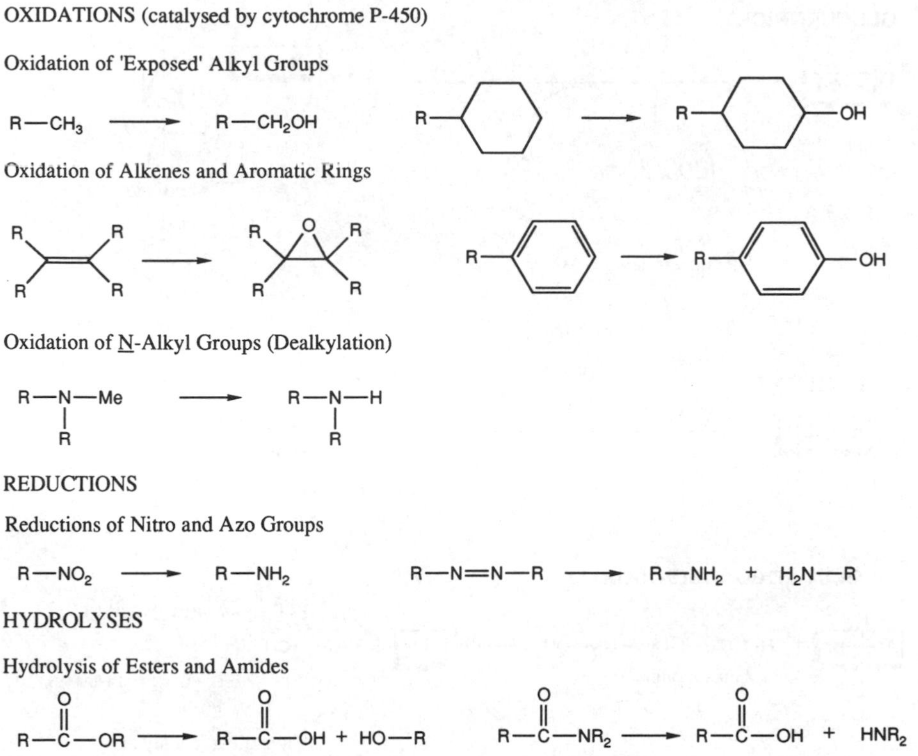

OXIDATIONS (catalysed by cytochrome P-450)

Oxidation of 'Exposed' Alkyl Groups

$R\text{—}CH_3 \longrightarrow R\text{—}CH_2OH$

Oxidation of Alkenes and Aromatic Rings

Oxidation of <u>N</u>-Alkyl Groups (Dealkylation)

REDUCTIONS

Reductions of Nitro and Azo Groups

$R\text{—}NO_2 \longrightarrow R\text{—}NH_2$

$R\text{—}N{=}N\text{—}R \longrightarrow R\text{—}NH_2 + H_2N\text{—}R$

HYDROLYSES

Hydrolysis of Esters and Amides

Fig. 8.1 Drug metabolism: phase I reactions.

8.1.3 Hydrophilic/hydrophobic balance

In order to cross hydrophobic cell membranes, a drug has to be reasonably lipophilic (fat loving). However, it cannot be too lipophilic. If it is, then it would very swiftly be extracted from an aqueous bloodstream and be stored away in the fatty tissues of the body. This fat solubility can lead to problems. For example, obese patients undergoing surgery require a larger then normal volume of anaesthetic since the gases used are particularly fat soluble. Unfortunately, once surgery is over and the patient has regained consciousness, the anaesthetics stored in the fat tissues will be released and may render the patient unconscious again.

Barbiturates such as thiopental were once seen as potential intravenous anaesthetics which could replace the anaesthetic gases. Unfortunately, they too are fat soluble and as a result it is extremely difficult to estimate a sustained safe dosage. The initial dose can be estimated safely enough to allow for barbiturate taken up by fat cells (and thus removed from the system). However, further doses eventually lead to saturation of the fat depot, and result in a sudden and perhaps fatal increase of barbiturate levels in the blood supply.

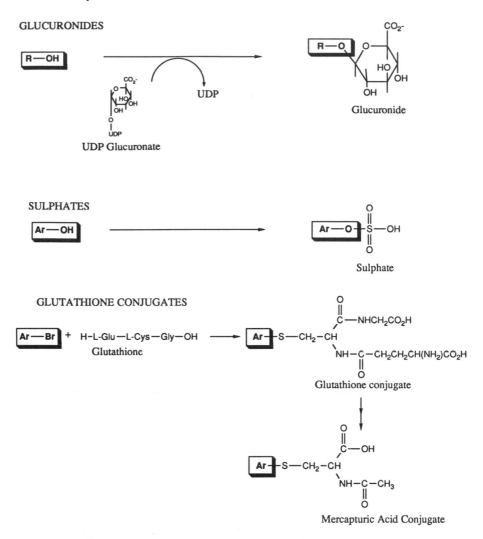

GLUCURONIDES

SULPHATES

GLUTATHIONE CONJUGATES

Glutathione

Glucuronide

UDP Glucuronate

UDP

Sulphate

Glutathione conjugate

Mercapturic Acid Conjugate

Fig. 8.2 Drug metabolism: phase II reactions (conjugation).

In general, fat-soluble drugs tend to get locked away in fat depots and consequently do not reach the target protein very efficiently.

If fat solubility is a problem, why not design drugs to be as hydrophilic as possible? After all, most drugs will need polar or ionic groups in order to bind to their receptor or enzyme.

Yet, how would such a molecule cross the fatty barriers of cell membranes?

We have here an apparent contradiction. The drugs which bind most strongly to the receptor are often very polar, ionized compounds, but have no chance of crossing the fatty cell membranes of the intestinal wall. On the other hand, the drugs which can easily negotiate the fatty cell membranes get mopped up by fat tissue or are too weak to bind to their receptor sites.

Consequently, the best drugs are usually a compromise. They are neither too lipophilic nor too hydrophilic. In general, it is found that the most effective drugs have a pK_a value in the range 6–8. In other words, they are drugs which are partially ionized at blood pH and can easily equilibrate between their ionized and non-ionized forms. This allows them to cross cell membranes in the non-ionized form and to bind to their receptor in the ionized form (Fig. 8.3).

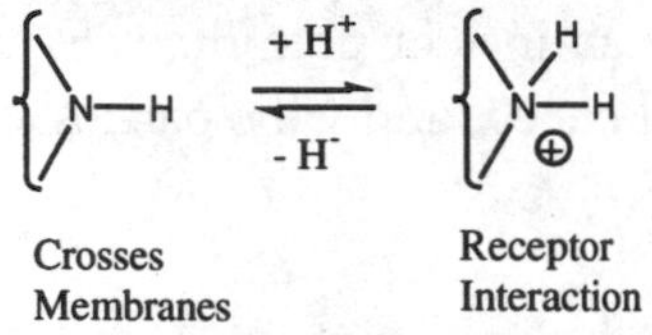

Fig. 8.3 Hydrophobic/hydrophilic balance.

On the other hand, it is sometimes useful to have a fully ionized drug which is incapable of crossing cell membranes. For example, highly ionized sulfonamides are used against gastrointestinal infections. They are incapable of crossing the gut wall and are therefore directed efficiently against the infection.

8.2 Drug dose levels

Estimating dose levels for certain drugs can be a problem with a further range of variables to be taken into account.

Ideally, the blood levels of any drug should be constant and controlled, but this would require a continuous, intravenous drip which is clearly impractical. Therefore, most drugs are taken at regular time intervals and the doses taken are designed to keep the blood levels of drug within a maximum and minimum level such that they are not too high to be toxic, yet not too low to be ineffective. This works well in most cases, but there are certain situations where it does not. The treatment of diabetes with insulin is a case in point. Insulin is normally secreted continuously by the pancreas and so the injection of insulin at timed intervals is unnatural and can lead to a whole range of physiological complications.

Other complications include differences of age, sex, and race. Diet, environment, and altitude also have an influence. Body weight is an important factor to be taken into account. Obese people present a particular problem since it can prove very difficult to estimate how much of a drug will be stored in fat tissue and how much will be free drug. The precise time when drugs are taken may be important since metabolic reaction rates can vary throughout the day.

Drugs can interact with other drugs. For example, some drugs used for diabetes are bound by plasma protein in the blood supply and are therefore not 'free' to react with receptors. However, they can be displaced from the plasma protein by aspirin and this can lead to a drug overdose. A similar phenomenon is observed between anticoagulents and aspirin.

Problems can also occur if a drug taken to inhibit a metabolic reaction is taken with a drug normally metabolized by that reaction. The latter would then be more slowly metabolized, increasing the risk of an overdose. For example, the antidepressant drug phenelzine (Fig. 8.4) inhibits the metabolism of amines and should not be taken with drugs such as amphetamines or pethidine. Even amine-rich foods can lead to adverse effects, implying that cheese and wine parties are hardly the way to cheer the victim of depression.

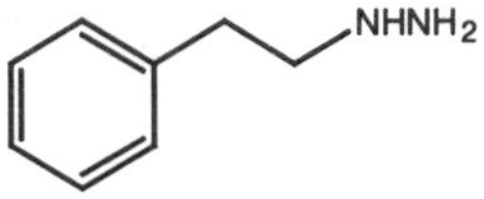

Fig. 8.4 Phenelzine.

When one considers all these complications, it is hardly surprising that individual variability to drugs can vary by as much as a factor of ten.

8.3 *Drug design for pharmacokinetic problems*

Drug design aimed at solving any or all of the above problems can involve a lot of trial and error, basically because of the many variables involved. However, there are some strategies which can be usefully employed.

8.3.1 Variation of substituents

Easily accessible substituents can often be varied to improve the pK_a and lipophilic properties of a compound (Chapter 7). Such studies are particularly open to a quantitative approach known as the quantitative structure–activity relationship (QSAR) approach, discussed in Chapter 9.

8.3.2 Stereoelectronic modifications

The development of the local anaesthetic lidocaine from procaine (Fig. 8.5) is a good example of how the use of steric and electronic effects can make a drug more stable, both chemically and metabolically. Procaine is a good local anaesthetic, but it is short-lasting due to the hydrolysis of the ester group. By changing the ester group to the less reactive amide group, chemical hydrolysis is reduced.

Furthermore, the presence of two *ortho*-methyl groups on the aromatic ring help to shield the carbonyl group from attack by enzymes.

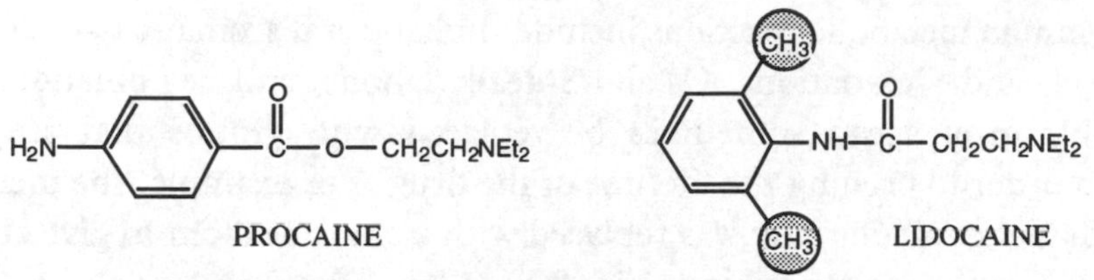

Fig. 8.5

A further example of these tactics is provided in the penicillin field with methicillin (Chapter 10).

8.3.3 Metabolic blockers

Some drugs are metabolized at particular positions in their skeleton. For example, the oral contraceptive megestrol acetate is oxidized at position 6 to give a hydroxyl group at that position. The introduction of a polar group such as this usually allows the formation of polar conjugates which can be quickly eliminated from the system.

The introduction of a stable group such as a methyl group at position 6 (Fig 8.6) can block metabolism and so prolong the activity of the drug.

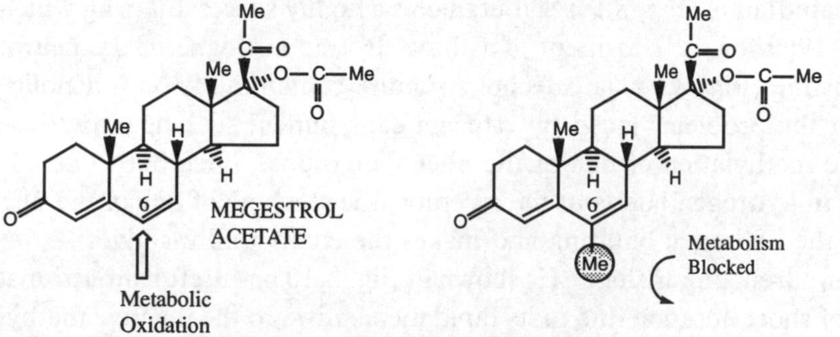

Fig. 8.6 Metabolic blockers.

8.3.4 Removal of susceptible metabolic groups

There are certain chemical groups which are particularly susceptible to metabolic enzymes. For example, methyl groups on aromatic rings can be oxidized to carboxylic acids (Fig. 8.7). These acids can then be quickly eliminated from the body.

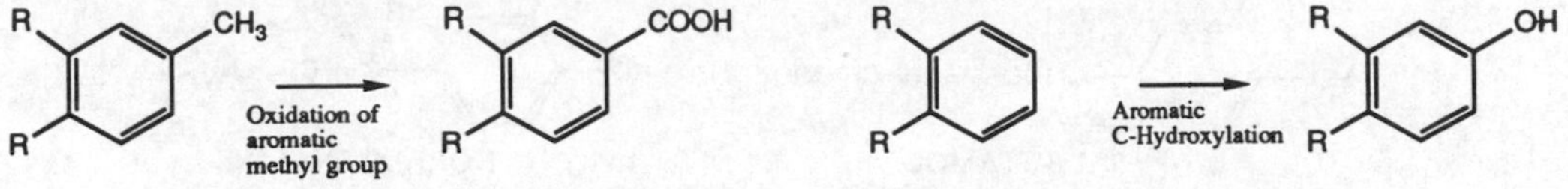

Fig. 8.7 Examples of chemical groups susceptible to metabolic enzymes.

Other common metabolic reactions include aliphatic and aromatic C-hydroxylations (Fig. 8.7), N- and S-oxidations, O- and S-dealkylations, and deamination.

Susceptible groups can sometimes be replaced with groups that are stable to oxidation in order to prolong the lifetime of the drug. For example, the methyl group of the antidiabetic tolbutamide was replaced with a chlorine atom to give chlorpropamide which is much longer lasting (Fig. 8.8).

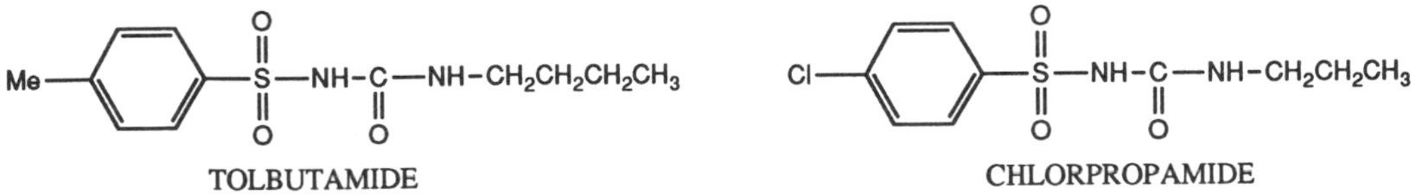

TOLBUTAMIDE   CHLORPROPAMIDE

Fig. 8.8

But suppose the vulnerable group is also crucial for activity? If we cannot replace it or remove it, what can we do?

There are two possible solutions. We can either mask the vulnerable group on a temporary basis by using a prodrug (see later) or we can try 'shifting' the vulnerable group away from whatever is making it vulnerable. The latter tactic was used in the development of salbutamol (Fig. 8.9). Salbutamol is a highly successful drug which was introduced in 1969 for the treatment of asthma. It is an analogue of the neurotransmitter noradrenaline (Fig. 8.9)—a catechol structure containing two *o*-phenolic groups

One of the problems faced by catechol compounds, such as noradrenaline, is the metabolic methylation of one of the phenolic groups. Since both phenol groups are involved in hydrogen bonds to the receptor, the masking of one of the phenol groups disrupts the hydrogen bonding and makes the compound inactive.

The noradrenaline analogue (I) shown in Fig. 8.10 has useful antiasthmatic activity, but it is of short duration due to its rapid metabolism to the inactive methyl ether (II).

Replacing the OH with something like a proton or a methyl group may prevent the metabolism but will also prevent the important hydrogen bonding. So how was the problem solved?

The answer was to move the vulnerable OH group out from the ring by one carbon unit. This was enough to make the compound unrecognizable to the metabolic enzyme.

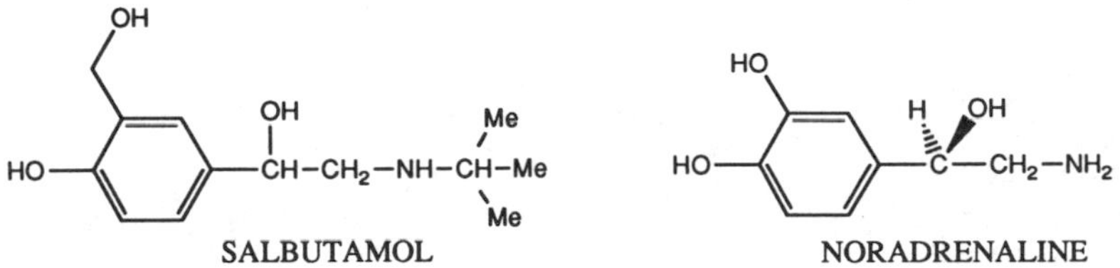

SALBUTAMOL   NORADRENALINE

Fig. 8.9

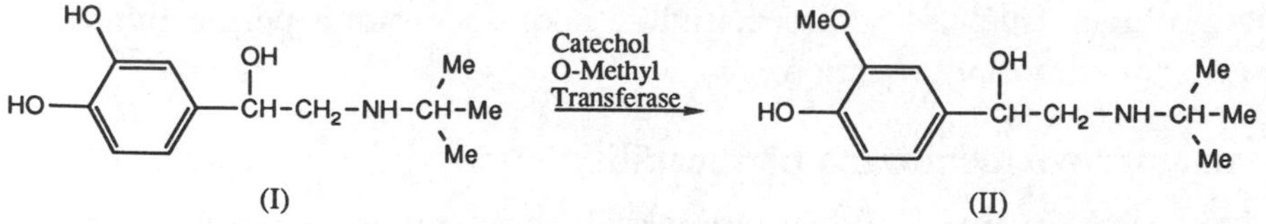

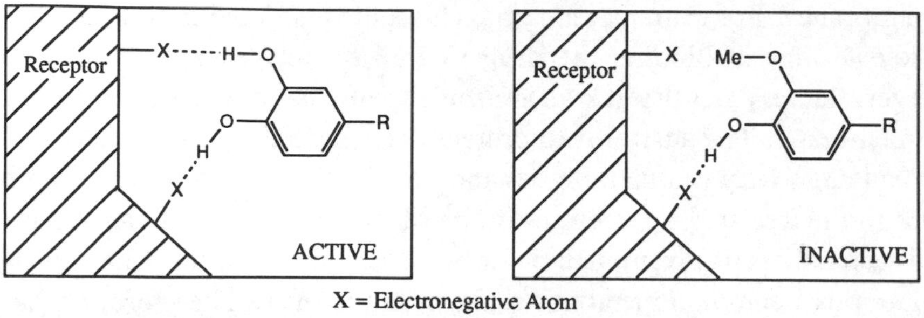

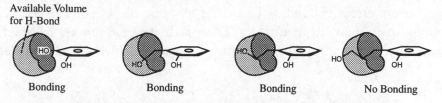

Fig. 8.10 Metabolic methylation of a noradrenaline analogue.

Fortunately, the receptor appears to be quite lenient over the position of this hydrogen bonding group and it is interesting to note that a hydroxyethyl group is also acceptable. Beyond that, activity is lost due to the OH being 'out of range' or being too large to fit. These results demonstrate that it is better to consider a receptor binding site as an available volume rather than imagining the binding site as fixed at one spot. A drug can then be designed such that the relevant binding group is positioned into any part of that available volume (Fig. 8.11).

The tactic worked, but there was no way of knowing beforehand whether the receptor itself would still recognize the structure. Several factors could well have prevented the necessary hydrogen bonding interaction. The CH_2OH group might have been too bulky to fit. The group may have been too far from the binding site. The fact that an acidic phenol had been replaced with a neutral alcohol group could have destroyed the bonding interaction.

Fig. 8.11 Receptor binding site as an available volume.

8.3.5 Prodrugs

Prodrugs are compounds which may be inactive in themselves, but which can be converted by chemical or enzymatic means to an active drug. They have been useful in tackling problems such as acid sensitivity, poor membrane permeability, drug toxicity, and short duration of action.

Prodrugs to improve membrane permeability

Prodrugs have proved very useful in temporarily masking an 'awkward' functional group which is important to receptor binding, but which hinders the drug from crossing cell membranes. For example, a carboxylic acid functional group may have an important role to play in binding the drug to a receptor via ionic or hydrogen bonding. However, the very fact that it is an ionizable group may prevent it from crossing a fatty cell membrane. The answer is to protect the acid function as an ester. The less polar ester can cross fatty cell membranes and once in the bloodstream, it will be hydrolysed back to the free acid by esterases in the blood. An example of such a prodrug is the antibacterial agent pivampicillin—a prodrug for ampicillin (Chapter 10).

N-Demethylation is a common metabolic reaction in the liver. Therefore, primary or secondary amines could be N-methylated to improve their membrane permeability. Several hypnotics and antiepileptics take advantage of this reaction (e.g. hexobarbitone (Fig. 8.12)).

Another way round the problem of membrane permeability is to design a prodrug which can take advantage of a carrier protein in the cell membrane, such as the one responsible for carrying amino acids into a cell. The best known example of such a prodrug is levodopa (Fig. 8.13).

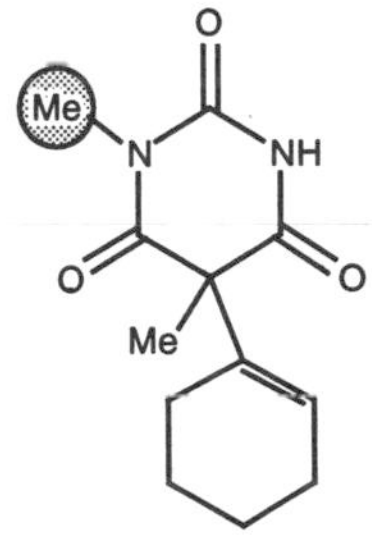

Levodopa is a prodrug for the neurotransmitter dopamine and as such has been used in the treatment of Parkinson's disease—a condition due primarily to a deficiency of the neurotransmitter dopamine. Dopamine itself cannot be used since it is too polar to cross the blood–brain barrier. Levodopa is even more polar and seems an unlikely

Fig. 8.12 Hexobarbitone.

prodrug. However, it is an amino acid and as such can make use of the special 'arrangements' made in order to move amino acids across the blood–brain barrier. Amino acids are essential building blocks for all cells, but are incapable of crossing hydrophobic membranes by themselves. There is, however, a process by which amino acids can be shuttled through membranes such as the blood–brain barrier. This

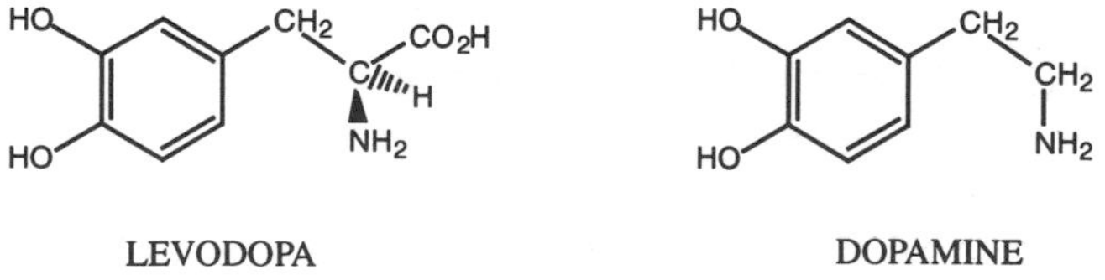

LEVODOPA   DOPAMINE

Fig. 8.13

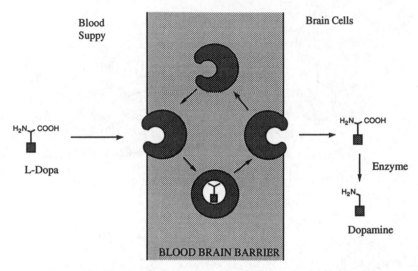

Fig. 8.14 Transport of levodopa across the blood-brain barrier.

involves a protein carrier system which is embedded in the membrane and 'smuggles' its passengers from one side to the other (Chapter 5). Once across the barrier, a decarboxylase enzyme removes the acid group and generates dopamine (Fig. 8.14).

Another means of taking advantage of the transport proteins is to attach the active drug to an amino acid or nucleic acid base such that the drug gets a 'piggyback' across the membrane. Uracil mustard (Fig. 8.23) is one such example.

Prodrugs for prolonged activity

6-Mercaptopurine (Fig. 8.15) suppresses the body's immune response and is therefore useful in protecting donor grafts. However, the drug tends to be eliminated from the body too quickly. The prodrug azathioprine (Fig 8.15) lasts far longer. Azathioprine is slowly converted to 6-mercaptopurine, allowing a more sustained activity. Since the conversion is chemical and unaffected by enzymes, the rate of conversion can be altered, depending on the electron withdrawing ability of the heterocyclic group. The greater the electron withdrawing power, the faster the breakdown. The NO_2 group is therefore present to ensure an efficient conversion to 6-mercaptopurine, since it is strongly electron withdrawing.

There is a belief that the well-known sedatives Valium (Fig. 8.16) and Librium might be prodrugs and are only active because they are metabolized by *N*-demethylation to nordazepam. Nordazepam itself has been used as a sedative, but loses activity quite quickly due to metabolism and excretion. Valium, if it is a prodrug for nordazepam, demonstrates again how a prodrug can be used to lead to a more sustained action.

One approach to maintaining a sustained level of drug over long periods is to

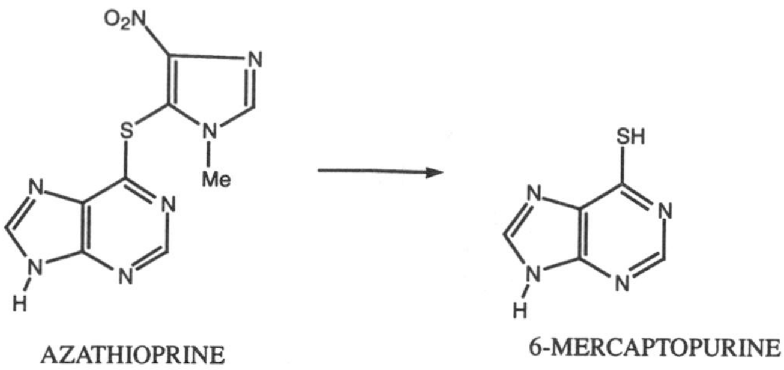

122   **Pharmacodynamics**

Fig. 8.15 Azathioprine acts as a prodrug for 6-mercaptopurine.

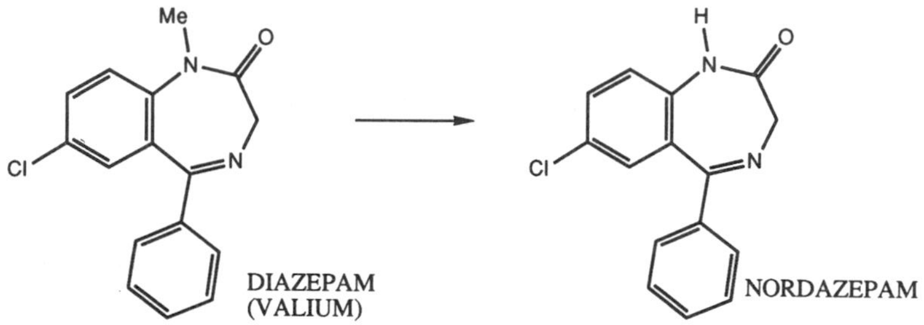

Fig. 8.16 Valium as a possible prodrug for nordazepam.

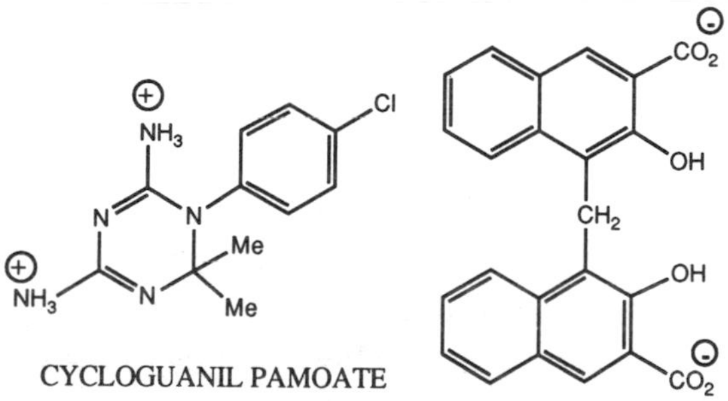

Fig. 8.17 CYCLOGUANIL PAMOATE

deliberately associate a very lipophilic group with an active drug. This means that the majority of the drug is stored in fat tissue, and if the lipophilic group is only slowly removed, then the drug is steadily released into the bloodstream over a long period of time. The antimalarial agent cycloguanil pamoate (Fig. 8.17) is one such agent. The active drug is bound ionically to an anion with a large lipophilic group.

Prodrugs masking drug toxicity and side-effects

Prodrugs can be used to mask the side-effects and toxicity of drugs. For example, salicylic acid is a good painkiller, but causes gastric bleeding due to the free phenolic group. This is overcome by masking the phenol as an ester (aspirin) (Fig. 8.18). The ester is later hydrolysed by esterases to free the active drug.

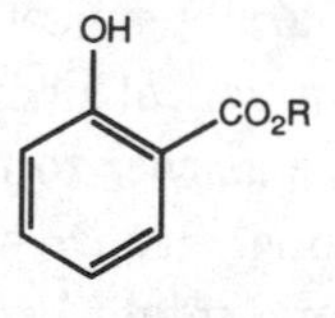

SALICYLIC ACID   R = H
ASPIRIN   R = Ac

Fig. 8.18

Prodrugs can be used to give a slow release of drugs which would be too toxic to give directly. Propiolaldehyde is useful in the aversion therapy of alcohol, but is not used itself since it is an irritant. However, the prodrug pargylene can be converted to propiolaldehyde by enzymes in the liver (Fig. 8.19).

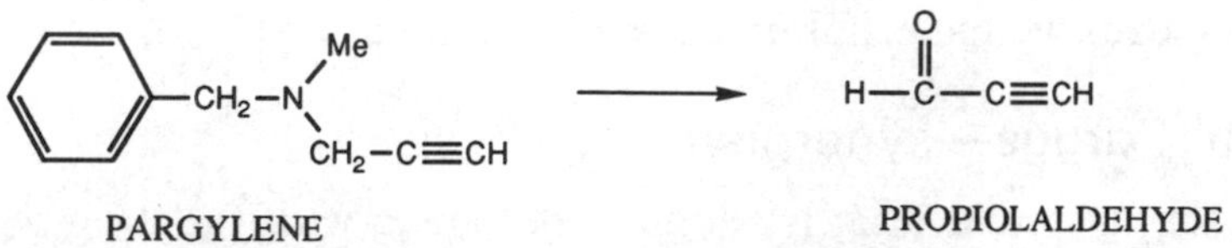

PARGYLENE   PROPIOLALDEHYDE

Fig. 8.19 Pargylene as a prodrug for propiolaldehyde.

An extension of this tactic is to design a prodrug such that it is converted to the active drug at the target site itself. If this can be achieved successfully, it will greatly reduce the side-effects of highly toxic drugs such as the anticancer agents. Cyclophosphamide is a successful anticancer drug which is not toxic itself, but which is converted in several steps to the toxic phosphoramide mustard (Fig. 8.20). This is a strong alkylating agent which will alkylate a cell's DNA and thus kill the cell. Since there is a high level of phosphoramidase enzyme in some tumour cells, it was hoped that the drug could be directed selectively against these cells. Some selectivity has indeed been observed and it is hoped that complete selectivity can eventually be achieved.

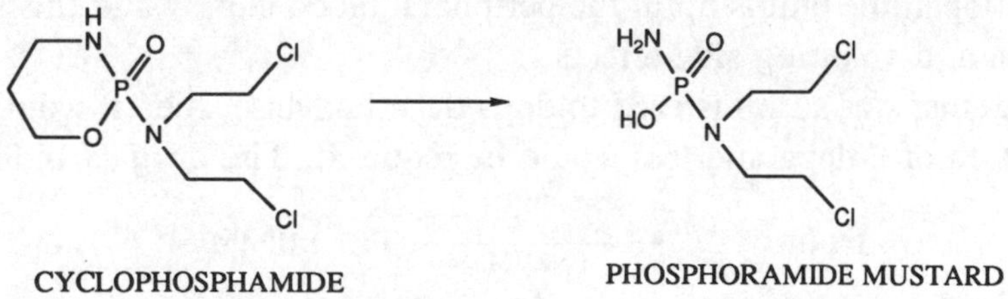

CYCLOPHOSPHAMIDE   PHOSPHORAMIDE MUSTARD

Fig. 8.20

8.3.6 Bioisosteres

A bioisostere is a chemical group which can replace another chemical group without affecting biological activity.

Many peptides and polypeptides are chemical messengers in the body, yet using such compounds as medicines is impractical since the body's own digestive enzymes can hydrolyse the peptide links. One answer to the problem has been to replace the peptide bond with another functional group which is stable to these hydrolytic enzymes. For example, a peptide bond might be replaced with a double bond. If the compound retains activity, then the double bond represents a bioisostere for the peptide link *in this particular case*. Note that bioisosteres are not general. They are specific for the drug and the protein with which it interacts. A successful bioisostere in one field of medicinal chemistry may be useless in a different field. Note also that bioisosteres are different from isosteres. It is the retention of important biological activity which determines whether a group is a bioisostere, not the valency.

An example of how bioisosteric groups have been used successfully is provided by the cholinergic drug bethanechol described in Chapter 11.

8.3.7 'Sentry' drugs—synergism

In this approach, a second drug is administered along with the drug which is 'going into action'. The role of the second drug is to guard or assist the principal drug. Usually, the second drug is an antagonist of an enzyme which metabolizes the principal drug.

For example, clavulanic acid inhibits the enzyme β-lactamase and is therefore able to protect penicillins which are labile to that particular enzyme (Chapter 10).

Another example is to be found in the drug therapy of Parkinson's disease. The use of L-dopa (levodopa) as a prodrug for dopamine has already been described. However, to be effective, large doses of L-dopa (3–8 g per day) are required, and over a period of time these dose levels lead to side-effects such as nausea and vomiting. L-Dopa is susceptible to the enzyme dopa decarboxylase and as a result, much of the L-dopa administered is decarboxylated to L-dopamine before it reaches the central nervous system (Fig. 8.21).

As stated earlier, dopamine is unable to cross the blood–brain barrier. As a result, an excess of dopamine builds up in the peripheral blood supply and this is what leads to the nausea and vomiting side-effects.

If an antagonist was administered to dopa decarboxylase, then it would inhibit the decarboxylation of L-dopa and less would be required. The drug carbidopa has been

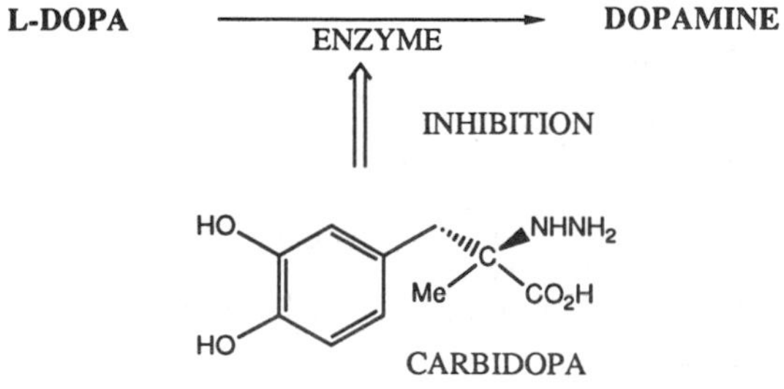

Fig. 8.21 Inhibition of L-dopa decarboxylation.

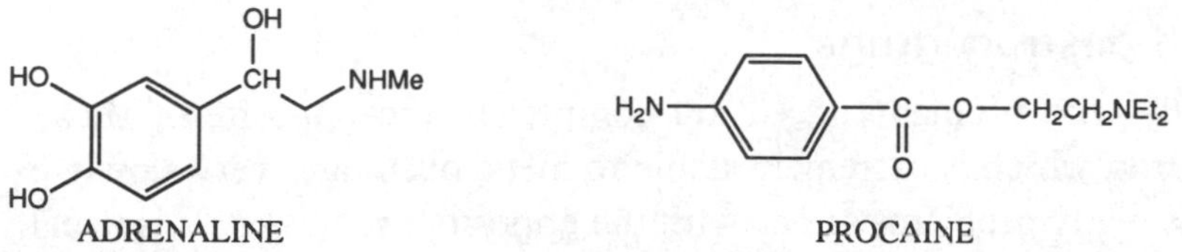

Fig. 8.22

used successfully in this respect and effectively inhibits dopa decarboxylase. Further-more, since it is a highly polar compound containing two phenolic groups, a hydrazine moiety, and an acidic group, it is unable to cross the blood–brain barrier and so cannot prevent the conversion of L-dopa to dopamine in the brain.

Adrenaline is an example of a 'sentry drug' which acts on a receptor rather than an enzyme. This drug is used along with the injectable local anaesthetic procaine to prolong its action (Fig. 8.22). Adrenaline constricts the blood vessels in the vicinity of the injection and so prevents procaine being 'washed away' by the blood supply.

8.3.8 'Search and destroy' drugs

A major goal in cancer chemotherapy is to target drugs efficiently against tumour cells rather than normal cells. One method of achieving this is to design a drug transport system. The idea is to attach the active drug to a molecule which is needed in large amounts by the rapidly dividing tumour cells. One approach has been to attach the active drug to an amino acid or a nucleic acid base, e.g. uracil mustard (Fig. 8.23).

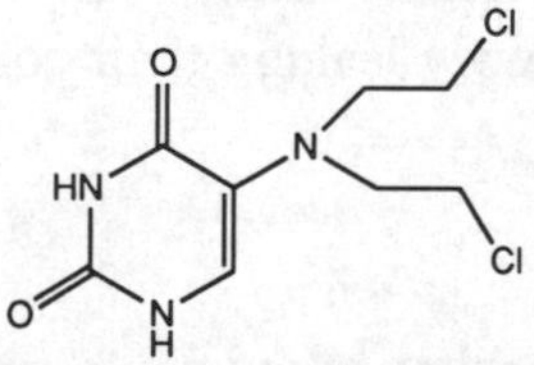

Fig. 8.23 Uracil mustard.

Of course, normal cells require these building blocks as well, but tumour cells often grow more quickly than normal cells and require the building blocks more urgently. Therefore, the uptake of these drugs should be greater in tumour cells than in normal cells. This approach has the added advantage that the drug can enter the cell more efficiently by using the transport proteins for the particular building block.

The tactic has been reasonably successful, but has not yet lived up to expectation.

A more recent idea has been to attach the active drug to monoclonal antibodies which can recognize antigens unique to the tumour cell. The difficulty is in finding suitable antigens and producing the antibodies in significant quantity. However, the approach has great promise for the future.

8.3.9 Self-destruct drugs

Occasionally, the problems faced are completely the opposite of those mentioned above. A drug which is extremely stable to metabolism and very slowly excreted can pose just as many problems as one with the opposite properties. It is usually desirable to have a drug which performs what it is meant to do, then stops doing it within a reasonable time. If not, the effects of the drug could last far too long and cause toxicity and lingering side-effects. Therefore, designing drugs with decreased chemical and metabolic stability can be useful on occasions.

The neuromuscular drug atracurium (Chapter 11) is a good example of this.

8.3.10 Delivery systems

Continuous minipumps have been developed which can release insulin at varying rates depending on blood-glucose levels. This appears to be the best answer to the problem of providing insulin at the correct levels at the correct times.

Some acid-sensitive drugs can be protected by the way they are formulated. For example, it is possible to coat pills with an acid-resistant polymer which protects the drug from the acids in the stomach. The polymer is designed to be removed under the slightly alkaline conditions of the large intestine. Unfortunately, absorption in the large intestine is not so efficient as the small intestine and so the applications are limited.

A physical way of protecting drugs from metabolic enzymes in the bloodstream is to inject small vesicles called liposomes filled with the drug. These vesicles or globules consist of a bilayer of fatty molecules in the same way as a cell membrane and will travel round the circulation, slowly leaking their contents.

8.4 *Testing of drugs*

It is unlikely that the thorny problem of animal testing will disappear for a long time. As can be seen already, there are so many variables involved in determining whether a drug will be effective or not that it is impossible to anticipate them all. One has also to take into account that the drug will be metabolized to other compounds, all with their own range of biological properties.

It appears impossible, therefore, to predict whether a potential drug will work or whether it will be safe by *in vitro* tests alone. Therein lies the importance of animal experiments. Only animal tests can test for the unexpected. Unless we are prepared to volunteer ourselves as guinea pigs, then animal experiments will remain an essential feature of drug development for many years to come. A more realistic ambition for the near future is to reduce the number of animal tests by developing more *in vitro* tests. For example, the metabolism of drugs is being studied initially by using liver cells grown in culture. If these tests are satisfactory, then animal studies can follow on. If the tests are unsatisfactory, then the animals are spared.

8.5 *Neurotransmitters as drugs?*

Before finishing this chapter, let us consider the body's own neurotransmitters. Why do we not use these as drugs? If the body is short of dopamine, why not administer more dopamine to make up the balance? After all, most neurotransmitters which have been identified are simple molecules, easily prepared in the laboratory.

Unfortunately, this is not possible for a number of reasons. Many are not chemically stable enough to survive the acid of the stomach and would have to be injected. Even if they were injected, there is little chance that they would survive to reach their target receptors. As mentioned already, the body has efficient mechanisms by which it inactivates its neurotransmitters as soon as they have passed on their message. Therefore, on injection, they would be swiftly inactivated by enzymes or by cell uptake.

Even if they were to survive, they could lead to undesirable side-effects. For example, the shortage of neurotransmitter may only be at one small area in the brain and be normal elsewhere. If we gave the natural neurotransmitter, how would we stop it producing an overdose of transmitter at these other sites? This of course is a problem with all drugs, but in recent years it has been discovered that receptors for a specific neurotransmitter are slightly different, depending on where they are in the body. The medicinal chemist can design synthetic drugs which take advantage of that difference, such that they 'ignore' receptors which the natural neurotransmitter would not. In this respect, the medicinal chemist has actually improved on nature.

We cannot even assume that the body's own neurotransmitters are perfectly safe and free from the horrors of tolerance and addiction associated with drugs such as heroin. It is quite possible to be addicted to one's own neurotransmitters and hormones. Some people are addicted to exercise and are compelled to exercise long hours each day in order to feel good. The very process of exercise leads to the release of hormones and neurotransmitters. This can produce a 'high', which drives susceptible people to exercise more and more. If they stop exercising, they suffer withdrawal symptoms such as deep depression.

The same phenomenum probably drives mountaineers into attempting feats which they know quite well may lead to their death. The thrill of danger produces hormones and neurotransmitters which in turn produce a 'high'. Perhaps this too is why war has such a fascination for mankind.

To conclude, many of the body's own neurotransmitters are known and can be easily synthesized, but they cannot be effectively used as medicines.

9 · Quantitative structure–activity relationships (QSAR)

9.1 *Introduction*

In Chapters 7 and 8 we studied the various strategies which can be used in the design of drugs. Several of these strategies involved a change in shape such that the new drug had a better 'fit' for its receptor. Other strategies involved a change in the physical properties of the drug such that its distribution, metabolism, or receptor binding interactions were affected. These latter strategies often involved the synthesis of analogues containing a range of substituents on aromatic/heteroaromatic rings or accessible functional groups. There are an infinite number of possible analogues which can be made, if we were to try and synthesize analogues with every substituent and combination of substituents possible. Therefore, it is clearly advantageous if a rational approach can be followed in deciding which substituents to use. The QSAR (quantitative structure–activity relationship) approach has proved extremely useful in tackling this problem.

The QSAR approach attempts to identify and quantify the physicochemical properties of a drug and to see whether any of these properties has an effect on the drug's biological activity. If such a relationship holds true, an equation can be drawn up which quantifies the relationship and allows the medicinal chemist to say with some confidence that the property (or properties) has an important role in the distribution or mechanism of the drug. It also allows the medicinal chemist some level of prediction. By quantifying physicochemical properties, it should be possible to calculate in advance what the biological activity of a novel analogue might be. There are two advantages to this. Firstly, it allows the medicinal chemist to target efforts on analogues which should have improved activity and thus cut down the number of analogues which have to be made. Secondly, if an analogue is discovered which does not fit the equation, it implies that some other feature is important and provides a lead for further development.

What are these physicochemical features which we have mentioned?

Essentially, they refer to any structural, physical, or chemical property of a drug. Clearly, any drug will have a large number of such properties and it would be a Herculean task to quantify and relate them all to biological activity at the same time. A simple, more practical approach is to consider one or two physicochemical proper-ties of the drug and to vary these while attempting to keep other properties constant. This is not as simple as it sounds, since it is not always possible to vary one property without affecting another. Nevertheless, there have been numerous examples where the approach has worked.

9.2 *Graphs and equations*

In the simplest situation, a range of compounds are synthesized in order to vary one physicochemical property (e.g. log P) and to test how this affects the biological activity (log $1/C$) (we will come to the meaning of log $1/C$ and log P in due course). A graph is then drawn to plot the biological activity on the y axis versus the physico-chemical feature on the x axis (Fig. 9.1).

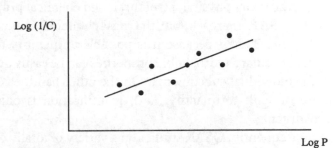

Fig. 9.1 Biological activity versus physicochemical property.

It is then necessary to draw the best possible line through the data points on the graph. This is done by a procedure known as 'linear regression analysis by the least squares method'. This is quite a mouthful and can produce a glazed expression on any chemist who is not mathematically orientated. In fact, the principle is quite straightforward.

If we draw a line through a set of data points, most of the points will be scattered on either side of the line. The best line will be the one closest to the data points. To measure how close the data points are, vertical lines are drawn from each point (Fig. 9.2). These verticals are measured and then squared in order to eliminate the negative values. The squares are then added up to give a total. The best line through the points will be the line where this total is a minimum.

The equation of the straight line will be $y = k_1x + k_2$ where k_1 and k_2 are constants. By varying k_1 and k_2, different equations are obtained until the best line is obtained. This whole process can be speedily done by computer programme.

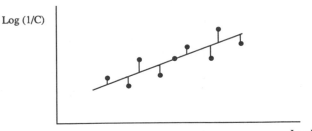

Fig. 9.2 Proximity of data points to line of best fit.

The next stage in the process is to see whether the relationship is significant. We may have obtained a straight line through points which are so random that it means nothing. The significance of the equation is given by a term known as the regression coefficient (r) This coefficient can again be calculated by computer. For a perfect fit, $r^2 = 1$. Good fits generally have r^2 values of 0.95 or above.

9.3 *Physicochemical properties*

There are many physical, structural, and chemical properties which have been studied by the QSAR approach, but the most commonly studied are hydrophobic, electronic, and steric. This is because it is possible to quantify these effects relatively easily.

In particular, hydrophobic properties can be easily quantified for complete molecules or for individual substituents. On the other hand, electronic and steric properties are more difficult to quantify, and quantification is only really feasible for individual substituents.

Consequently, QSAR studies on a variety of totally different structures are relatively rare and are limited to studies on hydrophobicity. It is more common to find QSAR studies being carried out on compounds of the same general structure, where substituents on aromatic rings or accessible functional groups are varied. The QSAR study then considers how the hydrophobic, electronic, and steric properties of the substituents affect biological activity.

The three most studied physicochemical properties will now be considered in some detail.

9.3.1 Hydrophobicity

The hydrophobic character of a drug is crucial to how easily it crosses cell membranes (see Section 8.1.3.) and may also be important in receptor interactions. Changing substituents on a drug may well have significant effects on its hydrophobic character and hence its biological activity. Therefore, it is important to have a means of predicting this quantitatively.

The partition coefficient (P)

The hydrophobic character of a drug can be measured experimentally by testing the drug's relative distribution in an octanol/water mixture. Hydrophobic molecules will prefer to dissolve in the octanol layer of this two-phase system, whereas hydrophilic molecules will prefer the aqueous layer. The relative distribution is known as the partition coefficient (P) and is obtained from the following equation:

$$P = \frac{\text{Concentration of drug in octanol}}{\text{Concentration of drug in aqueous solution}}$$

Hydrophobic compounds will have a high P value, whereas hydrophilic compounds will have a low P value.

Varying substituents on the lead compound will produce a series of analogues having different hydrophobicities and therefore different P values. By plotting these P values against the biological activity of these drugs, it is possible to see if there is any relationship between the two properties. The biological activity is normally expressed as $1/C$, where C is the concentration of drug required to achieve a defined level of biological activity. (The reciprocal of the concentration $(1/C)$ is used, since more active drugs will achieve a defined biological activity at lower concentration.)

The graph is drawn by plotting $\log(1/C)$ versus $\log P$. The scale of numbers involved in measuring C and P usually covers several factors of ten and so the use of logarithms allows the use of more manageable numbers.

In studies where the range of the $\log P$ values is restricted to a small range (e.g. $\log P = 1$–4), a straight-line graph is obtained (Fig. 9.1) showing that there is a relationship between hydrophobicity and biological activity. Such a line would have the following equation:

$$\log \left(\frac{1}{C}\right) = k_1 \log P + k_2.$$

For example, the binding of drugs to serum albumin is determined by their hydrophobicity and a study of 40 compounds resulted in the following equation:

$$\log \left(\frac{1}{C}\right) = 0.75 \log P + 2.30.$$

The equation shows that serum albumin binding increases as $\log P$ increases. In other words, hydrophobic drugs bind more strongly to serum albumin than hydrophilic drugs. Knowing how strongly a drug binds to serum albumin can be important in estimating effective dose levels for that drug. When bound to serum albumin, the drug cannot bind to its receptor and so the dose levels for the drug should be based on the amount of unbound drug present in the circulation. The equation above allows us to calculate how strongly drugs of similar structure will bind to serum albumin and gives an indication of how 'available' they will be for receptor interactions.

Despite such factors as serum albumin binding, it is generally found that increasing the hydrophobicity of a lead compound results in an increase in biological activity. This reflects the fact that drugs have to cross hydrophobic barriers such as cell membranes in order to reach their target. Even if no barriers are to be crossed (e.g. *in vitro* studies), the drug has to interact with a target system such as an enzyme or receptor where the binding site is usually hydrophobic. Therefore, increasing hydrophobicity aids the drug in crossing hydrophobic barriers or in binding to its target site.

This might imply that increasing $\log P$ should increase the biological activity *ad infinitum*. In fact, this does not happen. There are several reasons for this. For example, the drug may become so hydrophobic that it is poorly soluble in the aqueous phase. Alternatively, it may be 'trapped' in fat depots and never reach the intended site. Finally, hydrophobic drugs are often more susceptible to metabolism and subsequent elimination.

A straight-line relationship between $\log P$ and biological activity is observed in many QSAR studies because the range of $\log P$ values studied is often relatively narrow. For example, the study carried out on serum albumin binding was restricted to compounds having $\log P$ values in the range 0.78 to 3.82. If these studies were to be extended to include compounds with very high $\log P$ values then we would see a different picture. the graph would be parabolic, as shown in Fig. 9.3. Here, the biological activity increases as $\log P$ increases until a maximum value is obtained. The value of $\log P$ at the maximum ($\log P^0$) represents the optimum partition coefficient for biological activity. Beyond that point, an increase in $\log P$ results in a decrease in biological activity.

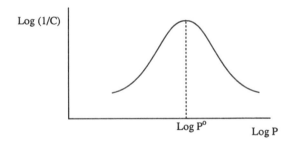

Fig. 9.3 Parabolic log (1/C) vs. log P curve.

In situations where the partition coefficient is the only factor influencing biological activity, the parabolic curve can be expressed by the mathematical equation:

$$\log \left(\frac{1}{C}\right) = -k_1(\log P)^2 + k_2 \log P + k_3.$$

Note that the $(\log P)^2$ term has a negative sign in front of it. When P is small, the $(\log P)^2$ term is very small and the equation is dominated by the $\log P$ term. This represents the first part of the graph where activity increases with increasing P. When

P is large, the $(\log P)^2$ term is more significant and eventually 'overwhelms' the $\log P$ term. This represents the last part of the graph where activity drops with increasing P. k_1, k_2, and k_3 are constants and can be determined by a suitable computer programme.

There are relatively few drugs where activity is related to the $\log P$ factor alone. Those that do, tend to operate in cell membranes where hydrophobicity is the dominant feature controlling their action. The best example of drugs which operate in cell membranes are the general anaesthetics. These are thought to function by entering the central nervous system and 'dissolving' into cell membranes where they affect membrane structure and nerve function. In such a scenario, there are no specific drug–receptor interactions and the mechanism of the drug is controlled purely by its ability to enter cell membranes (i.e. its hydrophobic character). The general anaesthetic activity of a range of ethers was found to fit the parabolic equation:

$$\log \left(\frac{1}{C}\right) = -0.22(\log P)^2 + 1.04 \log P + 2.16.$$

According to the equation, anaesthetic activity increases with increasing hydrophobicity (P), as determined by the $\log P$ factor. The negative $(\log P)^2$ factor shows that the relationship is parabolic and that there is an optimum value for $\log P$ ($\log P^0$) beyond which increasing hydrophobicity causes a decrease in anaesthetic activity.

With this equation, it is now possible to predict the anaesthetic activity of other ether structures, given their partition coefficients.

There are limitations to the use of this particular equation. For example, it is derived purely for anaesthetic ethers and is not applicable to other structural types of anaesthetics. This is generally true in QSAR studies. The procedure works best if it is applied to a series of compounds which have the same general structure.

QSAR studies have been carried out on other structural types of general anaesthetics and in each case a parabolic curve has been obtained. Although, the constants for each equation are different, it is significant that the optimum hydrophobicity (represented by $\log P^0$) for anaesthetic activity is close to 2.3, regardless of the class of anaesthetic being studied. This finding suggests that all general anaesthetics are operating in a similar fashion, controlled by the hydrophobicity of the structure.

Since different anaesthetics have similar $\log P^0$ values, the $\log P$ value of any compound can give some idea of its potential potency as an anaesthetic. For example, the $\log P$ values of the gaseous anaesthetics ether, chloroform, and halothane are 0.98, 1.97, and 2.3 respectively. Their anaesthetic activity increases in the same order.

Since general anaesthetics have a simple mechanism of action based on the efficiency with which they enter the central nervous system (CNS), it implies that $\log P$ values should give an indication of how easily any compound can enter the CNS. In other words, compounds having a log P value close to 2 should be capable of entering the CNS efficiently.

This is generally found to be true. For example, the most potent barbiturates for sedative and hypnotic activity are found to have log P values close to 2.

As a rule of thumb, drugs which are to be targeted for the CNS should have a log P value of approximately 2. Conversely, drugs which are designed to act elsewhere in the body should have log P values significantly different from 2 in order to avoid possible CNS side-effects (e.g. drowsiness).

As an example of this, the cardiotonic agent shown in Fig. 9.4(a) was found to produce 'bright visions' in some patients, which implied that it was entering the CNS. This was supported by the fact that the log P value of the drug was 2.59. In order to prevent the drug entering the CNS, the 4-OMe group was replaced with a 4-S(O)Me group. This particular group is approximately the same size as the methoxy group, but more hydrophilic. The log P value of the new drug (sulmazole) (Fig. 9.4(b)) was found to be 1.17. The drug was now too hydrophilic to enter the CNS and was free of CNS side-effects.

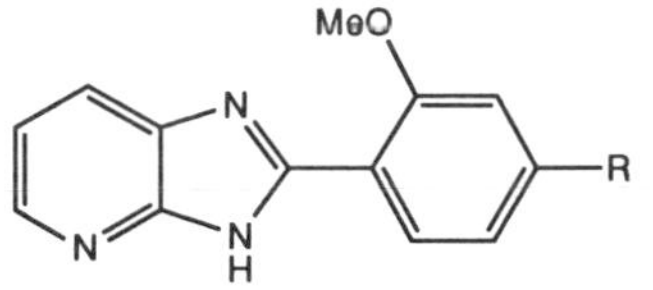

Fig. 9.4 Cardiotonic agents.

a) R= OMe
b) R= S(O)Me Sulmazole

The substituent hydrophobicity constant (π)

We have seen how the hydrophobicity of a compound can be quantified by using the partition coefficient P. However, in order to get P we have to measure it experimentally and that means that we have to synthesize the compounds. It would be much better if we could calculate P theoretically and decide in advance whether the compound is worth synthesizing. QSAR would then allow us to target the most promising looking structures. For example, if we were planning to synthesize a range of barbiturate structures, we could calculate log P values for them all and concentrate on the structures which had log P values closest to the optimum log P^0 value for barbiturates.

Fortunately, partition coefficients can be calculated by knowing the contribution that various substituents make to hydrophobicity. This contribution is known as the substituent hydrophobicity constant (π).

The substituent hydrophobicity constant is a measure of how hydrophobic a substituent is, relative to hydrogen. The value can be obtained as follows. Partition coefficients are measured experimentally for a standard compound with and without a substituent (X). The hydrophobicity constant (π_X) for the substituent (X) is then obtained using the following equation:

$$\pi_X = \log P_X - \log P_H$$

where P_H is the partition coefficient for the standard compound, and P_X is the partition coefficient for the standard compound with the substituent.

A positive value of π indicates that the substituent is more hydrophobic than hydrogen. A negative value indicates that the substituent is less hydrophobic. The π values for a range of substituents are shown in Fig. 9.5.

Group	CH_3	Bu^t	OH	OCH_3	CF_3	Cl	Br	F
π (Aliphatic Substituents)	0.50	1.68	−1.16	0.47	1.07	0.39	0.60	−0.17
π (Aromatic Substituents)	0.52	1.68	−0.67	−0.02	1.16	0.71	0.86	0.14

Fig. 9.5 π values for a range of substituents.

These π values are characteristic for the substituent and can be used to calculate how the partition coefficient of a drug would be affected by adding these substituents. The P value for the lead compound would have to be measured experimentally, but once that is known, the P value for analogues can be calculated quite simply.

As an example, consider the log P values for benzene (log P = 2.13), chlorobenzene (log P = 2.84), and benzamide (log P = 0.64) (Fig. 9.6). Since benzene is the parent compound, the substituent constants for Cl and $CONH_2$ are 0.71 and −1.49 respectively. Having obtained these values, it is now possible to calculate the theoretical log P value for *meta*-chlorobenzamide:

$$\begin{aligned} \log P_{(\text{chlorobenzamide})} &= \log P_{(\text{benzene})} + \pi_{Cl} + \pi_{CONH_2} \\ &= 2.13 + 0.71 + (-1.49) \\ &= 1.35. \end{aligned}$$

The observed log P value for this compound is 1.51.

It should be noted that π values for aromatic substituents are different from those used for aliphatic substituents. Furthermore, neither of these sets of π values are in

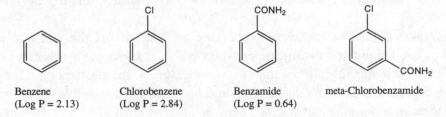

Benzene
(Log P = 2.13)

Chlorobenzene
(Log P = 2.84)

Benzamide
(Log P = 0.64)

meta-Chlorobenzamide

Fig. 9.6 Values for log P.

fact true constants and are accurate only for the structures from which they were derived. They can be used as good approximations when studying other structures, but it is possible that the values will have to be adjusted in order to get accurate results.

P vs. π

QSAR equations relating biological activity to the partition coefficient P have already been described, but there is no reason why the substituent hydrophobicity constant π cannot be used in place of P if only the substituents are being varied.

The equation obtained would be just as relevant as a study of how hydrophobicity affects biological activity. That is not to say that P and π are exactly equivalent— different equations would be obtained with different constants.

Apart from the fact that the constants would be different, the two factors have different emphases. The partition coefficient P is a measure of the drug's overall hydrophobicity and is therefore an important measure of how efficiently a drug is transported to its target site and bound to its receptor. The π factor measures the hydrophobicity of a specific region on the drug's skeleton. Thus, any hydrophobic bonding to a receptor involving that region will be more significant to the equation than the overall transport process. If the substituent is involved in hydrophobic bonding to a receptor, then the QSAR equation using the π factor will emphasize that contribution to biological activity more dramatically than the equation using P.

Most QSAR equations will have a contribution from P or from π or from both. However, there are examples of drugs which have only a slight contribution. For example, a study on antimalarial drugs showed very little relationship between anti-malarial activity and hydrophobic character. This finding lends support to the theory that these drugs are acting in red blood cells, since previous research has shown that the ease with which drugs enter red blood cells is not related to their hydrophobicity.

9.3.2 Electronic effects

The electronic effects of various substituents will clearly have an effect on a drug's ionization or polarity. This in turn may have an effect on how easily a drug can pass through cell membranes or how strongly it can bind to a receptor. It is therefore useful to have some measure of the electronic effect a substituent can have on a molecule.

As far as substituents on an aromatic ring are concerned, the measure used is known as the Hammett substitution constant which is given the symbol σ.

The Hammett substitution constant (σ) is a measure of the electron withdrawing or electron donating ability of a substituent and has been determined by measuring the dissociation of a series of substituted benzoic acids compared to the dissociation of benzoic acid itself.

Benzoic acid is a weak acid and only partially ionizes in water (Fig. 9.7).

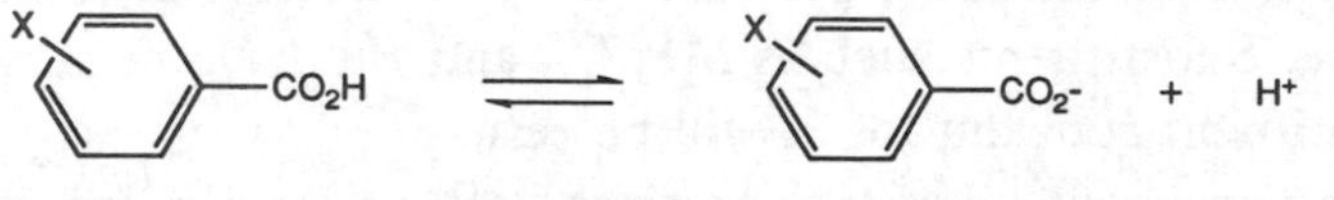

Fig. 9.7 Ionization of benzoic acid.

An equilibrium is set up between the ionized and non-ionized forms, where the relative proportions of these species is known as the equilibrium or dissociation constant K_H (the subscript H signifies that there are no substituents on the aromatic ring).

$$K_H = \frac{[PhCO_2^-]}{[PhCO_2H]}.$$

When a substituent is present on the aromatic ring, this equilibrium is affected. Electron withdrawing groups, such as a nitro group, result in the aromatic ring having a stronger electron withdrawing and stabilizing influence on the carboxylate anion. The equilibrium will therefore shift more to the ionized form such that the substituted benzoic acid is a stronger acid and has a larger K_X value (X represents the substituent on the aromatic ring) (Fig. 9.8).

If the substituent X is an electron donating group such as an alkyl group, then the aromatic ring is less able to stabilize the carboxylate ion. The equilibrium shifts to the left and a weaker acid is obtained with a smaller K_X value (Fig. 9.8).

The Hammett substituent constant (σ_X) for a particular substituent (X) is defined by the following equation:

$$\sigma_X = \log \frac{K_X}{K_H} = \log K_X - \log K_H.$$

Benzoic acids containing electron withdrawing substituents will have larger K_X values than benzoic acid itself (K_H) and therefore the value of σ_X for an electron withdrawing substituent will be positive. Substituents such as Cl, CN, or CF_3 have positive σ values.

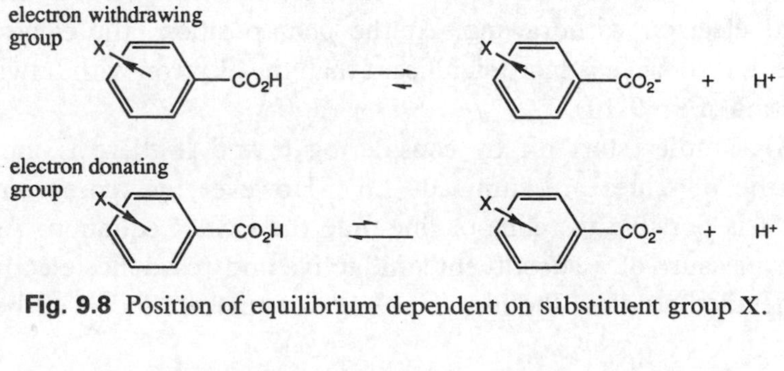

Fig. 9.8 Position of equilibrium dependent on substituent group X.

Benzoic acids containing electron donating substituents will have smaller K_X values than benzoic acid itself and hence the value of σ_X for an electron donating substituent will be negative. Substituents such as Me, Et, and Bu^t have negative σ values. The Hammett substituent constant for H will be zero.

The Hammett constant takes into account both resonance and inductive effects. Therefore, the value of σ for a particular substituent will depend on whether the substituent is *meta* or *para*. This is indicated by the subscript *m* or *p* after the σ symbol.

For example, the nitro substituent has $\sigma_p = 0.78$ and $\sigma_m = 0.71$. In the *meta* position, the electron withdrawing power is due to the inductive influence of the substituent, whereas at the *para* position inductive and resonance both play a part and so the σ_p value is greater (Fig. 9.9).

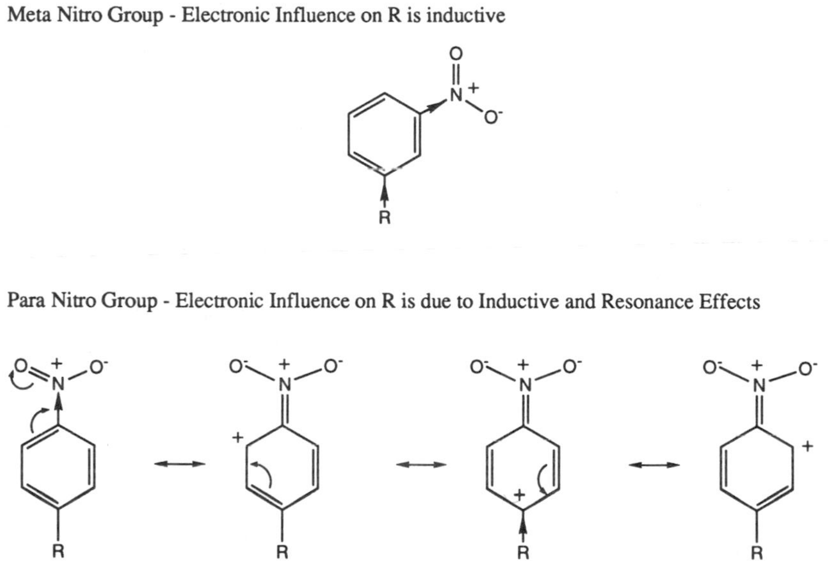

Meta Nitro Group - Electronic Influence on R is inductive

Para Nitro Group - Electronic Influence on R is due to Inductive and Resonance Effects

Fig. 9.9

For the OH group $\sigma_m = 0.12$ while $\sigma_p = -0.37$. At the *meta* position, the influence is inductive and electron withdrawing. At the *para* position, the electron donating influence due to resonance is more significant than the electron withdrawing influence due to induction (Fig. 9.10).

Most QSAR studies start off by considering σ and if there is more than one substituent, the σ values are summed ($\Sigma\sigma$). However, as more compounds are synthesized, it is possible to refine or fine-tune the QSAR equation. As mentioned above, σ is a measure of a substituent's inductive and resonance electronic effects. With more detailed studies, the inductive and resonance effects can be considered

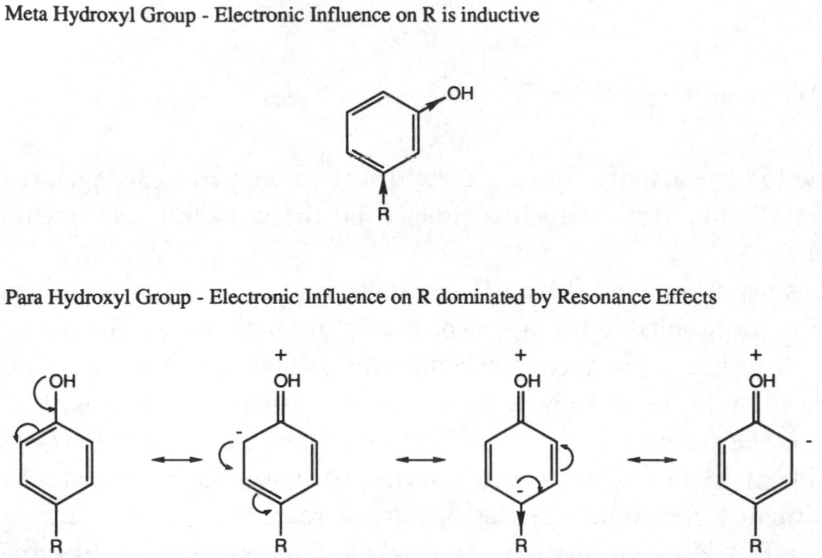

Meta Hydroxyl Group - Electronic Influence on R is inductive

Para Hydroxyl Group - Electronic Influence on R dominated by Resonance Effects

Fig. 9.10

separately. Tables of constants are available which quantify a substituent's inductive effect (F) and its resonance effect (R). In some cases, it might be found that a substituent's effect on activity is due to F rather than R, and vice versa. It might also be found that a substituent has a more significant effect at a particular position on the ring and this can also be included in the equation.

There are limitations to the electronic constants which we have described so far. For example, Hammett Substituent Constants cannot be measured for *ortho* substituents since such substituents have an important steric, as well as electronic, effect.

There are very few drugs whose activities are solely influenced by a substituent's electronic effect, since hydrophobicity usually has to be considered as well. Those that do are generally operating by a mechanism whereby they do not have to cross any cell membranes. Alternatively, *in vitro* studies on isolated enzymes may result in QSAR equations lacking the hydrophobicity factor, since there are no cell membranes to be considered.

The insecticidal activity of diethyl phenyl phosphates (Fig. 9.11) is one of the few examples where activity is related to electronic factors alone:

$$\log \left(\frac{1}{C}\right) = 2.282\sigma - 0.348.$$

The equation reveals that substituents with a positive value for σ (i.e. electron withdrawing groups) will increase activity. The fact that the π parameter is not significant is a good indication that the drugs do not have to pass into or through a cell

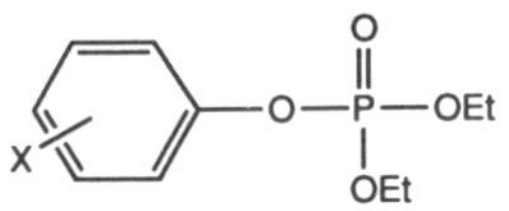

Fig. 9.11 Diethyl phenyl phosphate.

membrane to have activity. In fact, these drugs are known to act against an enzyme called acetylcholinesterase which is situated on the outside of cell membranes (see Chapter 11).

The above constants (σ, R, and F) can only be used for aromatic substituents and are therefore only suitable for drugs containing aromatic rings. However, a series of aliphatic electronic substituent constants are available. These were obtained by measuring the rates of hydrolysis for a series of aliphatic esters (Fig. 9.12). Methyl ethanoate is the parent ester and it is found that the rate of hydrolysis is affected by the substituent X. The extent to which the rate of hydrolysis is affected is a measure of the substituent's electronic effect at the site of reaction (i.e. the ester group). The electronic effect is purely inductive and is given the symbol σ_I. Electron donating groups reduce the rate of hydrolysis and therefore have negative values. For example, σ_I values for methyl, ethyl, and propyl are -0.04, -0.07, and -0.36 respectively. Electron withdrawing groups increase the rate of hydrolysis and have positive values. The σ_I values for NMe_3^+ and CN are 0.93 and 0.53 respectively.

Fig. 9.12 Hydrolysis of an aliphatic ester.

It should be noted that the inductive effect is not the only factor affecting the rate of hydrolysis. The substituent may also have a steric effect. For example, a bulky substituent may 'shield' the ester from attack and lower the rate of hydrolysis. It is therefore necessary to separate out these two effects. This can be done by measuring hydrolysis rates under basic conditions and also under acidic conditions. Under basic conditions, steric and electronic factors are important, whereas under acidic conditions only steric factors are important. By comparing the rates, values for the electronic effect (σ_I), and for the steric effect (E_S) (see below) can be determined.

9.3.3 Steric factors

In order for a drug to interact with an enzyme or a receptor, it has to approach, then bind to a binding site. The bulk, size, and shape of the drug may have an influence on this process. For example, a bulky substituent may act like a shield and hinder the ideal interaction between drug and receptor. Alternatively, a bulky substituent may help to orientate a drug properly for maximum receptor binding and increase activity.

Quantifying steric properties is more difficult than quantifying hydrophobic or electronic properties. Several methods have been tried and three are described here. It is highly unlikely that a drug's biological activity will be affected by steric factors alone, but these factors are frequently to be found in Hansch equations (Section 9.4.).

Taft's steric factor (E_s)

Attempts have been made to quantify the steric features of substituents by using Taft's steric factor (E_s). The value for E_s can be obtained as described in Section 9.3.2. However, the number of substituents which can be studied by this method is restricted.

Molar refractivity (MR)

Another measure of the steric factor is provided by a parameter known as molar refractivity (MR). This is a measure of the volume occupied by an atom or group of atoms. The molar refractivity is obtained from the following equation:

$$MR = \frac{(n^2 - 1)}{(n^2 + 2)} \times \frac{MW}{d}$$

where n is the index of refraction, MW is the molecular weight, and d is the density. The term MW/d defines a volume, while the $(n^2 - 1)/(n^2 + 2)$ term provides a correction factor by defining how easily the substituent can be polarized. this is particularly significant if the substituent has pi electrons or lone pairs of electrons.

Verloop steric parameter

Another approach to measuring the steric factor involves a computer programme called STERIMOL which calculates steric substituent values (Verloop steric parameters) from standard bond angles, van der Waals radii, bond lengths, and possible conformations for the substituent. Unlike E_s, the Verloop steric parameter can be measured for any substituent.

9.3.4 Other physicochemical parameters

The physicochemical properties most commonly studied by the QSAR approach have been described above, but other properties have also been studied. These include dipole moments, hydrogen bonding, conformation, and interatomic distances. However, difficulties in quantifying these properties limit the use of these parameters.

9.4 *Hansch equation*

In Section 9.3. we looked at the physicochemical properties commonly used in QSAR studies and how it is possible to quantify them. In a simple situation where biological activity is related to only one such property, a simple equation can be drawn up. However, the biological activity of most drugs is related to a combination of physico-chemical properties. In such cases, simple equations involving only one parameter are relevant only if the other parameters are kept constant. In reality, this is not easy to achieve and equations which relate biological activity to more than one parameter are more common. These equations are known as Hansch equations and they usually relate biological activity to the most commonly used physicochemical properties (P and/or π, σ, and a steric factor). If the range of hydrophobicity values is limited to a small range then the equation will be linear as follows:

$$\log \left(\frac{1}{C}\right) = k_1 \log P + k_2 \sigma + k_3 E_s + k_4.$$

If the P values are spread over a large range then the equation will be parabolic for the same reasons described in Section 9.3.1.

$$\log \left(\frac{1}{C}\right) = -k(\log P)^2 + k_2 \log P + k_3 \sigma + k_4 E_s + k_5.$$

The constants k_1–k_5 are determined by computer in order to get the best fitting line.

Not all the parameters will necessarily be significant. For example, the adrenergic blocking activity of β-halo-β-arylamines (Fig. 9.13) was related to π and σ and did not include a steric factor:

$$\log \left(\frac{1}{C}\right) = 1.22\pi - 1.59\sigma + 7.89.$$

This equation tells us that biological activity increases if the substituents have a positive π value and a negative σ value. In other words, the substituents should be hydrophobic and electron donating.

Since the P value and the π factor are not necessarily correlated, it is possible to have Hansch equations containing both of these factors. For example, a series of 102 phenanthrene aminocarbinols (Fig. 9.14) were tested for antimalarial activity and found to fit the following equation:

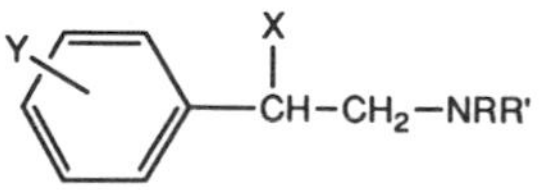

Fig. 9.13 β-Halo-β-arylamines.

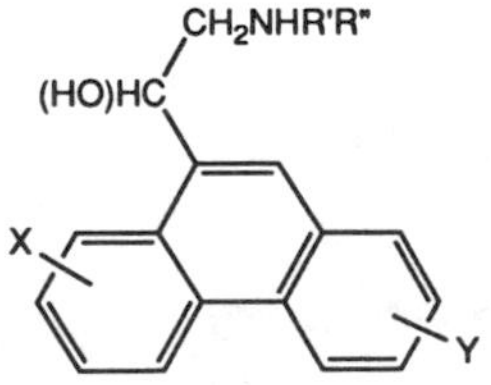

Fig. 9.14 Phenanthrene amino-carbinol structure.

$$\log \left(\frac{1}{C}\right) = -0.015(\log P)^2 + 0.14 \log P + 0.27\Sigma\pi_X + 0.40\Sigma\pi_Y + 0.65\Sigma\sigma_X +$$

$$0.88\Sigma\sigma_Y + 2.34.$$

This equation tells us that antimalarial activity increases very slightly as the hydrophobicity of the molecule (P) increases. The constant of 0.14 is low and shows that the increase is slight. The $(\log P)^2$ term shows that there is an optimum P value for activity. The equation also shows that activity increases significantly if hydrophobic substituents are present on ring X and in particular on ring Y. This could be taken to imply that some form of hydrophobic interaction is involved at these sites. Electron withdrawing substituents on both rings are also beneficial to activity, more so on ring Y than ring X.

When carrying out a Hansch analysis, it is important to choose the substituents carefully to ensure that the change in biological activity can be attributed to a particular parameter. There are plenty of traps for the unwary. Take, for example, drugs which contain an amine group. One of the most frequently carried out studies on amines is to synthesize analogues containing a homologous series of alkyl substituents on the nitrogen atom (i.e. Me, Et, Pr^n, Bu^n). If activity increases with the chain length of the substituent, is it due to increasing hydrophobicity or to increasing size or to both? If we look at the π and MR values of these substituents, then we find that both increase in a similar fashion across the series and we would not be able to distinguish between them (Fig. 9.15).

Substituent	H	Me	Et	Pr^n	Bu^n	OMe	$NHCONH_2$	I	CN
π	0.00	0.56	1.02	1.50	2.13	-0.02	-1.30	1.12	-0.57
MR	0.10	0.56	1.03	1.55	1.96	0.79	1.37	1.39	0.63

Fig. 9.15 Values for π and MR for a series of substituents.

In this example, a series of substituents would have to be chosen where π and MR are not related. The substituents H, Me, OMe, $NHCOCH_2$, I, and CN would be more suitable.

9.5 The Craig plot

Although tables of π and σ factors are readily available for a large range of substituents, it is often easier to visualize the relative properties of different substituents by considering a plot where the y axis is the value of the σ factor and the x axis is the value of the π factor. Such a plot is known as a Craig plot. The example shown in Fig.

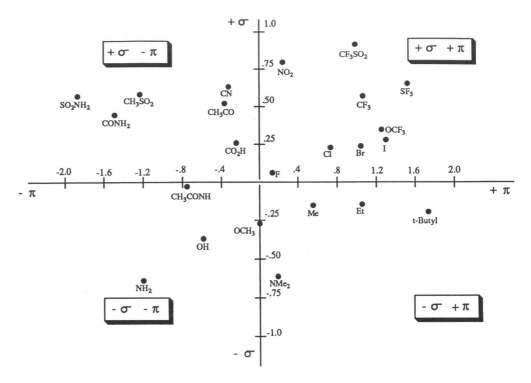

Fig. 9.16 Craig plot.

9.16 is the Craig plot for the σ and π factors of *para*-aromatic substituents. There are several advantages to the use of such a Craig plot.

- The plot shows clearly that there is no overall relationship between π and σ. The various substituents are scattered around all four quadrants of the plot.
- It is possible to tell at a glance which substituents have positive π and σ parameters, which substituents have negative π and σ parameters, and which substituents have one positive and one negative parameter.
- It is easy to see which substituents have similar π values. For example, the ethyl, bromo, trifluoromethyl, and trifluoromethylsulfonyl groups are all approximately on the same vertical line on the plot. In theory, these groups could be interchangeable on drugs where the principal factor affecting biological activity is the π factor. Similarly, groups which form a horizontal line can be identified as being isoelectronic or having similar σ values (e.g. CO_2H, Cl, Br, I).
- The Craig plot is useful in planning which substituents should be used in a QSAR study. In order to derive the most accurate equation involving π and σ, analogues should be synthesized with substituents from each quadrant. For example, halide substituents are useful representatives of substituents with increased hydrophobicity

and electron withdrawing properties (positive π and positive σ), whereas an OH substituent has more hydrophilic and electron donating properties (negative π and negative σ). Alkyl groups are examples of substituents with positive π and negative σ values, whereas acyl groups have negative π and positive σ values.

• Once the Hansch equation has been derived, it will show whether π or σ should be negative or positive in order to get good biological activity. Further developments would then concentrate on substituents from the relevant quadrant. For example, if the equation shows that positive π and positive σ values are necessary, then further substituents should only be taken from the top right quadrant.

Craig plots can also be drawn up to compare other sets of physicochemical parameters, such as hydrophobicity and *MR*.

9.6 *The Topliss scheme*

In certain situations, it might not be feasible to make the large range of structures required for a Hansch equation. For example, the synthetic route involved might be difficult and only a few structures can be made in a limited time. In these circumstances, it would be useful to test compounds for biological activity as they are synthesized and to use these results to determine the next analogue to be synthesized.

A Topliss scheme is a 'flow diagram' which allows such a procedure to be followed. There are two Topliss schemes, one for aromatic substituents (Fig. 9.17) and one for aliphatic side-chain substituents (Fig. 9.18). The schemes were drawn up by considering the hydrophobicity and electronic factors of various substituents and are designed such that the optimum substituent can be found as efficiently as possible. However, they are not meant to be a replacement for a full Hansch analysis. Such an analysis

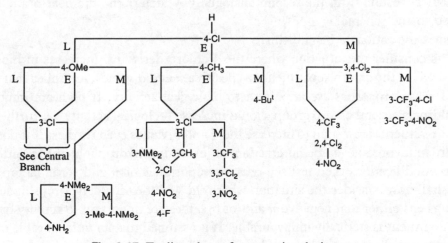

Fig. 9.17 Topliss scheme for aromatic substituents.

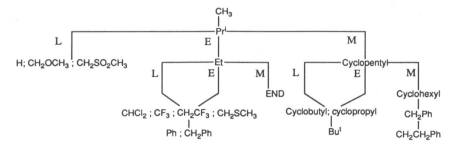

Fig. 9.18 Topliss scheme for aliphatic side chain substituents.

would be carried out in due course, once a suitable number of structures have been synthesized.

The Topliss scheme for aromatic substituents (Fig. 9.17) assumes that the lead compound has been tested for biological activity and contains a monosubstituted aromatic ring. The first analogue in the scheme is the 4-chloro derivative, since this derivative is usually easy to synthesize. The chloro substituent is more hydrophobic and electron withdrawing than hydrogen and therefore, π and σ are positive.

Once the chloro analogue has been synthesized, the biological activity is measured. There are three possibilities. The analogue will have less activity (L), equal activity (E), or more activity (M). The type of activity observed will determine which branch of the Topliss scheme is followed next.

If the biological activity increases, then the (M) branch is followed and the next analogue to be synthesized is the 3,4-dichloro-substituted analogue. If, on the other hand, the activity stays the same, then the (E) branch is followed and the 4-methyl analogue is synthesized. Finally, if activity drops, the (L) branch is followed and the next analogue is the 4-methoxy analogue.

Biological results from the second analogue now determine the next branch to be followed in the scheme.

What is the rationale behind this?

Let us consider the situation where the 4-chloro derivative increases in biological activity. Since the chloro substituent has positive π and σ values, it implies that one or both of these properties are important to biological activity. If both are important, then adding a second chloro group should increase biological activity yet further. If it does, substituents are varied to increase the π and σ values even further. If it does not, then an unfavourable steric interaction or excessive hydrophobicity is indicated. Further modifications then test the relative importance of π and steric factors.

We shall now consider the situation where the 4-chloro analogue drops in activity. This suggests either that negative π and/or σ values are important to activity or that a *para* substituent is sterically unfavourable. It is assumed that an unfavourable σ effect is the most likely reason for the reduced activity and so the next substituent is one with a negative σ factor (i.e. 4-OMe). If activity improves, further changes are suggested to test the relative importance of the σ and π factors. If, on the other hand, the 4-OMe group does not improve activity, it is assumed that an unfavourable steric factor is at work and the next substituent is a 3-chloro group. Modifications of this group would then be carried out in the same way as shown in the centre branch of Fig. 9.17.

The last scenario is where the activity of the 4-chloro analogue is little changed from the lead compound. This could arise from the drug requiring a positive π value and a negative σ value. Since both values for the chloro group are positive, the beneficial effect of the positive π value might be cancelled out by the detrimental effects of a positive σ value. The next substituent to try in that case is the 4-methyl group which has the necessary positive π value and negative σ value. If this still has no beneficial effect, then it is assumed that there is an unfavourable steric interaction at the *para* position and the 3-chloro substituent is chosen next. Further changes continue to vary the relative values of the π and σ factors.

The validity of the Topliss scheme was tested by looking at structure–activity results for various drugs which had been reported in the literature. For example, the biological activity of nineteen substituted benzenesulfonamides (Fig. 9.19) have been reported. The second most active compound was the nitro-substituted analogue which would have been the fifth compound synthesized if the Topliss scheme had been followed.

Another example comes from the anti-inflammatory activities of substituted aryl-tetrazolylalkanoic acids (Fig. 9.20). Twenty-eight of these were synthesized. Using the Topliss scheme, three out of the four most active structures would have been synthesized from the first eight compounds synthesized.

The Topliss scheme for aliphatic side-chains (Fig. 9.18) was set up following a similar rationale to the aromatic scheme, and is used in the same way for side-groups attached to a carbonyl, amino, amide, or similar functional group. The scheme only attempts to differentiate between the hydrophobic and electronic effects of substituents and not the steric properties. Thus, the substituents involved have been chosen to try and minimize any steric differences. It is assumed that the lead com-

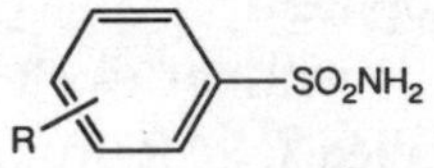

Order of Synthesis	R	Biological Activity	High Potency
1	H	-	
2	4-Cl	M	
3	3,4-Cl$_2$	L	
4	4-Br	E	
5	4-NO$_2$	M	*

M = More Activity
L = Less Activity
E = Equal Activity

Fig. 9.19 Biological activity of substituted benzenesulfonamides.

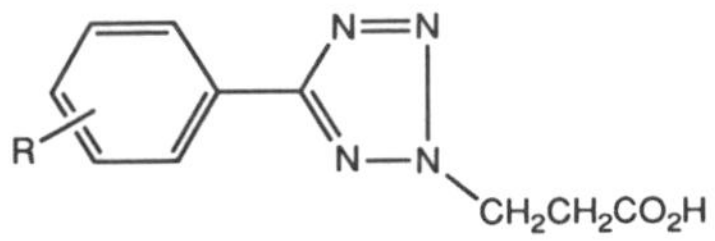

Order of Synthesis	R	Biological Activity	High Potency
1	H	-	
2	4-Cl	L	
3	4-MeO	L	
4	3-Cl	M	*
5	3-CF$_3$	L	
6	3-Br	M	*
7	3-I	L	
8	3,5-Cl$_2$	M	*

M= More Activity
L= Less Activity
E = Equal Activity

Fig. 9.20 Anti-inflammatory activities of substituted aryltetrazolylakonoic acids.

pound has a methyl group. The first analogue suggested is the isopropyl analogue. This has an increased π value and in most cases would be expected to increase activity, since it has been found from experience that the hydrophobicity of most lead compounds are less than the optimum hydrophobicity required for activity.

Let us concentrate first of all on the situation where activity rises. Following this branch, a cyclopentyl group is now used. A cyclic structure is used since it has a larger π value, but keeps any increase in steric factor to a minimum. If activity rises again, more hydrophobic substituents are tried. If activity does not rise, then there could be two explanations. Either the optimum hydrophobicity has been passed or there is an electronic effect (σ_I) at work. Further substituents are then used to determine which is the correct explanation.

Let us now look at the situation where the activity of the isopropyl analogue stays much the same. The most likely explanation is that the methyl group and the isopropyl group are on either side of the hydrophobic optimum. Therefore, an ethyl group is used next, since it has an intermediate π value. If this does not lead to an improvement, it is possible that there is an unfavourable electronic effect. The groups used have been electron donating, and so electron withdrawing groups with similar π values are now suggested.

Finally, we shall look at the case where activity drops for the isopropyl group. In this case, hydrophobic and/or electron donating groups could be bad for activity and the groups suggested are suitable choices for further development.

The Topliss scheme has proved useful many times, but it will not work in every case, and it is not meant to be a replacement for more detailed QSAR studies.

9.7 Bioisosteres

Tables of substituent constants are available for various physicochemical properties. A knowledge of these constants allows the medicinal chemist to identify substituents

Substituent	$-\overset{O}{\underset{\|}{C}}-CH_3$	$\overset{NC\diagdown\diagup CN}{\underset{\|}{\overset{C}{C}}}-CH_3$	$-\overset{O}{\underset{\|}{S}}-CH_3$	$-\overset{O}{\underset{\overset{\|}{\underset{O}{\|}}}{S}}-CH_3$	$-\overset{O}{\underset{\overset{\|}{\underset{O}{\|}}}{S}}-NHCH_3$	$-\overset{O}{\underset{\|}{C}}-NMe_2$
π	-0.55	0.40	-1.58	-1.63	-1.82	-1.51
σ_p	0.50	0.84	0.49	0.72	0.57	0.36
σ_m	0.38	0.66	0.52	0.60	0.46	0.35
MR	11.2	21.5	13.7	13.5	16.9	19.2

Fig. 9.21 Physicochemical parameters for six substituents.

which may be potential bioisosteres. Thus, the substituents CN, NO_2, and COMe have similar hydrophobic, electronic, and steric factors, and might be interchangeable. Such interchangeability was observed in the development of cimetidine (Chapter 13). The important thing to notice is that groups can be bioisosteric in some situations, but not others. Consider for example the table shown in Fig. 9.21.

This table shows some physicochemical parameters for six different substituents. If the most important physicochemical parameter for biological activity is σ_p, then the $COCH_3$ group (0.50) would be a reasonable bioisostere for the $SOCH_3$ group (0.49). If, on the other hand, the dominant parameter is π, then a more suitable bioisostere for $SOCH_3$ (−1.58) would be SO_2CH_3 (−1.63).

9.8 Planning a QSAR study

When starting a QSAR study it is important to decide which physicochemical parameters are going to be studied and to plan the analogues such that the parameters under study are suitably varied. For example, it would be pointless to synthesize analogues where the hydrophobicity and steric volume of the substituents are correlated, if these two parameters are to go into the equation.

It is also important to make enough structures to make the results statistically meaningful. As a rule of thumb, five structures should be made for every parameter studied. Typically, the initial QSAR study would involve the two parameters π and σ, and possibly E_s. Craig plots could be used in order to choose suitable substituents.

Certain substituents are worth avoiding in the initial study since they may have properties other than those being studied. For example, substituents which might ionize (CO_2H, NH_2, SO_2H) should be avoided. Groups which might easily be metabolized should be avoided if possible (e.g. esters or nitro groups).

If there are two or more substituents, then the initial equation usually considers the total π and σ contribution.

As more analogues are made, it is often possible to consider the hydrophobic and electronic effect of substituents at specific positions of the molecule. Furthermore, the electronic parameter σ can be split into its inductive and resonance components (F and R). Such detailed equations may show up a particular localized requirement for activity. For example, a hydrophobic substituent may be favoured in one part of the skeleton, while an electron withdrawing substituent is favoured at another. This in turn gives clues about the binding interactions involved between drug and receptor.

9.9 *Case study*

An example of how the QSAR equation can change and become more specific as a study develops is demonstrated from a study carried out by workers at Smith, Kline, & French on the antiallergic activity of a series of pyranenamines (Fig. 9.22). In this study, substituents were varied on the aromatic ring, and the remainder of the molecule was kept constant. Nineteen compounds were synthesized and the first QSAR equation was obtained by only considering π and σ:

$$\log\left(\frac{1}{C}\right) = -0.14\Sigma\pi - 1.35(\Sigma\sigma)^2 - 0.72$$

where $\Sigma\pi$ and $\Sigma\sigma$ are the total π and σ values for all substituents present.

The negative coefficient for the π term shows that activity is inversely proportional to hydrophobicity, which is quite unusual. The $(\Sigma\sigma)^2$ term is also quite unusual. It was chosen since there was no simple relationship between activity and σ. In fact, it was observed that activity dropped if the substituent was electron withdrawing *or* electron donating. Activity was best with neutral substituents. To take account of this, the $(\Sigma\sigma)^2$ term was introduced. Since the coefficient in the equation is negative, activity is lowered if σ is anything other than zero.

A further range of compounds were synthesized with hydrophilic substituents to test this equation, making a total of 61 structures. This resulted in the following inconsistencies.

* The activities for the substituents 3-NHCOMe, 3-NHCOEt, and 3-NHCOPr were all similar. However, according to the equation, the activities should have dropped as the alkyl group became larger due to increasing hydrophobicity.
* Activity was greater than expected if there was a substituent such as OH, SH, NH_2, or NHCOR, at position 3, 4, or 5.
* The substituent $NHSO_2R$ was bad for activity.
* The substituents 3,5-$(CF_3)_2$ and 3,5-$(NHCOMe)_2$ had much greater activity than expected.

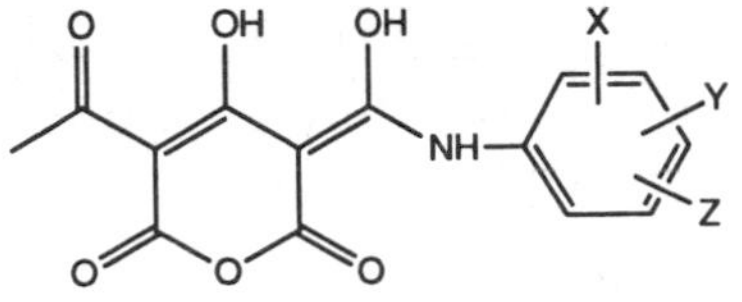

Fig. 9.22 Structure of pyranenamine.

- An acyloxy group at the 4-position resulted in an activity five times greater than predicted by the equation.

These results implied that the initial equation was too simple and that properties other than π and σ were important to activity. At this stage, the following theories were proposed to explain the above results.

- The similar activities for 3-NHCOMe, 3-NHCOEt, and 3-NHCOPr could be due to a steric factor. The substituents had increasing hydrophobicity which is bad for activity, but they were also increasing in size and it was proposed that this was good for activity. The most likely explanation is that the size of the substituent is forcing the drug into the correct orientation for optimum receptor interaction.
- The substituents which unexpectedly increased activity when they were at positions 2, 3, or 4 are all capable of hydrogen bonding. This suggests an important hydrogen bonding interaction with the receptor. For some reason, the $NHSO_2R$ group is an exception, which implies there is some other unfavourable steric or electronic factor peculiar to this group.
- The increased activity for 4-acyloxy groups was explained by suggesting that these analogues are acting as prodrugs. The acyloxy group is less polar than the hydroxyl group and so these analogues would be expected to cross cell membranes and reach the receptor more efficiently than analogues bearing a free hydroxyl group. At the receptor, the ester group could be hydrolysed to reveal the hydroxyl group which would then take part in hydrogen bonding with the receptor.
- The structures having substituents $3,5\text{-}(CF_3)_2$ and $3,5\text{-}(NHCOMe)_2$ are the only disubstituted structures where a substituent at position 5 has an electron withdrawing effect, so this feature was also introduced into the next equation.

The revised QSAR equation was as follows:

$$\log \left(\frac{1}{C}\right) = -0.30\Sigma\pi - 1.5(\Sigma\sigma)^2 + 2.0(F\text{-}5) + 0.39(345\text{-}HBD) - 0.63(NHSO_2) +$$
$$0.78(M\text{-}V) + 0.72(4\text{-}OCO) - 0.75.$$

The π and σ parameters are still present, but a number of new parameters have now appeared.

- The $F\text{-}5$ term represents the inductive effect of a substituent at position 5. Since the coefficient is positive and large, it shows that an electron withdrawing group substantially increases activity. However, since only 2 compounds in the 61 synthesized had a 5-substituent, there might be quite an error in this result.
- The advantage of having hydrogen bonding substituents at position 3,4, or 5 is accounted for by including a hydrogen bonding term ($345\text{-}HBD$). The value of this term depends on the number of hydrogen bonding substituents present. If one such

group is present, the *345-HBD* term is 1. If two such groups were present, the parameter is 2. Therefore, for each hydrogen bonding substituent present at positions 3, 4, or 5, $\log(1/C)$ increases by 0.39.

- The $NHSO_2$ term was introduced since this group was poor for activity despite being capable of hydrogen bonding. The negative coefficient indicates the drop in activity. A figure of 1 is used for any $NHSO_2R$ substituent present.
- The *M-V* term represents the volume of any *meta* substituent, and since the coefficient is positive, it indicates that substitutents with a large volume at the *meta* position increase activity.
- The *4-OCO* term is either 0 or 1 and is only present if an acyloxy group is present at position 4, and so $\log(1/C)$ is increased by 0.72 if the acyl group is present.

The most important parameters in the above equation are the hydrophobic parameter and the 4-OCO parameter.

A further 37 structures were synthesized to test steric and *F-5* parameters as well as exploring further groups capable of hydrogen bonding. Since hydrophilic substituents were good for activity, a range of very hydrophilic substituents were also tested to see if there was an optimum value for hydrophilicity.

The results obtained highlighted one more anomaly in that two hydrogen bonding groups *ortho* to each other were bad for activity. This was attributed to the groups hydrogen bonding with each other rather than to the receptor.

A revised equation was obtained as follows:

$$\log\left(\frac{1}{C}\right) = -0.034(\Sigma\pi)^2 - 0.33(\Sigma\pi) + 4.3(F\text{-}5) + 1.3(R\text{-}5) - 1.7(\Sigma\sigma)^2 +$$

$$0.73(345\text{-}HBD) - 0.86(HB\text{-}INTRA) - 0.69(NHSO_2) + 0.72(4\text{-}OCO) - 0.59.$$

The main points of interest from this equation are as follows.

- Increasing the hydrophilicity of substituents allowed the identification of an optimum value for hydrophobicity ($\Sigma\pi = -5$) and introduced the $(\Sigma\pi)^2$ parameter into the equation. The value of -5 is remarkably low and indicates that the receptor site is hydrophilic.
- As far as electronic effects are concerned, it is revealed that the resonance effects of substituents at the 5-position also have an influence on activity.
- The unfavourable situation where two hydrogen bonding groups are *ortho* to each other is represented by the *HB-INTRA* parameter. This parameter is given the value 1 if such an interaction is possible and the negative constant (-0.86) shows that such interactions decrease activity.
- It is interesting to note that the steric parameter is no longer significant and has disappeared from the equation.

The compound having the greatest activity has two $NHCOCH(OH)CH_2OH$ substituents at the 3- and 5-positions and is 1000 times more active than the original lead compound. The substituents are very polar and are not ones which would normally be used. They satisfy all the requirements determined by the QSAR study. They are highly polar groups which can take part in hydrogen bonding. They are *meta* with respect to each other, rather than *ortho*, to avoid undesirable intramolecular hydrogen bonding. One of the groups is at the 5-position and has a favourable *F-5* parameter. Together the two groups have a negligible $(\Sigma\sigma)^2$ value. Such an analogue would certainly not have been obtained by trial and error and this example demonstrates the strengths of the QSAR approach.

All the evidence from this study suggests that the aromatic ring of this series of compounds is fitting into a hydrophilic pocket in the receptor which contains polar groups capable of hydrogen bonding.

It is further proposed that a positively charged residue such as arginine, lysine, or histidine might be present in the pocket which could interact with an electronegative substituent at position 5 of the aromatic ring. (Fig. 9.23).

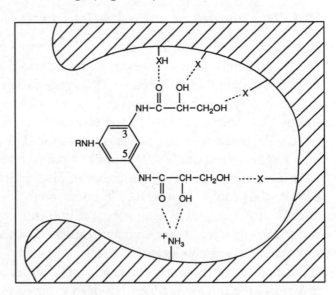

Fig. 9.23 Hypothetical receptor binding interactions of a pyranenamine.

This example demonstrates that QSAR studies and computers are powerful tools in medicinal chemistry. However, it also shows that the QSAR approach is a long way from replacing the human factor. One cannot put a series of facts and figures into a computer and expect it to magically produce an instant explanation of how a drug works. The medicinal chemist still has to interpret results, propose theories, and test those theories by incorporating the correct parameters into the QSAR equation. Imagination and experience still count for a great deal.

10 ▪ Antibacterial agents

The fight against bacterial infection is one of the great success stories of medicinal chemistry. The topic is a large one and there are terms used in this chapter which are unique to this particular field. Rather than clutter the text with explanations and definitions, Appendices 4 and 5 contain explanations of such terms as antibacterial, antibiotic, Gram-positive, Gram-negative, cocci, bacilli, streptococci, and staphylococci.

10.1 *The history of antibacterial agents*

Bacteria were first identified in the 1670s by van Leeuwenhoek, following his invention of the microscope. However, it was not until the nineteenth century that their link with disease was appreciated. This appreciation followed the elegant experiments carried out by the French scientist Pasteur, who demonstrated that specific bacterial strains were crucial to fermentation and that these and other microorganisms were far more widespread than was previously thought. The possibility that these microorganisms might be responsible for disease began to take hold.

An early advocate of a 'germ theory of disease' was the Edinburgh surgeon Lister. Despite the protests of several colleagues who took offence at the suggestion that they might be infecting their own patients, Lister introduced carbolic acid as an antiseptic and sterilizing agent for operating theatres and wards. The improvement in surgical survival rates was significant.

During that latter half of the nineteenth century, scientists such as Koch were able to identify the microorganisms responsible for diseases such as tuberculosis, cholera, and typhoid. Methods such as vaccination for fighting infections were studied. Research was also carried out to try and find effective antibacterial agents or antibiotics. However, the scientist who can lay claim to be the father of chemotherapy—the use of chemicals against infection—was Paul Ehrlich. Ehrlich spent much of his career studying histology, then immunochemistry, and won a Nobel prize for his contributions to immunology. However, in 1904 he switched direction and entered a field which he defined as chemotherapy. Ehrlich's 'Principle of Chemotherapy' was that a chemical could directly interfere with the proliferation of microorganisms, at concentrations tolerated by the host. This concept was popularly known as the 'magic bullet', where the chemical was seen as a bullet which could search out and destroy the invading microorganism without adversely affecting the host. The process is one of selective toxicity, where the chemical shows greater toxicity to the target micro-organism than to the host cells. Such selectivity can be represented by a 'chemo-therapeutic index', which compares the minimum effective dose of a drug with the maximum dose which can be tolerated by the host. This measure of selectivity was eventually replaced by the currently used therapeutic index (see Glossary).

By 1910, Ehrlich had successfully developed the first example of a purely synthetic antimicrobial drug. This was the arsenic-containing compound salvarsan (Fig. 10.1). Although it was not effective against a wide range of bacterial infections, it did prove effective against the protozoal disease sleeping sickness (trypanosomiasis), and the spirochaete disease of syphilis. The drug was used until 1945 when it was replaced by penicillin.

Over the next twenty years, progress was made against a variety of protozoal diseases, but little progress was made in finding antibacterial agents, until the intro-duction of proflavine in 1934.

Proflavine (Fig. 10.2) is a yellow-coloured aminoacridine structure which is particu-

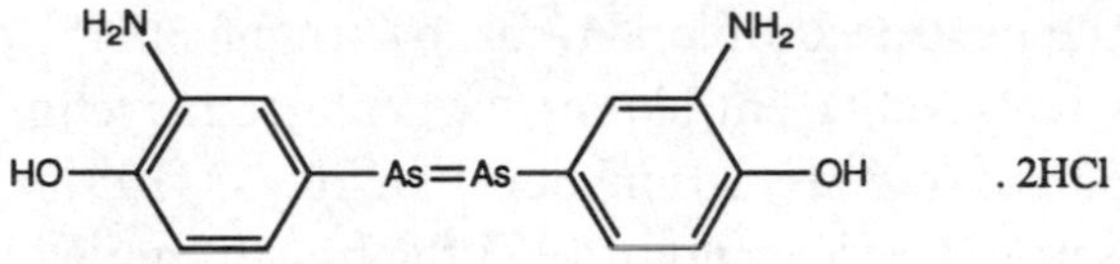

Fig. 10.1 Salvarsan.

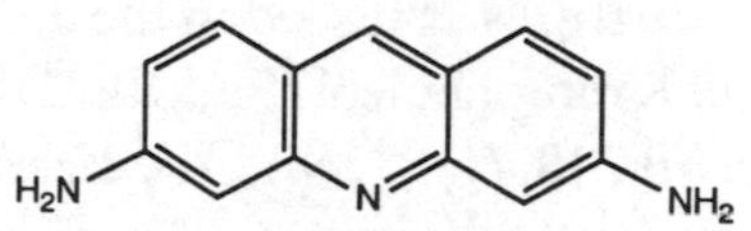

Fig. 10.2 Prolavine.

larly effective against bacterial infections in deep surface wounds, and was used to great effect during the Second World War. It is an interesting drug since it targets bacterial DNA rather than protein (see Chapter 6). Despite the success of this drug, it was not effective against bacterial infections in the bloodstream and there was still an urgent need for agents which would fight these infections.

This need was answered in 1935 with the discovery that a red dye called prontosil (Fig. 10.3) was effective against streptococci infections *in vivo*. As discussed later,

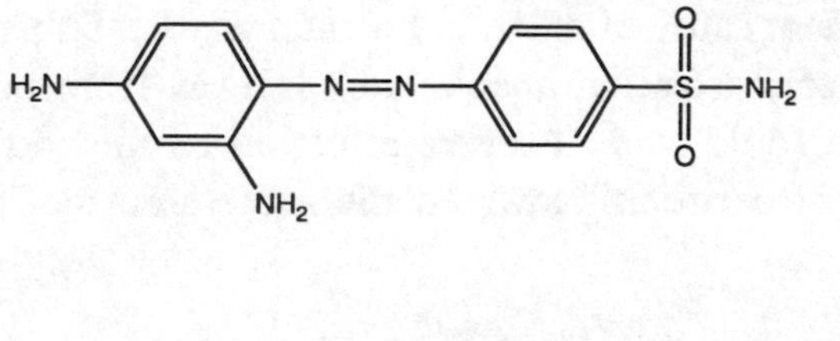

Fig. 10.3 Prontosil.

prontosil was eventually recognized as being a prodrug for a new class of antibacterial agents—the sulfa drugs (sulfonamides). The discovery of these drugs was a real breakthrough, since they represented the first drugs to be effective against bacterial infections carried in the bloodstream. They were the only effective drugs until penicillin became available in the early 1940s.

Although penicillin (Fig. 10.18) was discovered in 1928, it was not until 1940 that effective means of isolating it were developed by Florey and Chain. Society was then rewarded with a drug which revolutionized the fight against bacterial infection and proved even more effective than the sulfonamides.

Despite penicillin's success, it was not effective against all types of infection and the need for new antibacterial agents still remained. Penicillin is an example of a toxic chemical produced by a fungus to kill bacteria which might otherwise compete with it for nutrients. The realization that fungi might be a source for novel antibiotics spurred scientists into a huge investigation of microbial cultures, both known and unknown.

In 1944, the antibiotic streptomycin (Fig. 10.70) was discovered from a systematic search of soil organisms. It extended the range of chemotherapy to *Tubercle bacillus* and a variety of Gram-negative bacteria. This compound was the first example of a series of antibiotics known as the aminoglycoside antibiotics.

After the Second World War, the effort continued to find other novel antibiotic structures. This led to the discovery of the peptide antibiotics (e.g. bacitracin (1945)), chloramphenicol (Fig. 10.72) (1947), the tetracycline antibiotics (e.g. chlortetracycline (Fig. 10.71) (1948)), the macrolide antibiotics (e.g. erythromycin (Fig. 10.73) (1952)), the cyclic peptide antibiotics (e.g. cycloserine (1955)), and in 1955 the first example of a second major group of β-lactam antibiotics, cephalosporin C (Fig. 10.41).

As far as synthetic agents were concerned, isoniazid (a pyridine hydrazide structure) was found to be effective against human tuberculosis in 1952, and in 1962 nalidixic acid (Fig. 10.74) (the first of the quinolone antibacterial agents) was discovered. A second generation of this class of drugs was introduced in 1987 with ciprofloxacin (Fig. 10.74).

Many antibacterial agents are now available and the vast majority of bacterial diseases have been brought under control (e.g. syphilis, tuberculosis, typhoid, bubonic plague, leprosy, diphtheria, gas gangrene, tetanus, gonorrhoea).

This represents a great achievement for medicinal chemistry and it is perhaps sobering to consider the hazards which society faced in the days before penicillin.

Septicaemia was a risk faced by mothers during childbirth and could lead to death. Ear infections were common especially in children and could lead to deafness. Pneumonia was a frequent cause of death in hospital wards. Tuberculosis was a major problem, requiring special isolation hospitals built away from populated centres. A simple cut or a wound could lead to severe infection requiring the amputation of a limb, while the threat of peritonitis lowered the success rates of surgical operations.

These were the days of the thirties—still within living memory for many. Perhaps those of us born since the Second World War take the success of antibacterial agents too much for granted.

10.2 *The bacterial cell*

The success of antibacterial agents owes much to the fact that they can act selectively against bacterial cells rather than animal cells. This is largely due to the fact that bacterial cells and animal cells differ both in their structure and in the biosynthetic pathways which proceed inside them. Let us consider some of the differences between the bacterial cell (Fig. 10.4) and the animal cell.

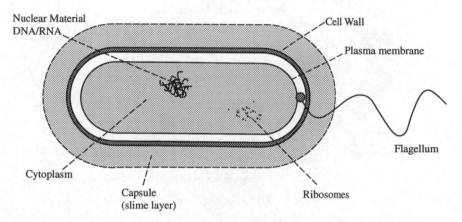

Fig. 10.4 The bacterial cell.

Differences between bacterial and animal cells

- The bacterial cell has a cell wall, as well as a cell membrane, whereas the animal cell has only a cell membrane. The cell wall is crucial to the bacterial cell's survival. Bacteria have to survive a wide range of environments and osmotic pressures, whereas animal cells do not. If a bacterial cell lacking a cell wall was placed in an aqueous environment containing a low concentration of salts, water would freely enter the cell due to osmotic pressure. This would cause the cell to swell and eventually 'burst'. The cell wall does not stop water flowing into the cell directly, but it does prevent the cell from swelling and so indirectly prevents water entering the cell.
- The bacterial cell does not have a defined nucleus, whereas the animal cell does.
- Animal cells contain a variety of structures called organelles (e.g. mitochondria, etc.), whereas the bacterial cell is relatively simple.

- The biochemistry of a bacterial cell differs significantly from that of an animal cell. For example, bacteria may have to synthesize essential vitamins which animal cells can acquire intact from food. The bacterial cells must have the enzymes to catalyse these reactions. Animal cells do not, since the reactions are not required.

10.3 *Mechanisms of antibacterial action*

There are four main mechanisms by which antibacterial agents act.

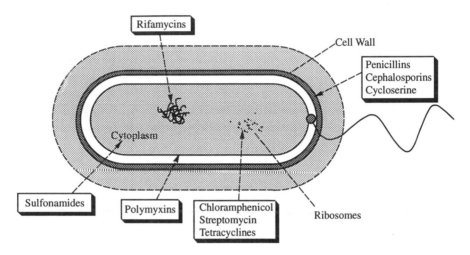

Fig. 10.5 Sites of antibacterial action.

- Inhibition of cell metabolism.
  Antibacterial agents which inhibit cell metabolism are called antimetabolites. These compounds inhibit the metabolism of a microorganism, but not the metabolism of the host. They do this by inhibiting an enzyme-catalysed reaction which is present in the bacterial cell, but not in animal cells. The best known examples of antibacterial agents acting in this way are the sulfonamides.
- Inhibition of bacterial cell wall synthesis.
  Inhibition of cell wall synthesis leads to bacterial cell lysis (bursting) and death. Agents operating in this way include penicillins and cephalosporins. Since animal cells do not have a cell wall, they are unaffected by such agents.
- Interactions with the plasma membrane.
  Some antibacterial agents interact with the plasma membrane of bacterial cells to affect membrane permeability. This has fatal results for the cell. Polymyxins and tyrothricin operate in this way.

- Disruption of protein synthesis.
  Disruption of protein synthesis means that essential enzymes required for the cell's survival can no longer be made. Agents which disrupt protein synthesis include the rifamycins, aminoglycosides, tetracyclines, and chloramphenicol.
- Inhibition of nucleic acid transcription and replication.
  Inhibition of nucleic acid function prevents cell division and/or the synthesis of essential enzymes. Agents acting in this way include nalidixic acid and proflavin.

We shall now consider these mechanisms in more detail.

10.4 *Antibacterial agents which act against cell metabolism (antimetabolites)*

10.4.1 Sulfonamides

The history of sulfonamides

The best example of antibacterial agents acting as antimetabolites are the sulfonamides (sometimes called the sulfa drugs).

The sulfonamide story began in 1935 when it was discovered that a red dye called prontosil had antibacterial properties *in vivo* (i.e. when given to laboratory animals). Strangely enough, no antibacterial effect was observed *in vitro*. In other words, prontosil could not kill bacteria grown in the test tube. This remained a mystery until it was discovered that prontosil was not in fact the antibacterial agent.

Instead, it was found that the dye was metabolized by bacteria present in the small intestine of the test animal, and broken down to give a product called sulfanilamide (Fig. 10.6). It was this compound which was the true antibacterial agent. Thus, prontosil was the first example of a prodrug (see Chapter 8). Sulfanilamide was synthesized in the laboratory and became the first synthetic antibacterial agent active against a wide range of infections. Further developments led to a range of sulfonamides which proved effective against Gram-positive organisms, especially pneumococci and meningococci.

Despite their undoubted benefits, sulfa drugs have proved ineffective against infections such as *Salmonella*—the organism responsible for typhoid. Other problems have resulted from the way these drugs are metabolized, since toxic products

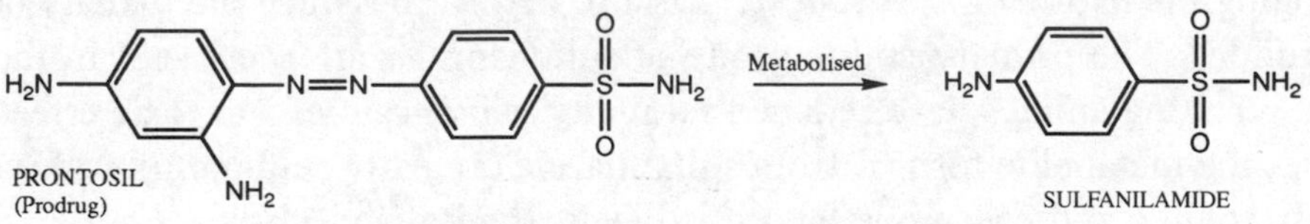

Fig. 10.6 Metabolism of prontosil.

are frequently obtained. This led to the sulfonamides mainly being superseded by penicillin.

Structure–activity relationships (SAR)

The synthesis of a large number of sulfonamide analogues (Fig. 10.7) led to the following conclusions.

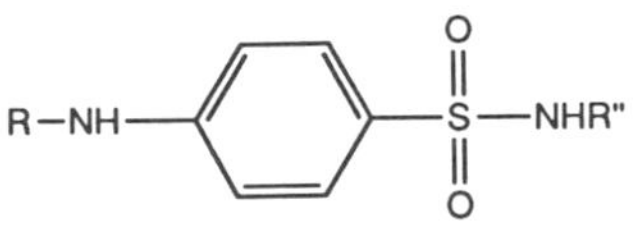

Fig. 10.7 Sulfonamide analogues.

- The *p*-amino group is essential for activity and must be unsubstituted (i.e. R = H). The only exception is when R = acyl (i.e. amides). The amides themselves are inactive but can be metabolized in the body to regenerate the active compound (Fig. 10.8). Thus amides can be used as sulfonamide prodrugs (see later).
- The aromatic ring and the sulfonamide functional group are both required.
- The aromatic ring must be *para*-substituted only.
- The sulfonamide nitrogen must be secondary.
- R″ is the only possible site that can be varied in sulfonamides.

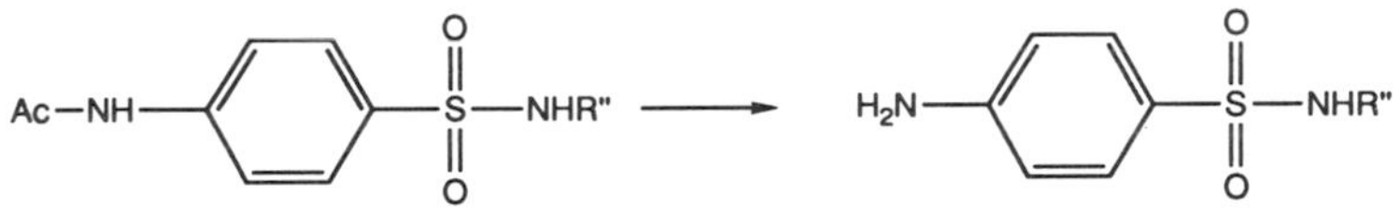

Fig. 10.8 Metabolism of acyl group to regenerate active compound.

Sulfanilamide analogues

R″ can be varied by incorporating a large range of heterocyclic or aromatic structures which affects the extent to which the drug binds to plasma protein. This in turn controls the blood levels of the drug such that it can be short acting or long acting. Thus, a drug which binds strongly to plasma protein will be slowly released into the blood circulation and will be longer lasting.

Changing the nature of the group R″ has also helped to reduce the toxicity of some sulfonamides. The primary amino group of sulfonamides are acetylated in the body and the resulting amides have reduced solubility which can lead to toxic effects. For example, the metabolite formed from sulfathiazole (an early sulfonamide) (Fig. 10.9) is poorly soluble and can prove fatal if it blocks the kidney tubules.

It is interesting to note that certain nationalities are more susceptible to this than others. For example, the Japanese and Chinese metabolize sulfathiazole more quickly than the Americans and are therefore more susceptible to its toxic effects.

It was discovered that the solubility problem could be overcome by replacing the thiazole ring in sulfathiazole with a pyrimidine ring to give sulfadiazine. The reason for the improved solubility lies in the acidity of the sulfonamide NH proton (Fig. 10.10). In sulfathiazole, this proton is not very acidic (high pK_a). Therefore, sulfathiazole and its metabolite are mostly un-ionized at blood pH. Replacing the thiazole ring with a more electron withdrawing pyrimidine ring increases the acidity of the NH proton by stabilizing the anion which results. Therefore, sulfadiazine and its metabolite are significantly ionized at blood pH. As a consequence, they are more soluble and less toxic.

Sulfadiazine was also found to be more active than sulfathiazole and soon replaced it in therapy.

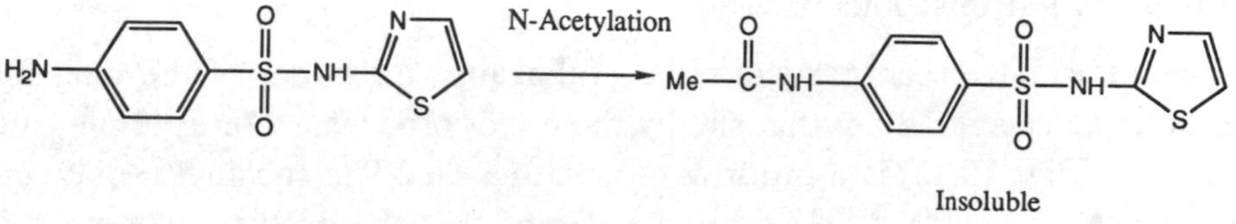

Fig. 10.9 Metabolism of sulfathiazole.

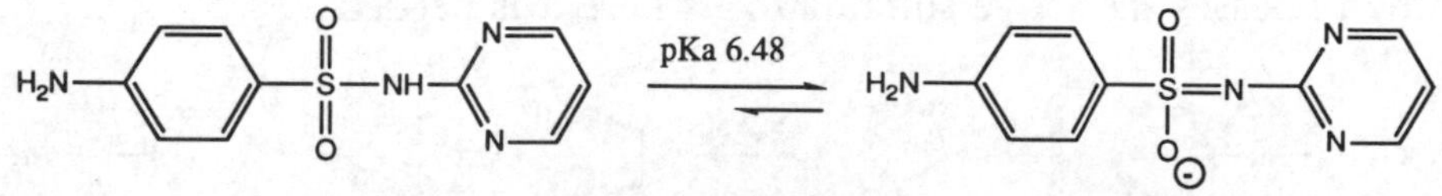

Fig. 10.10 Sulfadiazine.

To conclude, varying R″ can affect the solubility of sulfonamides or the extent to which they bind to plasma protein. These variations are therefore affecting the pharmacodynamics of the drug, rather than its mechanism of action.

Applications of sulfonamides

Before the appearance of penicillin, the sulfa drugs were the drugs of choice in the treatment of infectious diseases. Indeed, they played a significant part in world history by saving Winston Churchill's life during the Second World War. Whilst visiting North Africa, Churchill became ill with a serious infection and was bedridden for several weeks. At one point, his condition was deemed so serious that his daughter was flown out from Britain to be at his side. Fortunately, he responded to the novel sulfonamide drugs of the day.

Penicillins largely superseded sulfonamides in the fight against bacterial infections

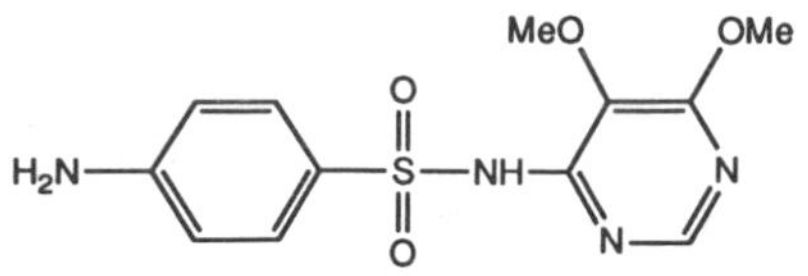

Fig. 10.11 Sulfamethoxine.

and for a long time sulfonamides were relegated backstage. However, there has been a revival of interest with the discovery of a new 'breed' of longer lasting sulfonamides. One example of this new generation is sulfamethoxine (Fig. 10.11) which is so stable in the body that it need only be taken once a week.

The sulfa drugs presently have the following applications in medicine:

- treatment of urinary tract infections
- eye lotions
- treatment of infections of mucous membranes
- treatment of gut infections

Sulfonamides have been particularly useful against infections of the intestine and can be targeted specifically to that site by the use of prodrugs. For example, succinyl sulfathiazole (Fig. 10.12) is a prodrug of sulfathiazole. The succinyl group converts the basic sulfathiazole into an acid which means that the prodrug is ionized in the slightly alkaline conditions of the intestine. As a result, it is not absorbed into the bloodstream and is retained in the intestine. Slow enzymatic hydrolysis of the succinyl group then releases the active sulfathiazole where it is needed.

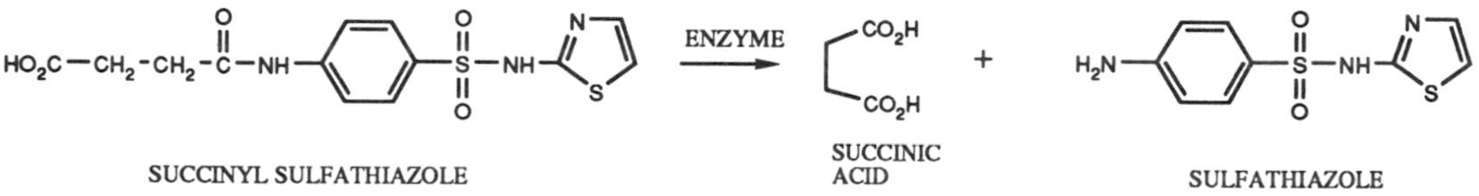

Fig. 10.12 Succinyl sulfathiazole is a prodrug of sulfathiazole.

Substitution on the aniline nitrogen with benzoyl groups (Fig. 10.13) has also given useful prodrugs which are poorly absorbed through the gut wall and can be used in the same way.

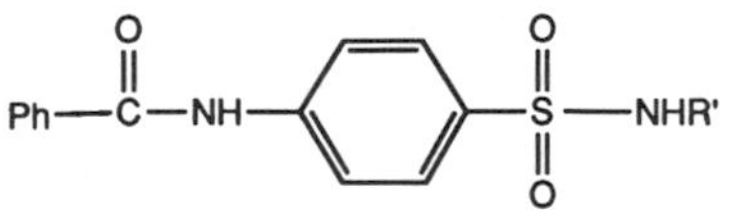

Fig. 10.13 Substitution on the aniline nitrogen with benzoyl groups.

Mechanism of action

The sulfonamides act as competitive enzyme inhibitors and block the biosynthesis of the vitamin folic acid in bacterial cells (Fig. 10.14). They do this by inhibiting the

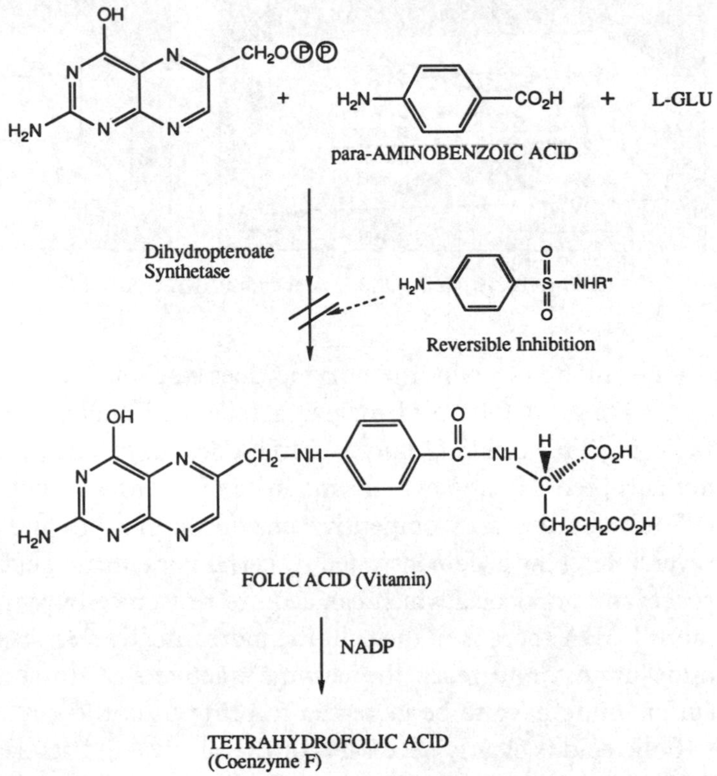

Fig. 10.14 Mechanism of action of sulfonamides.

enzyme responsible for linking together the component parts of folic acid. The consequences of this are disastrous for the cell. Under normal conditions, folic acid is the precursor for tetrahydrofolate—a compound which is crucial to cell biochemistry since it acts as the carrier for one-carbon units, necessary for many biosynthetic pathways. If tetrahydrofolate is no longer synthesized, then any biosynthetic pathway requiring one-carbon fragments is disrupted. The biosynthesis of nucleic acids is particularly disrupted and this leads to the cessation of cell growth and division.

Note that sulfonamides do not actively kill bacterial cells. They do, however, prevent the cells dividing and spreading. This gives the body's own defense systems enough time to gather their resources and wipe out the invader. Antibacterial agents which inhibit cell growth are classed as bacteriostatic, whereas agents which can actively kill bacterial cells (e.g. penicillin) are classed as bactericidal.

Sulfonamides act as inhibitors by mimicking *p*-aminobenzoic acid (PABA) (Fig. 10.14)—one of the normal constituents of folic acid. The sulfonamide molecule is similar enough in structure to PABA that the enzyme is fooled into accepting it into its active site (Fig. 10.15). Once it is bound, the sulfonamide prevents PABA from binding. As a result, folic acid is no longer synthesized. Since folic acid is essential to cell growth, the cell will stop dividing.

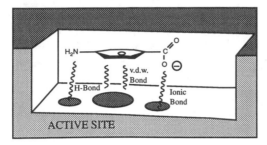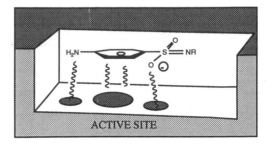

Fig. 10.15 Sulfonamide prevents PABA from binding by mimicking PABA.

One might ask why the enzyme does not join the sulfonamide to the other two components of folic acid to give a folic acid analogue containing the sulfonamide skeleton. This can in fact occur, but it does the cell no good at all since the analogue is not accepted by the next enzyme in the biosynthetic pathway.

Sulfonamides are competitive enzyme inhibitors and as such the effect can be reversible. This is demonstrated by certain organisms such as staphylococci, pneumococci, and gonococci which can acquire resistance by synthesizing more PABA. The more PABA there is in the cell, the more effectively it can compete with the sulfonamide inhibitor to reach the enzyme's active site. In such cases, the dose levels of sulfonamide have to be increased to bring back the same level of inhibition.

Folic acid is clearly necessary for the survival of bacterial cells. However, folic acid is also vital for the survival of human cells, so why do the sulfa drugs not affect human cells as well? The answer lies in the fact that human cells cannot make folic acid. They lack the necessary enzymes and so there is no enzyme for the sulfonamides to attack. Human cells acquire folic acid as a vitamin from the diet. Folic acid is brought through the cell membrane by a transport protein and this process is totally unaffected by sulfonamides.

We could now ask, 'If human cells can acquire folic acid from the diet, why can't bacterial cells infecting the human body do the same?' In fact, it is found that bacterial cells are unable to acquire folic acid since they lack the necessary transport protein required to carry it across the cell membrane. Therefore, they are forced to make it from scratch.

To sum up, the success of sulfonamides is due to two metabolic differences between mammalian and bacterial cells. In the first place, bacteria have a susceptible enzyme which is not present in mammalian cells. In the second place, bacteria lack the transport protein which would allow them to acquire folic acid from outside the cell.

10.4.2 Examples of other antimetabolites

There are other antimetabolites in medical use apart from the sulfonamides. Two examples are trimethoprim and a group of compounds known as sulfones (Fig. 10.16).

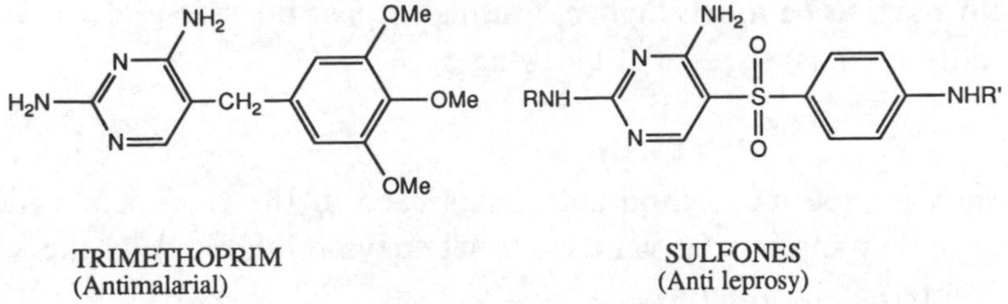

TRIMETHOPRIM
(Antimalarial)

SULFONES
(Anti leprosy)

Fig. 10.16 Examples of antimetabolites in medical use.

Trimethoprim

Trimethoprim is a diaminopyrimidine structure which has proved to be a highly selective, orally active, antibacterial, and antimalarial agent. Unlike the sulfonamides, it acts against dihydrofolate reductase—the enzyme which carries out the conversion of folic acid to tetrahydrofolate. The overall effect, however, is the same as with sulfonamides—the inhibition of DNA synthesis and cell growth.

Dihydrofolate reductase is present in mammalian cells as well as bacterial cells, so we might wonder why trimethoprim does not affect our own cells. The answer is that trimethoprim is able to distinguish between the enzymes in either cell. Although this enzyme is present in both types of cell and carries out the same reaction, mutations over millions of years have resulted in a significant difference in structure between the two enzymes such that trimethoprim recognizes and inhibits the bacterial enzyme, but does not recognize the mammalian enzyme.

Trimethoprim is often given in conjunction with the sulfonamide sulfamethoxazole (Fig. 10.17). The latter inhibits the incorporation of PABA into folic acid, while the former inhibits dihydrofolate reductase. Therefore, two enzymes in the one biosynthetic route are inhibited. This is a very effective method of inhibiting a biosynthetic route and has the advantage that the doses of both drugs can be kept down to safe levels. To get the same level of inhibition using a single drug, the dose level of that

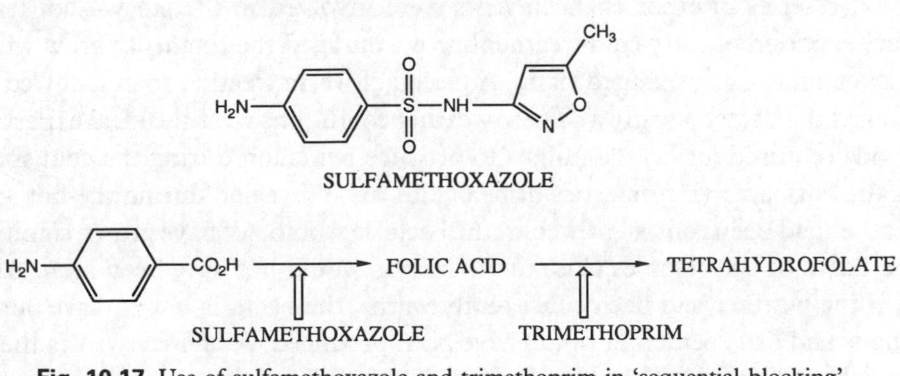

SULFAMETHOXAZOLE

Fig. 10.17 Use of sulfamethoxazole and trimethoprim in 'sequential blocking'.

drug would have to be much higher, leading to possible side-effects. This approach has been described as 'sequential blocking'.

Sulfones

The sulfones are the most important drugs used in the treatment of leprosy. It is believed that they inhibit the same bacterial enzyme inhibited by the sulfonamides, i.e. dihydropteroate synthetase.

10.5 *Antibacterial agents which inhibit cell wall synthesis*

There are two major classes of drug which act in this fashion—penicillins and cephalosporins. We shall consider penicillins first.

10.5.1 Penicillins

History of penicillins

In 1877, Pasteur and Joubert discovered that certain moulds could produce toxic substances which killed bacteria. Unfortunately, these substances were also toxic to humans and of no clinical value. However, they did demonstrate that moulds could be a potential source of antibacterial agents.

In 1928, Fleming noted that a bacterial culture which had been left several weeks open to the air had become infected by a fungal colony. Of more interest was the fact that there was an area surrounding the fungal colony where the bacterial colonies were dying. He correctly concluded that the fungal colony was producing an antibacterial agent which was spreading into the surrounding area. Recognizing the significance of this, he set out to culture and identify the fungus and showed it to be a relatively rare species of *Penicillium*. It has since been suggested that the *Penicillium* spore responsible for the fungal colony originated from another laboratory in the building and that the spore was carried by air currents and eventually blown through the window of Fleming's laboratory. This in itself appears a remarkable stroke of good fortune. However, a series of other chance events were involved in the story—not least the weather! A period of early cold weather had encouraged the fungus to grow while the bacterial colonies had remained static. A period of warm weather then followed which encouraged the bacteria to grow. These weather conditions were the ideal experimental conditions required for (a) the fungus to produce penicillin during the cold spell and (b) for the antibacterial properties of penicillin to be revealed during the hot spell. If the weather had been consistently cold, the bacteria would not have grown significantly and the death of cell colonies close to the fungus would not have been seen. Alternatively, if the weather had been consistently warm, the bacteria would have outgrown the fungus and little penicillin would have been produced. As a final twist to the story, the crucial agar plate had been stacked in a bowl of disinfectant prior to washing up, but was actually placed above the surface of the disinfectant. It says much for Fleming's observational powers that he bothered to take any notice of a culture plate which had been so discarded and that he spotted the crucial area of inhibition.

Fleming spent several years investigating the novel antibacterial substance and showed it to have significant antibacterial properties and to be remarkably non-toxic to humans. Unfortunately, the substance was also unstable and Fleming was unable to isolate and purify the compound. He therefore came to the conclusion that penicillin was too unstable to be used clinically.

The problem of isolating penicillin was eventually solved in 1938 by Florey and Chain by using a process known as freeze-drying which allowed isolation of the antibiotic under much milder conditions than had previously been available. By 1941, Florey and Chain were able to carry out the first clinical trials on crude extracts of penicillin and achieved spectacular success. Further developments aimed at producing the new agent in large quantities were developed in the United States such that by 1944, there was enough penicillin for casualties arising from the D-Day landings.

Although the use of penicillin was now widespread, the structure of the compound was still not settled and was proving to be a source of furious debate due to the unusual structures being proposed. The issue was finally settled in 1945 when Dorothy Hodgkins established the structure by X-ray analysis (Fig. 10.18).

The synthesis of such a highly strained molecule presented a huge challenge—a challenge which was met successfully by Sheehan who completed a full synthesis of penicillin by 1957. The full synthesis was too involved to be of commercial use, but the following year Beechams isolated a biosynthetic intermediate of penicillin called 6-aminopenicillanic acid (6-APA) which provided a readily accessible biosynthetic intermediate of penicillin. This revolutionized the field of penicillins by providing the starting material for a huge range of semisynthetic penicillins.

Penicillins were used widely and often carelessly, so that the evolution of penicillin-resistant bacteria became more and more of a problem. The fight against these penicillin-resistant bacteria was promoted greatly when, in 1976, Beechams discovered a natural product called clavulanic acid which has proved highly effective in protecting penicillins from the bacterial enzymes which attack penicillin.

Structure of penicillin

As mentioned above, the structure of penicillin (Fig. 10.18) is so unusual that many scientists remained sceptical until an X-ray analysis was carried out.

Penicillin contains a highly unstable-looking bicyclic system consisting of a four-membered β-lactam ring fused to a five-membered thiazolidine ring. The skeleton of the molecule suggests that it is derived from the amino acids cysteine and valine (Fig. 10.19), and this has been established.

The overall shape of the molecule is like a half-open book, as shown in Fig. 10.20.

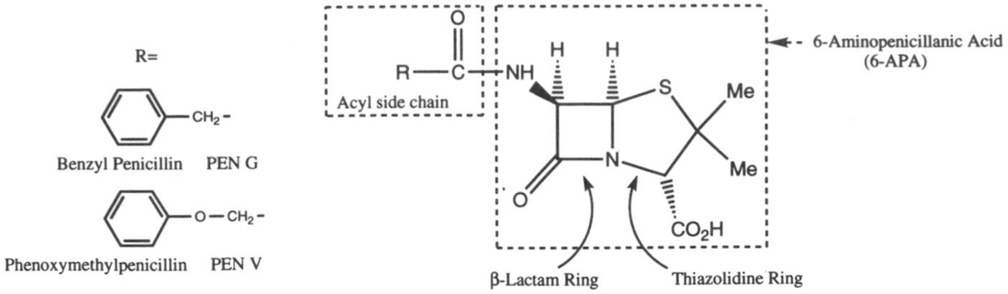

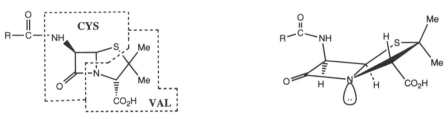

Fig. 10.18 The structure of penicillin.

Fig. 10.19 Penicillin appears to be derived from cysteine and valine.

Fig. 10.20 Shape of penicillin.

The acyl side-chain (R) varies, depending on the make up of the fermentation media. For example, corn steep liquor was used as the medium when penicillin was first mass-produced in the United States and this gave penicillin G (R=benzyl). This was due to high levels of phenylacetic acid ($PhCH_2CO_2H$) present in the medium.

Penicillin analogues

One method of varying the side-chain is to add different carboxylic acids to the fermentation medium; for example, adding phenoxyacetic acid ($PhOCH_2CO_2H$) gives penicillin V (Fig. 10.18).

However, there is a limitation to the sort of carboxylic acid one can add to the medium (i.e. only acids of general formula RCH_2CO_2H), and this in turn restricts the variety of analogues which can be obtained.

The other major disadvantage in obtaining analogues in this way is that it is a tedious and time-consuming business.

In 1957, Sheehan succeeded in synthesizing penicillin, and obtained penicillin V in 1% yield using a multistep synthetic route. Clearly, a full synthesis was not an efficient way of making penicillin analogues.

In 1958–60, Beechams managed to isolate a biosynthetic intermediate of penicillin which was also one of Sheehan's synthetic intermediates. The compound was 6-APA and it allowed the synthesis of a huge number of analogues by a semisynthetic

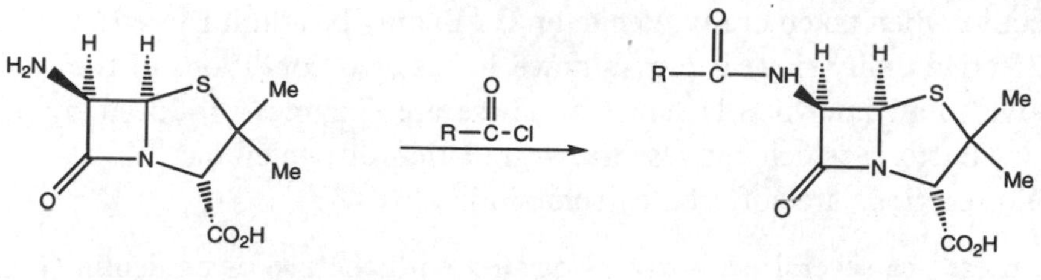

Fig. 10.21 Penicillin analogues achieved by acylating 6-APA.

method; thus, fermentation yielded 6-APA which could then be treated synthetically to give penicillin analogues. This was achieved by acylating the 6-APA with a range of acid chlorides (Fig. 10.21).

6-APA is now produced by hydrolysing penicillin G or penicillin V with an enzyme (penicillin acylase) (Fig. 10.22) or by chemical methods (see later). These are more efficient procedures than fermentation.

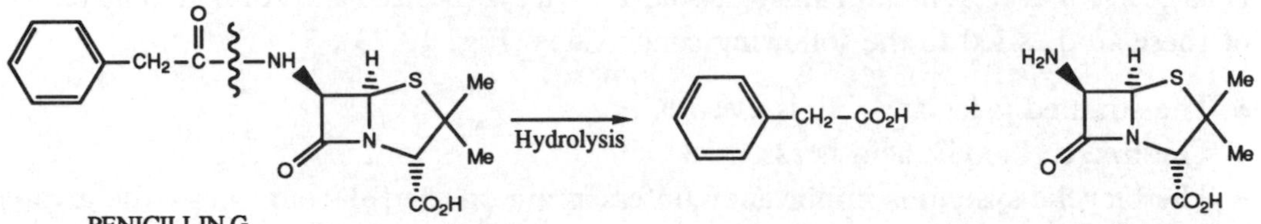

Fig. 10.22 Production of 6-APA.

We have emphasized the drive to make penicillin analogues with varying acyl side-chains. No doubt, the question could be asked—why bother? Is penicillin not good enough? Furthermore, what is so special about the acyl side-chain? Could changes not be made elsewhere in the molecule as well?

In order to answer these questions we need to look at penicillin G (the first penicillin to be isolated) in more detail and to consider its properties. Just how good an antibiotic is penicillin G?

Properties of penicillin G

The properties of benzyl penicillin are summarized below.

- Active versus Gram-positive bacilli (e.g. staphylococci, meningitis, and gonorrhoea) and many (but not all) Gram-negative cocci.
- **Non-toxic!**
  This point is worth emphasizing. The penicillins are amongst the safest drugs known to medicine.
- Not active over a wide range (or spectrum) of bacteria.

- Ineffective when taken orally. Penicillin G can only be administered by injection. It is ineffective orally since it breaks down in the acid conditions of the stomach.
- Sensitive to all known β-lactamases. These are enzymes produced by penicillin-resistant bacteria which catalyse the degradation of penicillins.
- Allergic reactions are suffered by some individuals.

Clearly, there are several problems associated with the use of penicillin G, the most serious being acid sensitivity, sensitivity to penicillinase, and a narrow spectrum of activity. The purpose of making semisynthetic penicillin analogues is therefore to find compounds which do not suffer from these disadvantages.

However, before launching into such a programme, a structure–activity study is needed to find out what features of the penicillin molecule are important to its activity. These features would then be retained in any analogues which are made.

Structure–activity relationships of penicillins

A large number of penicillin analogues have been synthesized and studied. The results of these studies led to the following conclusions (Fig. 10.23).

- The strained β-lactam ring is essential.
- The free carboxylic acid is essential.
- The bicyclic system is important (confers strain on the β-lactam ring—the greater the strain, the greater the activity, but the greater the instability of the molecule to other factors).
- The acylamino side-chain is essential (except for thienamycin, see later).
- Sulfur is usual but not essential.
- The stereochemistry of the bicyclic ring with respect to the acylamino side-chain is important.

The results of this analysis lead to the inevitable conclusion that very little variation

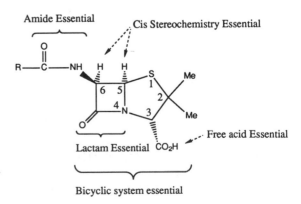

Fig. 10.23 Structure activity relationships of penicillins.

is tolerated by the penicillin nucleus and that any variation which can be made is restricted to the acylamino side-chain.

We can now look at the three problems mentioned earlier and see how they can be tackled.

The acid sensitivity of penicillins

Why is penicillin G acid sensitive? If we know the answer to that question, we might be able to plan how to solve the problem.

There are three reasons for the acid sensitivity of penicillin G.

- Ring strain.
  The bicyclic system in penicillin consists of a four-membered ring and a five-membered ring. As a result, penicillin suffers large angle and torsional strains. Acid-catalysed ring opening relieves these strains by breaking open the more highly strained four-membered lactam ring (Fig. 10.24).

Fig. 10.24 Ring opening.

- A highly reactive β-lactam carbonyl group.
  The carbonyl group in the β-lactam ring is highly susceptible to nucleophiles and as such does not behave like a normal tertiary amide which is usually quite resistant to nucleophilic attack. This difference in reactivity is due mainly to the fact that stabilization of the carbonyl is possible in the tertiary amide, but impossible in the β-lactam ring (Fig. 10.25).

  The β-lactam nitrogen is unable to feed its lone pair of electrons into the carbonyl group since this would require the bicyclic rings to adopt an impossibly strained flat system. As a result, the lone pair is localized on the nitrogen atom and the carbonyl group is far more electrophilic than one would expect for a tertiary amide. A normal tertiary amide is far less susceptible to nucleophiles since the resonance structures above reduce the electrophilic character of the carboxyl group.
- Influence of the acyl side-chain (neighbouring group participation).
  Figure 10.26 demonstrates how the neighbouring acyl group can actively participate in a mechanism to open up the lactam ring. Thus, penicillin G has a self-destruct mechanism built into its structure.

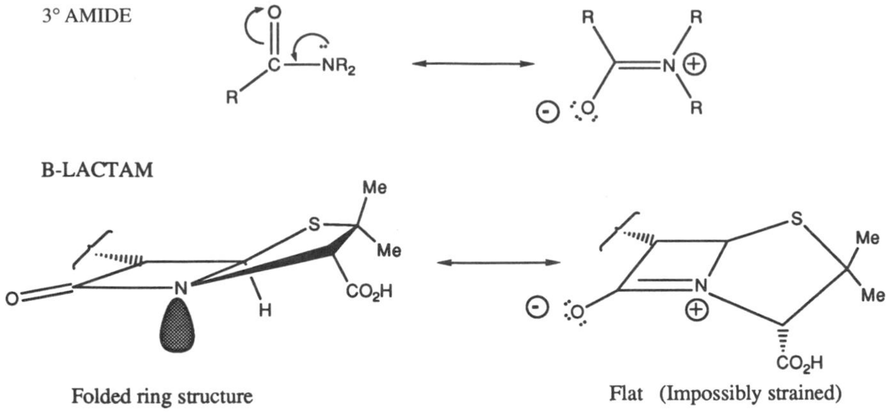

3° AMIDE

B-LACTAM

Folded ring structure   Flat   (Impossibly strained)

Fig. 10.25 Highly reactive β-lactam carbonyl group.

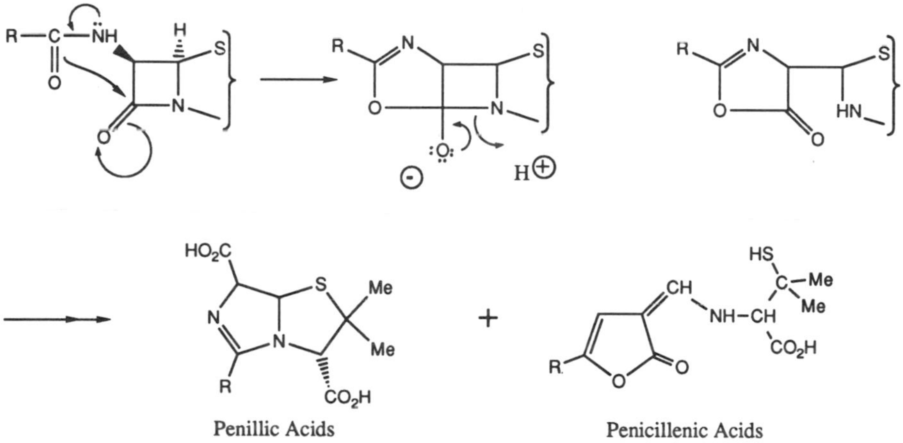

Penillic Acids   Penicillenic Acids

Fig. 10.26 Influence of the acyl side chain on acid sensitivity.

Tackling the problem of acid sensitivity

It can be seen that countering acid sensitivity is a difficult task. Nothing can be done about the first two factors since the β-lactam ring is vital for antibacterial activity. Without it, the molecule has no useful biological activity at all.

Therefore, only the third factor can be tackled. The task then becomes one of reducing the amount of neighbouring group participation to make it difficult, if not impossible, for the acyl carbonyl group to attack the β-lactam ring. Fortunately, such an objective is feasible. If a good electron withdrawing group is attached to the carbonyl group, then the inductive pulling effect should draw electrons away from the carbonyl oxygen and reduce its tendency to act as a nucleophile (Fig. 10.27).

Penicillin V (Fig. 10.28) has an electronegative oxygen on the acyl side-chain with

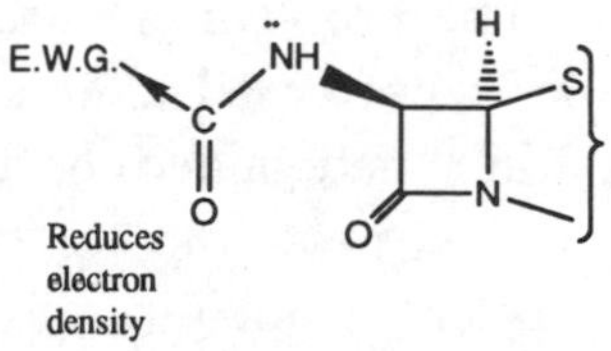

Fig. 10.27 Reduction of Neighbouring Group Participation with electron withdrawing group.

the electron withdrawing effect required. The molecule has better acid stability than penicillin G and is stable enough to survive the acid in the stomach. Thus, it can be given orally. However, Penicillin V is still sensitive to penicillinases and is slightly less active than penicillin G. It also shares with penicillin G the problem of allergic sensitivity in some individuals.

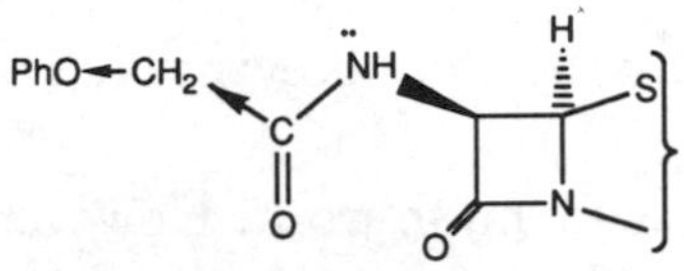

Fig. 10.28 Penicillin V.

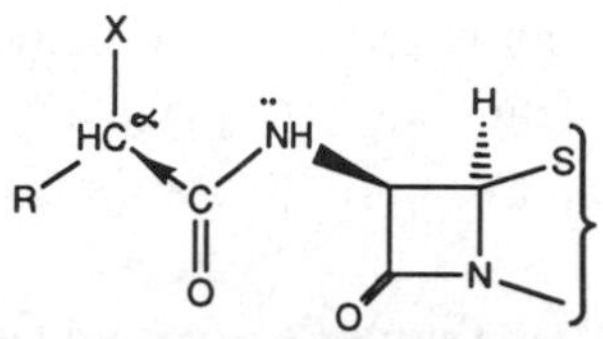

X = NH₂, Cl, PhOCONH, Heterocycles

Fig. 10.29 Penicillin analogues.

A range of penicillin analogues which have been very successful are penicillins which are disubstituted on the alpha-carbon next to the carbonyl group (Fig. 10.29). As long as one of the groups is electron withdrawing, these compounds are more resistant to acid hydrolysis and can be given orally (e.g. ampicillin (Fig. 10.36) and oxacillin (Fig. 10.33)).

To conclude, the problem of acid sensitivity is fairly easily solved by having an electron withdrawing group on the acyl side-chain.

Penicillin sensitivity to β-lactamases

β-Lactamases are enzymes produced by penicillin-resistant bacteria which can catalyse the reaction shown in Fig. 10.30—i.e. the same ring opening and deactivation of penicillin which occurred with acid hydrolysis.

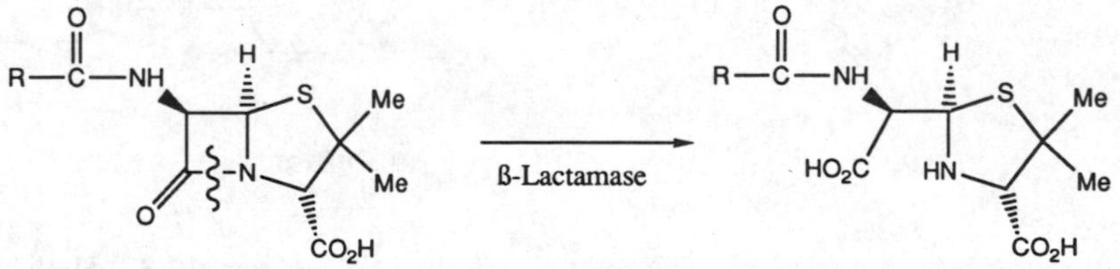

Fig. 10.30 β-Lactamase deactivation of penicillin.

The problem of β-lactamases became critical in 1960 when the widespread use of penicillin G led to an alarming increase of *Staph. aureus* infections. These problem strains had gained the lactamase enzyme and had thus gained resistance to the drug. At one point, 80 per cent of all *Staph. aureus* infections in hospitals were due to virulent, penicillin-resistant strains. Alarmingly, these strains were also resistant to all other available antibiotics.

Fortunately, a solution to the problem was just around the corner—the design of penicillinase-resistant penicillins. We say design, which implies that some sort of plan was used to counter the effects of the penicillinase enzyme. How then does one tackle a problem of this sort?

Tackling the problem of β-lactamase sensitivity

The strategy is to block the penicillin from reaching the penicillinase active site. One way of doing that is to place a bulky group on the side-chain. This bulky group can then act as a 'shield' to ward off the penicillinase and therefore prevent binding (Fig. 10.31).

Several analogues were made and the strategy was found to work. However, there was a problem. If the side-chain was made too bulky, then the steric shield also prevented the penicillin from attacking the enzyme responsible for bacterial cell wall synthesis. Therefore, a great deal of work had to be done to find the ideal 'shield' which would be large enough to ward off the lactamase enzyme, but would be small enough to allow the penicillin to do its duty. The fact that it is the β-lactam ring which is interacting with both enzymes highlights the difficulty in finding the ideal 'shield'.

Fortunately, 'shields' were found which could make that discrimination.

Methicillin (Fig. 10.32) was the first semisynthetic penicillin unaffected by penicillinase and was developed just in time to treat the *Staph. aureus* problem already mentioned.

The principle of the steric shield can be seen by the presence of two *ortho*-methoxy groups on the aromatic ring. Both of these are important in shielding the lactam ring.

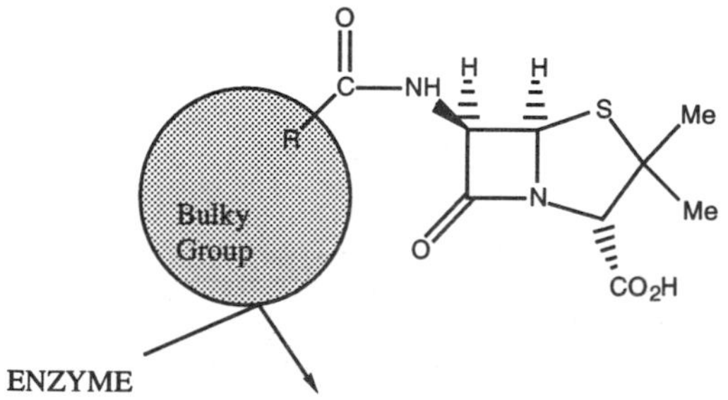

Fig. **10.31** Blocking penicillin from reaching the penicillinase active site.

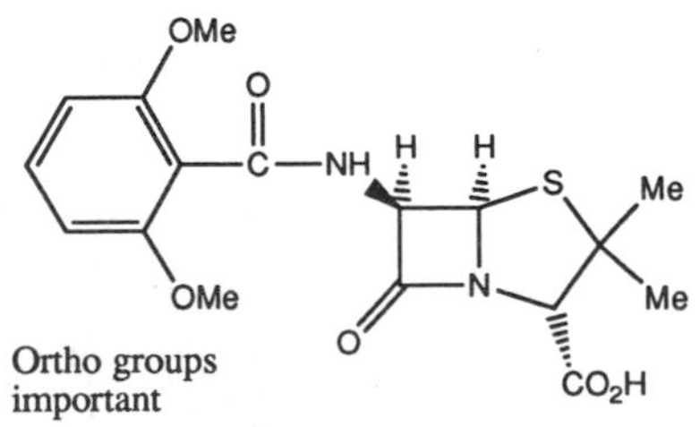

Fig. **10.32** Methicillin.

However, methicillin is by no means an ideal drug. Since there is no electron withdrawing group on the side-chain, it is acid sensitive, and so has to be injected. It is only one-fiftieth the activity of penicillin G against penicillin G sensitive organisms, it shows poor activity against some streptococci, and it is inactive against Gram-negative bacteria.

Further work eventually got round the problem of acid sensitivity by incorporating into the side-chain a five-membered heterocycle which was designed to act as a steric shield and also to be electron withdrawing (Fig. 10.33).

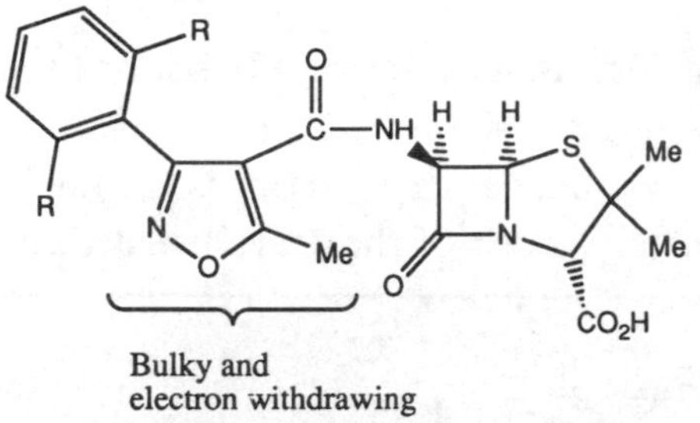

OXACILLIN   R = R' = H
CLOXACILLIN   R = Cl, R' = H
FLUCLOXACILLIN R = Cl, R' = F

Bulky and
electron withdrawing

Fig. 10.33 Incorporation of a five-membered heterocycle.

These compounds (oxacillin, cloxacillin, and flucloxacillin) are acid-resistant and penicillinase-resistant, and are also useful against *Staph. aureus* infections.

The only difference between the above three compounds is the type of halogen substitution on the aromatic ring. The influence of these groups is found to be pharmacodynamic, that is, they influence such factors as absorption of the drug and plasma protein binding. For example, cloxacillin is better absorbed through the gut wall than oxacillin, whereas flucloxacillin is less bound to plasma protein, resulting in higher levels of the free drug in the blood supply.

Having pointed out the advantages of these drugs over methicillin, it is worth putting things into context by pointing out that these three penicillins have inferior activity to the original penicillins when they are used against bacteria without the penicillinase enzyme. They also prove to be inactive against Gram-negative bacteria.

To sum up, acid-resistant penicillins would be the first choice of drug against an infection. However, if the bacteria proved resistant due to the presence of a penicillinase enzyme, then the therapy would be changed to a penicillinase-resistant penicillin.

Narrow spectrum of activity

One problem has cropped up in everything described so far; most penicillins show a poor activity against Gram-negative bacteria. There are several reasons for this resistance.

- Permeability barrier.

It is difficult for penicillins to invade a Gram-negative bacterial cell due to the make up of the cell wall. Gram-negative bacteria have a coating on the outside of their cell wall which consists of a mixture of fats, sugars, and proteins (Fig. 10.34). This coating can act as a barrier in various ways. For example, the outer surface may have an overall negative or positive charge depending on its constituent triglycerides. An excess of phosphatidylglycerol would result in an overall anionic charge, whereas an excess of lysylphosphatidylglycerol would result in an overall cationic charge. Penicillin has a free carboxylic acid which if ionized would be repelled by the former type of cell membrane.

Alternatively, the fatty portion of the coating may act as a barrier to the polar, hydrophilic penicillin molecule.

The only way in which penicillin can negotiate such a barrier is through protein channels in the outer coating. Unfortunately, most of these are usually closed.

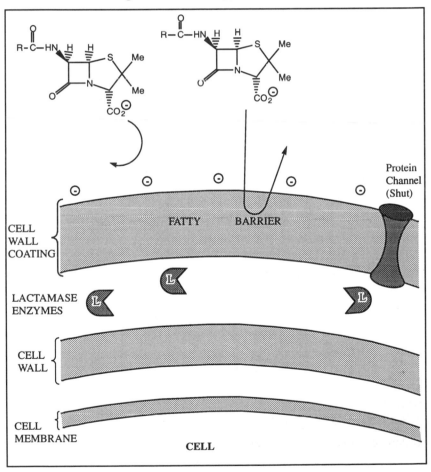

Fig. 10.34 Permeability barrier of a Gram-negative bacterial cell.

- High levels of transpeptidase enzyme produced.
  The transpeptidase enzyme is the enzyme attacked by penicillin. In some gram-negative bacteria, a lot of transpeptidase enzyme is produced, and the penicillin is incapable of inactivating all the enzyme molecules present.
- Modification of the transpeptidase enzyme.
  A mutation may occur which allows the bacterium to produce a transpeptidase enzyme which is not antagonized by penicillin.
- Presence of β-lactamase.
  We have already seen that β-lactamases are enzymes which degrade penicillin. They are situated between the cell wall and its outer coating.
- Transfer of the β-lactamase enzyme.
  Bacteria can transfer small portions of DNA from one cell to another through structures called plasmids. These are small pieces of circular bacterial DNA. If the transferred DNA contains the code for the β-lactamase enzyme, then the recipient cell acquires immunity.

Tackling the problem of narrow activity spectrum

One, some, or all of these factors might be at work, and therefore it is impossible to come up with a sensible strategy to completely solve the problem. The search for broad-spectrum antibiotics has been one of trial and error which involved making a huge variety of analogues. These changes were again confined to variations in the side-chain and gave the following results.

- Hydrophobic groups on the side-chain (e.g. penicillin G) favour activity against Gram-positive bacteria, but result in poor activity against Gram-negative bacteria.
- If the hydrophobic character is increased, there is little effect on the Gram-positive activity, but what activity there is against Gram-negative bacteria drops even more.
- Hydrophilic groups on the side-chain have either little effect on Gram-positive activity (e.g. penicillin T) or cause a reduction of activity (e.g. penicillin N) (Fig. 10.35). However, they lead to an increase in activity against Gram-negative bacteria.
- Enhancement of Gram-negative activity is found to be greatest if the hydrophilic group (e.g. NH_2, OH, CO_2H) is attached to the carbon, alpha to the carbonyl group on the side-chain.

Those penicillins having useful activity against both Gram-positive and Gram-negative bacteria are known as broad-spectrum antibiotics. There are two classes of broad-spectrum antibiotics. Both have an alpha-hydrophilic group. However, in one class the hydrophilic group is an amino function as in ampicillin or amoxycillin (Fig. 10.36), while in the other the hydrophilic group is an acid group as in carbenicillin (Fig. 10.39).

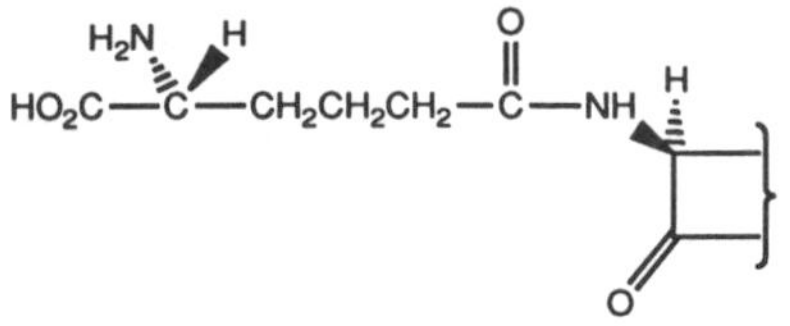

Penicillin N 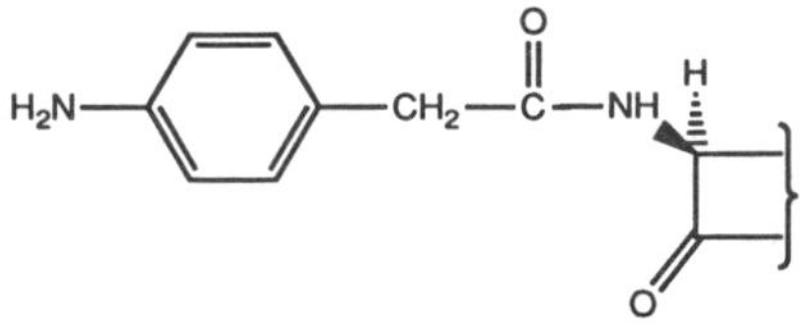   Penicillin T

Antibacterial Activities with respect to Pen G

Gram +ve	Gram -ve	Gram +ve	Gram -ve
1 %	Greater	@ same	2-4 times greater

Fig. 10.35 Effect of hydrophilic groups on the side chain on antibacterial activity.

Class I broad-spectrum antibiotics—ampicillin and amoxycillin (Beechams 1964)

Ampicillin is the second most used penicillin in medical practice. Amoxycillin differs merely in having a phenolic group. It has similar properties, but is better absorbed through the gut wall.

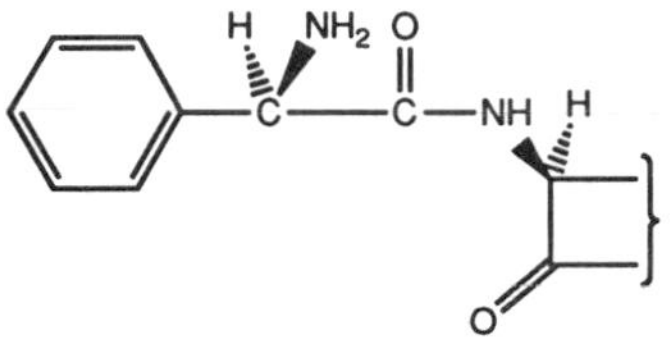

AMPICILLIN   AMOXYCILLIN
(Penbritin)   (Amoxil)

Fig. 10.36 Class I broad spectrum antibiotics.

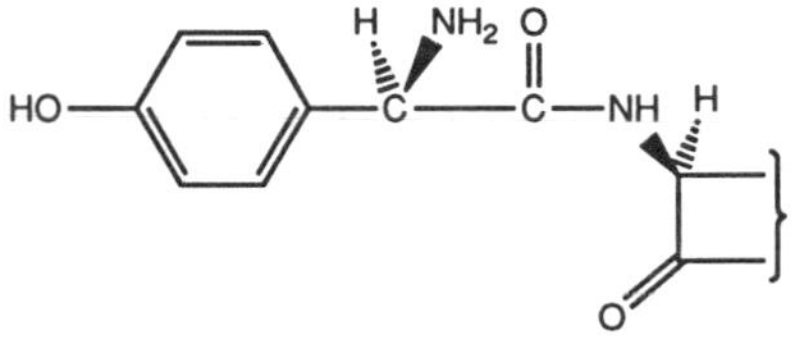

Properties:

• Active versus Gram-positive bacteria and against those Gram-negative bacteria which do not produce penicillinase.
• Acid-resistant due to the NH_2 group, and is therefore orally active.
• Non-toxic.
• Sensitive to penicillinase (no 'shield').
• Inactive against *Pseudomonas aeruginosa* (a particularly resistant species).
• Can cause diarrhoea due to poor absorption through the gut wall leading to disruption of gut flora.

The last problem of poor absorption through the gut wall is due to the dipolar nature of the molecule since it has both a free amino group and a free carboxylic acid function. This problem can be alleviated by using a prodrug where one of the polar groups is masked with a protecting group. This group is removed metabolically once the prodrug has been absorbed through the gut wall. Three examples are shown in Fig. 10.37.

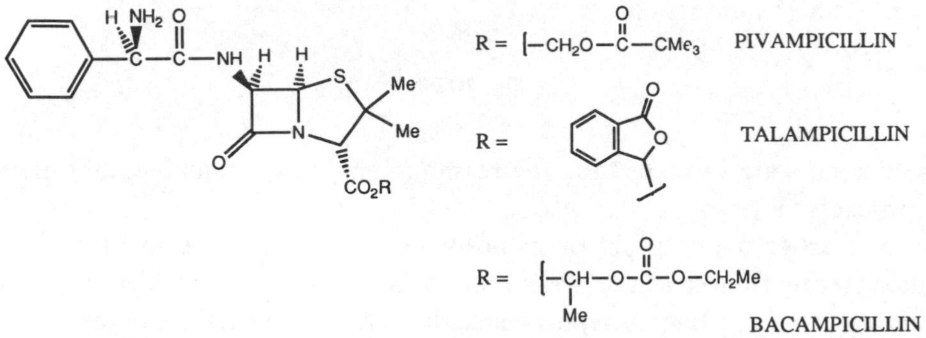

Fig. 10.37 Prodrugs used to aid absorption of antibiotic through gut wall.

These three compounds are all prodrugs of ampicillin. In all three examples, the esters used to mask the carboxylic acid group seem rather elaborate and one may ask why a simple methyl ester is not used. The answer is that methyl esters of penicillins are not metabolized in man. Perhaps the bulkiness of the penicillin skeleton being so close to the ester functional group prevents the esterases from binding the penicillin.

Fortunately, it is found that acyloxymethyl esters are susceptible to non-specific esterases. These esters contain a second ester group further away from the penicillin nucleus which is more exposed to attack. The products which are formed from hydrolysis are inherently unstable and decompose spontaneously to reveal the free carboxylic acid (Fig. 10.38).

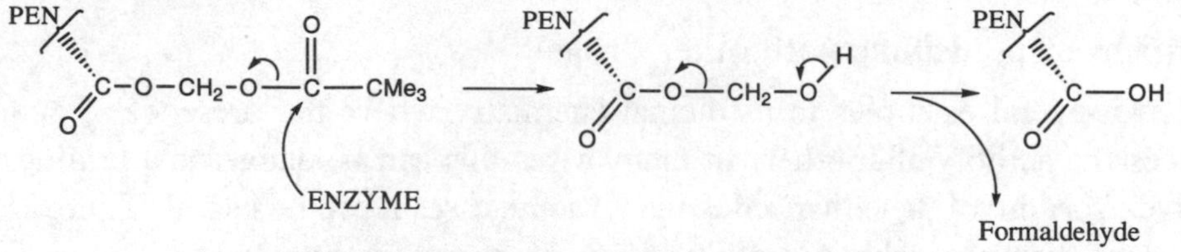

Fig. 10.38 Decomposition of acyloxymethyl esters.

Class II broad-spectrum antibiotics—carbenicillin

Carbenicillin has an activity against a wider range of Gram-negative bacteria than ampicillin. It is resistant to most penicillinases and is also active against the stubborn *Pseudomonas aeruginosa*.

This particular organism is known as an 'opportunist' pathogen since it strikes patients when they are in a weakened condition. The organism is usually present in the body, but is kept under control by the body's own defence mechanisms. However,

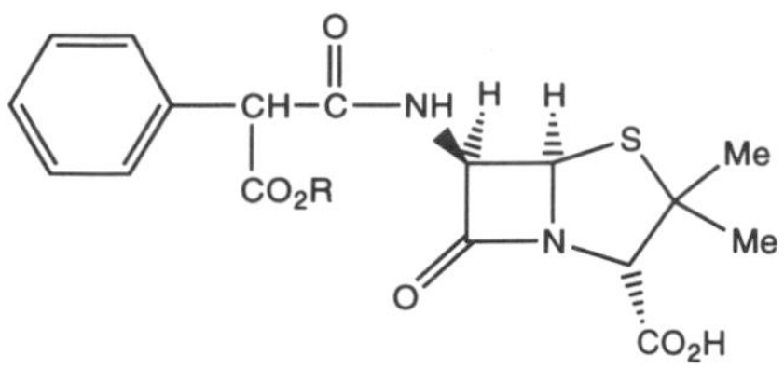

R = H   CARBENICILLIN

R = Ph   CARFECILLIN

Fig. 10.39

if these defences are weakened for any reason (e.g. shock, chemotherapy), then the organism can strike.

This can prove a real problem in hospitals where there are many susceptible patients suffering from cancer or cystic fibrosis. Burn victims are particularly prone to infection and this can lead to septicaemia which can be fatal. The organism is also responsible for serious lung infections. Carbenicillin represents one of the few penicillins which is effective against this organism.

However, there are drawbacks to carbenicillin. It shows a marked reduction in activity against Gram-positive bacteria (note the hydrophilic acid group). It is also acid sensitive and has to be injected.

In general, carbenicillin is used against penicillin-resistant Gram-negative bacteria. The broad activity against Gram-negative bacteria is due to the hydrophilic acid group (ionized at pH 7) on the side-chain. It is particularly interesting to note that the stereochemistry of this group is important. The alpha-carbon is chiral and only one of the two enantiomers is active. This implies that the acid group is involved in some sort of binding interaction with the target enzyme.

Carfecillin (Fig. 10.39) is the prodrug for carbenicillin and shows an improved absorption through the gut wall.

Synergism of penicillins with other drugs

There are several examples in medicinal chemistry where the presence of one drug enhances the activity of another. In many cases this can be dangerous, leading to an effective overdose of the enhanced drug. In some cases it can be useful. There are two interesting examples whereby the activity of penicillin has been enhanced by the presence of another drug.

One of these is the effect of clavulanic acid, described in Section 10.5.3.

The other is the administration of penicillins with a compound called probenecid (Fig. 10.40). Probenecid is a moderately lipophilic carboxylic acid and as such is similar to penicillin. It is found that probenecid can block facilitated transport of penicillin through the kidney tubules. In other words,

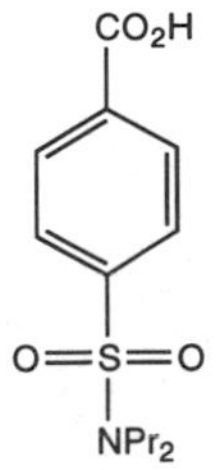

Fig. 10.40 Probenicid.

probenicid slows down the rate at which penicillin is excreted by competing with it in the excretion mechanism. As a result, penicillin levels in the bloodstream are enhanced and the antibacterial activity increases—a useful tactic if faced with a particularly resistant bacterium.

10.5.2 Cephalosporins

Discovery and structure of cephalosporin C

The second major group of β-lactam antibiotics to be discovered were the cephalosporins. The first cephalosporin was cephalosporin C—isolated in 1948 from a fungus obtained from sewer waters on the island of Sardinia. Although its antibacterial properties were recognized at the time, it was not until 1961 that the structure was established.

It is perhaps hard for modern chemists to appreciate how difficult and painstaking structure determination could be, even in the post-war period. The advent of NMR spectroscopy in the sixties and seventies has revolutionized the field so that if a new fungal metabolite is discovered today, its structure can be worked out in a matter of days rather than a matter of years.

The structure of cephalosporin C (Fig. 10.41) has similarities to that of penicillin in that it has a bicyclic system containing a four-membered β-lactam ring. However, this time the β-lactam ring is fused with a six-membered dihydrothiazine ring. This larger ring relieves the strain in the bicyclic system to some extent, but it is still a reactive system.

A study of the cephalosporin skeleton reveals that cephalosporins can be derived from the same biosynthetic precursors as penicillin, i.e. cysteine and valine (Fig. 10.42).

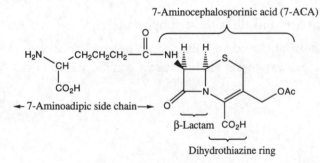

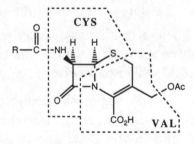

Fig. 10.41 Cephalosporin C.   **Fig. 10.42** Cephalosporin skeleton.

Properties of cephalosporin C

The properties of cephalosporin C can be summarized as follows.

- Difficult to isolate and purify due to a highly polar side-chain.
- Low potency (one-thousandth of penicillin G).
- Not absorbed orally.

- Non-toxic.
- Low risk of allergenic reactions.
- Relatively stable to acid hydrolysis compared to penicillin G.
- More stable than penicillin G to penicillinase (equivalent to oxacillin).
- Good ratio of activity against Gram-negative bacteria and Gram-positive bacteria.

Cephalosporin C has few clinical uses, is not particularly potent and at first sight seems rather uninteresting. However, its importance lies in its potential as a lead compound to something better. This potential resides in the last property mentioned above. Cephalosporin C may have low activity, but the antibacterial activity which it *does* have is more evenly directed against Gram-negative and Gram-positive bacteria than is the case with penicillins. By modifying cephalosporin C we might be able to increase the potency whilst retaining the breadth of activity against both Gram-positive and Gram-negative bacteria. Another in-built advantage of cephalosporin C over penicillin is that it is already resistant to acid hydrolysis and to penicillinase enzymes.

Cephalosporin C has been used in the treatment of urinary tract infections since it is found to concentrate in the urine and survive the body's hydrolytic enzymes.

Structure–activity relationships of cephalosporin C

Many analogues of cephalosporin C have been made and the structure–activity relationship (SAR) conclusions are as follows.

- The β-lactam ring is essential.
- A free carboxyl group is needed at position 4.
- The bicyclic system is essential.
- The stereochemistry of the side-groups and the rings is important.

These results tally closely with those obtained for the penicillins and once again there are only a limited number of places where modifications can be made (Fig. 10.43). Those places are:

- the 7-acylamino side-chain;
- the 3-acetoxymethyl side-chain;
- substitution at carbon 7.

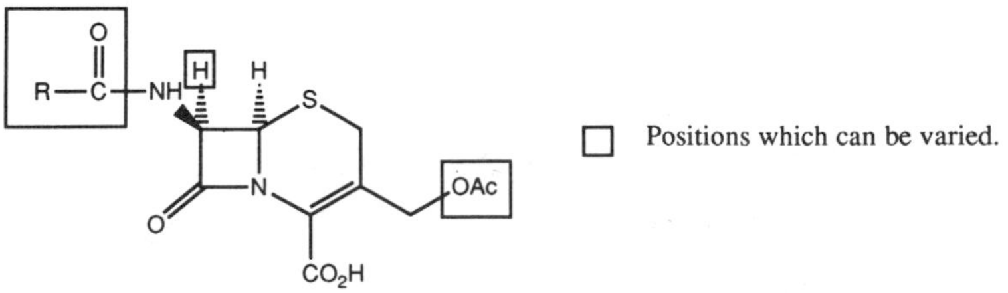

☐ Positions which can be varied.

Fig. 10.43 Positions for possible modification of cephalosporin C.

Analogues of cephalosporin C by variation of the 7-acylamino side-chain

Access to analogues with varied side-chains at the 7-position initially posed a problem. Unlike penicillins, it proved impossible to obtain cephalosporin analogues by fermentation. Similarly, it was not possible to obtain the 7-ACA (7-aminocephalosporinic acid) skeleton (Fig. 10.44) either by fermentation or by enzymic hydrolysis of cephalosporin C, thus preventing the semisynthetic approach analogous to the preparation of penicillins from 6-APA.

Therefore, a way had to be found of obtaining 7-ACA from cephalosporin C by chemical hydrolysis. This is not an easy task. After all, a secondary amide has to be hydrolysed in the presence of a highly reactive β-lactam ring. Normal hydrolytic procedures are not suitable and so a special method had to be worked out as shown in Fig. 10.44.

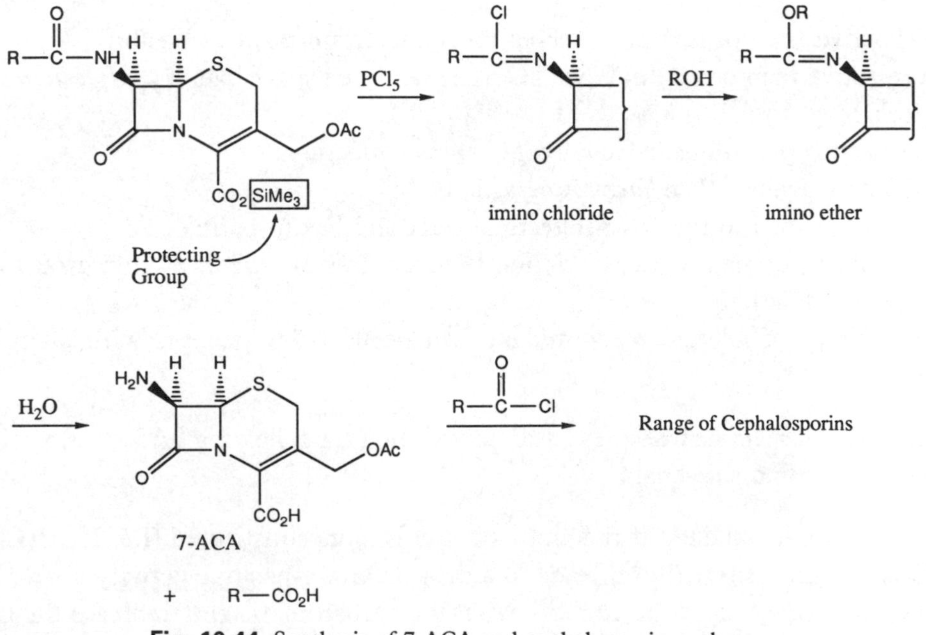

Fig. 10.44 Synthesis of 7-ACA and cephalosporin analogues.

The strategy used takes advantage of the fact that the β-lactam nitrogen is unable to share its lone pair of electrons with its neighbouring carbonyl group.

The first step of the procedure requires the formation of a double bond between the nitrogen on the side-chain and its neighbouring carbonyl group. This is only possible for the secondary amide group since ring constraints prevent the β-lactam nitrogen forming a double bond with the β-lactam ring (see Section 10.5.1.).

A chlorine atom is now introduced to form an imino chloride which can then be

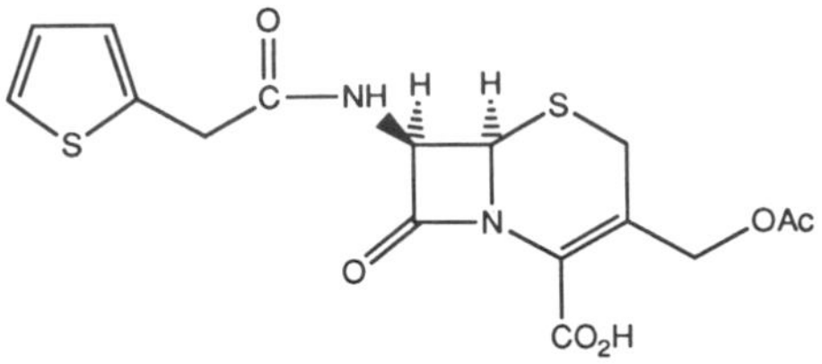

Fig. 10.45 Cephalothin.

reacted with an alcohol to give an imino ether. This product is now more susceptible to hydrolysis than the β-lactam ring and so treatment with aqueous acid successfully gives the desired 7-ACA which can then be acylated to give a range of analogues.

The most commonly used of these cephalosporin analogues is cephalothin (Fig. 10.45).

Properties of cephalothin:

- Less active than penicillin G versus cocci and Gram-positive bacilli.
- More active than penicillin G versus some Gram-negative bacilli (*Staph. aureus* and *E. coli*).
- Resistant to penicillinase from *Staph. aureus* infections.
- Not active against *Pseudomonas aeruginosa*.
- Poorly absorbed in the gastrointestinal tract and has to be injected.
- Metabolized in man by deacetylation to give a free 3-hydroxymethyl group which has reduced activity.
- Less chance of allergic reactions and can be used for patients with allergies to penicillin.

The study of several analogues has demonstrated the following SAR results relevant to the 7-acylamino side-chain.

- Best activity is obtained if the alpha-carbon is monosubstituted (i.e. RCH$_2$CO–7 ACA). Further substitution leads to a drop in Gram-positive activity.
- Lipophilic substituents on the aromatic or heteroaromatic ring increase the Gram positive activity and decrease the Gram-negative activity.

Analogues of cephalosporin C by variation of the 3-acetoxymethyl side-chain

The first observation which can be made about this area of the molecule is that losing the 3-acetyl group releases the free alcohol group and results in a drop of activity. This hydrolysis occurs metabolically and therefore it would be useful if this process was blocked to prolong the activity of cephalosporins.

An example is cephaloridine (Fig. 10.46) which contains a pyridinium group in place of the acetoxy group.

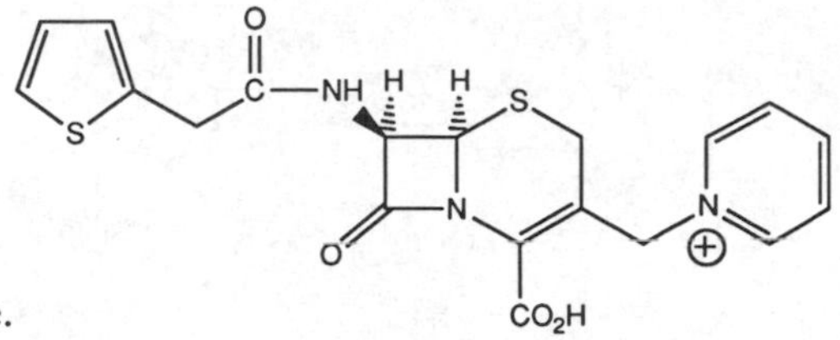

Fig. 10.46 Cephaloridine.

Properties of cephaloridine:

- Stable to metabolism.
- Soluble in water because of the positive charge.
- Low serum protein binding leads to good levels of free drug in the circulation.
- Excellent activity against Gram-positive bacteria.
- Same activity as cephalothin against Gram-negative bacteria.
- Slightly lower resistance than cephalothin to penicillinase.
- Some kidney toxicity at high doses.
- Poorly absorbed through gut wall and has to be injected.

A second example is cephalexin (Fig. 10.47) which has no substitution at position 3. This is one of the few cephalosporins which is absorbed through the gut wall and can be taken orally. This better absorption appears to be related to the presence of the 3-methyl group. Usually, the presence of such a group lowers the activity of cephalosporins, but if the correct 7-acylamino group is present as in cephalexin, then activity can be retained. The mechanism of the absorption through the gut wall is poorly understood and therefore it is not clear why the 3-methyl group is so advantageous.

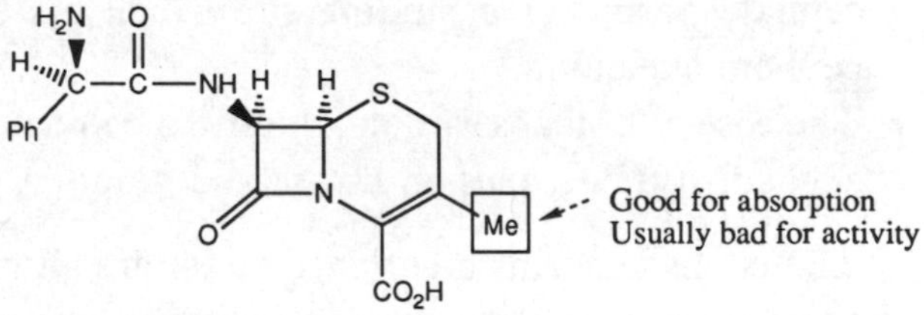

Fig. 10.47 Cephalexin.

The activity of cephalexin against Gram-positive bacteria is lower than injectable cephalosporins, but it is still useful. The activity versus Gram-negative bacteria is similar to the injectable cephalosporins.

Synthesis of 3-methylated cephalosporins

The synthesis of 3-methylated cephalosporins from cephalosporins is very difficult and it is easier to start from the penicillin nucleus as shown in Fig. 10.48. The

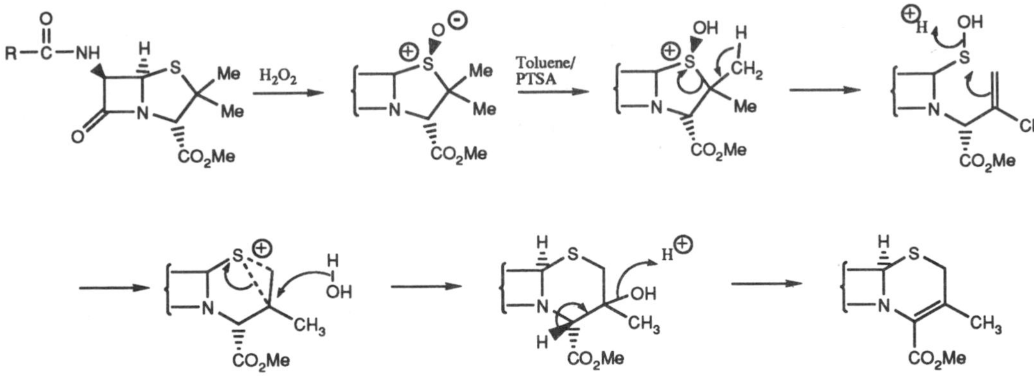

Fig. 10.48 Synthesis of 3-methylated cephalosporins.

synthesis, which was first demonstrated by Eli Lilly, involves a ring expansion, where the five-membered thiazolidine ring in penicillin is converted to the six-membered dihydrothiazine ring in cephalosporin.

Summary of properties of cephalosporins

The following conclusions can be drawn on the analogues studied to this point.

- Injectable cephalosporins of clinical use have a high activity against a large number of Gram-positive and Gram-negative organisms including the penicillin-resistant staphylococci.
- Most cephalosporins are poorly absorbed through the gut wall.
- In general, cephalosporins have lower activity than comparable penicillins, but a better range. This implies that the enzyme which is attacked by penicillin and cephalosporin has a binding site which fits the penam skeleton better than the cephem skeleton.
- The ease of oral absorption appears to be related to an alpha-amino group on the 7-acyl substituent, plus an uncharged group at position 3.

The cephalosporins mentioned so far are all useful agents, but as with penicillins, the appearance of resistant organisms has posed a problem. Gram-negative organisms, in particular, appear to have a β-lactamase which can degrade even those cephalosporins which are resistant to β-lactamase enzymes in Gram-positive species. Attempts to introduce some protection against these lactamases by means of steric shields (compare Section 10.5.1) were successful, but led to inactive compounds. Clearly the introduction of such groups in cephalosporins not only prevents access to the β-lactamase enzyme, but also to the target transpeptidase enzyme.

The next advance came when it was discovered that cephalosporins substituted at the 7-position were active.

Analogues of cephalosporin C by substitution at position 7

The only substitution which has been useful at position 7 has been the introduction of the 7-alpha-methoxy group to give a class of compounds known as the cephamycins (Fig. 10.49).

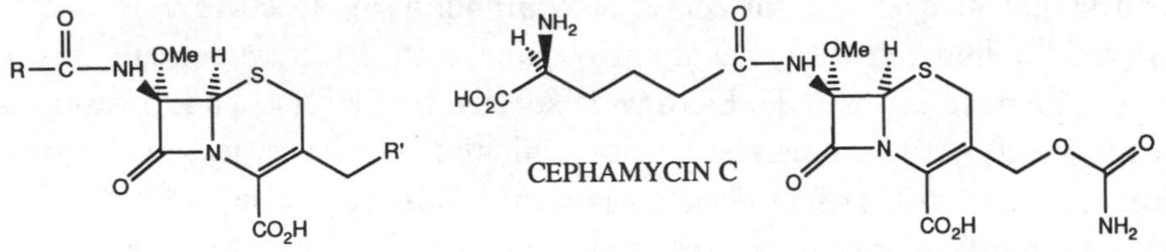

Fig. 10.49 Cephamycin C and analogues.

The parent compound cephamycin C was isolated from a culture of *Streptomyces clavuligerus* and was the first β-lactam to be isolated from a bacterial source. Modification of the side-chain gave cefoxitin (Fig. 10.50) which showed a broader spectrum of activity than most cephalosporins, due to greater resistance to penicillinase enzymes. This increased resistance is thought to be due to the steric hindrance provided by the extra methoxy group. However, it is interesting to note that introduction of the methoxy group at the corresponding 6-alpha-position of penicillins results in loss of activity.

Modifications of the cephamycins are aimed at increasing Gram-positive activity whilst retaining Gram-negative activity, as in cefoxitin (Fig. 10.50).

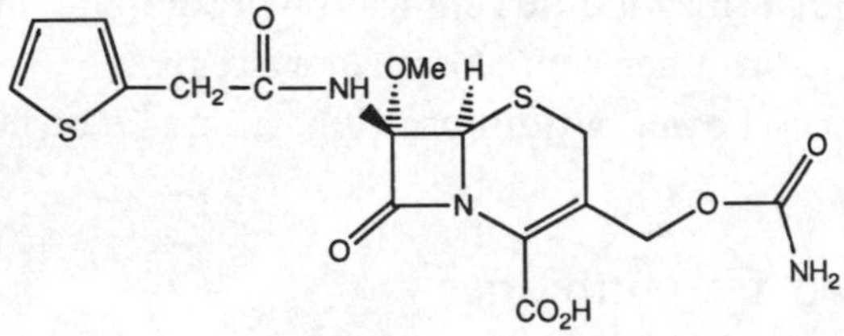

Fig. 10.50 Cefoxitin.

Properties of cefoxitin:

- Stable to β-lactamases.
- Stable to mammalian hydrolytic enzymes (due to NH_2 in place of CH_3—compare Section 11.9.2).
- Broader spectrum of activity than previous cephalosporins.
- Poor absorption through the gut wall and therefore administered by injection.
- Painful at injection site and therefore administered with a local anaesthetic.
- Poor activity against *Pseudomonas aeruginosa*.

Second- and third-generation cephalosporins—oximinocephalosporins

Research is continually being carried out to try and discover cephalosporins with an improved spectrum of activity or which are active against particularly resistant bacteria. One group of cephalosporins which has resulted from this effort has been the oximinocephalosporins.

The first useful agent in this class of compounds was cefuroxime (Fig. 10.51) (Glaxo) which, like cefoxitin, has good resistance to β-lactamases and mammalian esterases. The drug is very safe, has a wide spectrum of activity, and is useful against organisms which have become resistant to penicillin. However, it is not active against 'difficult' bacteria such as *Pseudomonas aeruginosa* and it also has to be injected.

Various modifications have resulted in another injectable cephalosporin—ceftazidime (Fig. 10.52).

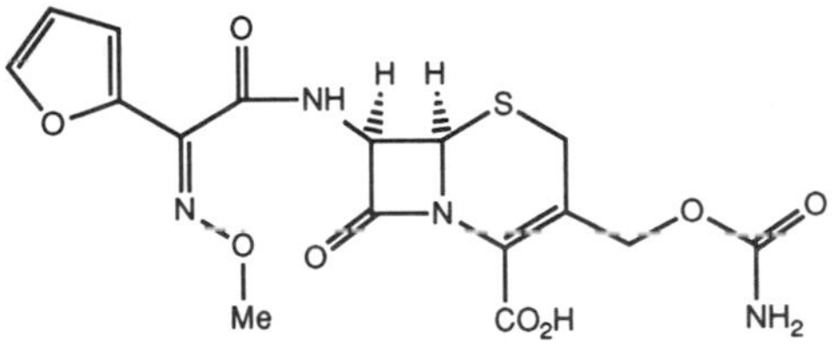

Fig. 10.51 Cefuroxime.

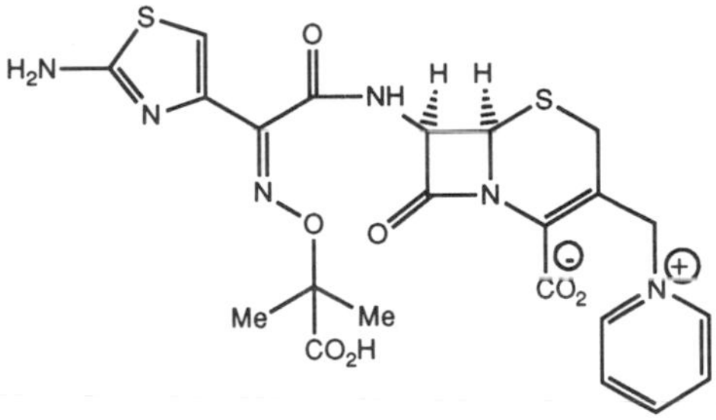

Fig. 10.52 Ceftazidime.

This drug is particularly useful since it is effective against *Pseudomonas aeruginosa*. The new five-membered thiazolidine ring was incorporated, since the literature shows that it is advantageous in other cephalosporin systems.

We have already seen how a pyridinium ring can make cephalosporins more stable to metabolism.

10.5.3 Novel β-lactam antibiotics

Although penicillins and cephalosporins are the best known and most researched β-lactams, there are other β-lactam structures which are of great interest in the antibacterial field.

Clavulanic acid (Beechams 1976)

Clavulanic acid (Fig. 10.53) was isolated from *Streptomyces clavuligerus* by Beechams (1976). It has weak and unimportant antibiotic activity. However, it is a powerful and irreversible inhibitor of most β-lactamases[1] and as such is now used in combination

---

[1] It must be realized that there are various types of β-lactamases. Clavulanic acid is effective against most but not all.

with traditional penicillins such as amoxy-cillin (Augmentin). This allows the amount of amoxycillin to be reduced and also increases the spectrum of activity.

The structure of clavulanic acid proved quite a surprise once it was determined, since it was the first example of a naturally occurring β-lactam ring which was not fused to a sulfur-containing ring. It is instead fused to an oxazolidine ring structure.

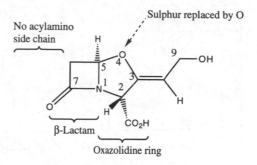

Fig. 10.53 Clavulanic acid.

It is also unusual in that it does not have an acylamino side-chain.

Many analogues have now been made and the essential requirements for β-lactamase activity are:

- The β-lactam ring.
- The double bond.
- The double bond has the *Z* configuration. (Activity is reduced but not eliminated if the double bond is *E*.)
- No substitution at C6.
- (*R*)-stereochemistry at positions 2 and 5.
- The carboxylic acid group.

The variability allowed is therefore strictly limited to the 9-hydroxyl group. Small hydrophilic groups appear to be ideal, suggesting that the original hydroxyl group is involved in a hydrogen bonding interaction with the active site of the β-lactamase.

Clavulanic acid is a mechanism-based irreversible inhibitor and could be classed as a suicide substrate (Chapter 4). The drug fits the active site of β-lactamase and the β-lactam ring is opened by a serine residue in the same manner as penicillin. However, the acyl-enzyme intermediate then reacts further with another enzymic nucleophilic group (possibly NH_2) to bind the drug irreversibly to the enzyme (Fig. 10.54). The mechanism requires the loss or gain of protons at various stages and an amino acid such as histidine present in the active site would be capable of acting as a proton donor/acceptor (compare the mechanism of acetylcholinesterase in Chapter 11).

Thienamycin (Merck 1976)

Thienamycin (Fig. 10.55) was isolated from *Streptomyces cattleya*. It is potent with an extraordinarily broad range of activity against Gram-positive and Gram-negative bacteria (including *P. aeruginosa*). It has low toxicity and shows a high resistance to β-lactamases. This resistance has been ascribed to the presence of the hydroxyethyl side-chain.

However, it shows poor metabolic and chemical stability, and is not absorbed from

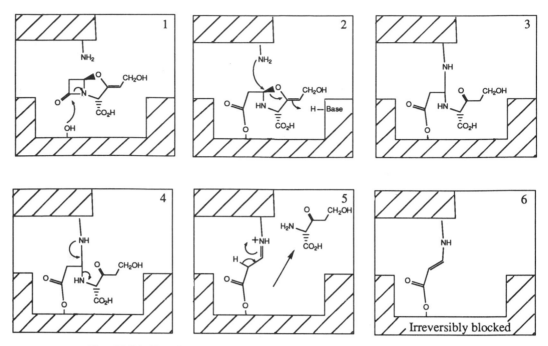

Fig. 10.54 Clavulanic acid as an irreversible mechanism based inhibitor.

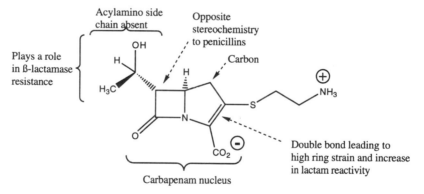

Fig. 10.55 Thienamycin.

the gastrointestinal tract. Therefore, analogues with increased chemical stability and oral activity would be useful.

The big surprise concerning the structure of thienamycin is the missing sulfur atom and acylamino side-chain, both of which were thought to be essential to antibacterial activity. Furthermore, the stereochemistry of the side-chain at substituent 6 is opposite from the usual stereochemistry in penicillins.

Olivanic acids

The olivanic acids (e.g. MM13902) (Fig. 10.56) were isolated from strains of *Strepto-myces olivaceus* and are carbapenam structures like thienamycin. They have very strong β-lactamase activity, in some cases 1000 times more potent than clavulanic acid. They are also effective against the β-lactamases which can break down cephalo-sporins. These β-lactamases are unaffected by clavulanic acid.

Unfortunately, these compounds are susceptible to metabolic degradation in the kidney.

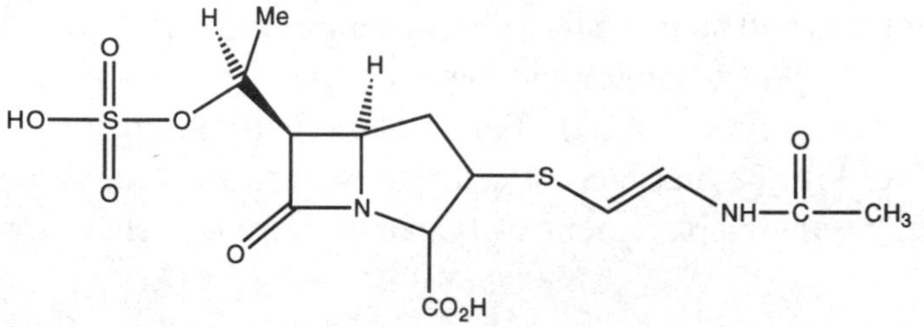

Fig. 10.56 MM 13902.

Nocardicins

At least seven nocardicins (e.g. nocardicin A (Fig. 10.57)) have been isolated from natural sources by the Japanese company Fujisawa. They show moderate activity *in vitro* against a narrow group of Gram-negative bacteria including *Pseudomonas aeruginosa*. However, it is surprising that they should show any activity at all since they contain a single β-lactam ring unfused to any other ring system. The presence of a fused second ring has always been thought to be essential in order to strain the β-lactam ring sufficiently for antibacterial activity.

One explanation for the surprising activity of the nocardicins is that they operate via a different mechanism from penicillins and cephalosporins. There is some evidence supporting this in that the nocardicins are inactive against Gram-positive bacteria and generally show a different spectrum of activity from the other β-lactam antibiotics. It

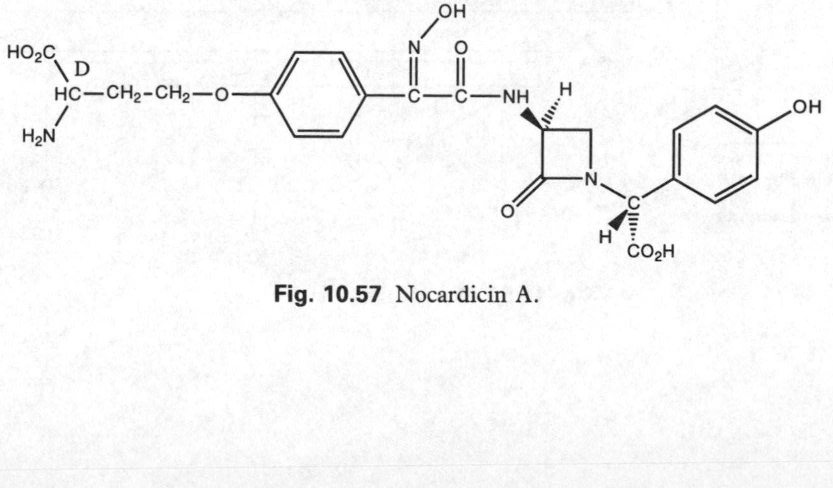

Fig. 10.57 Nocardicin A.

is possible that these compounds act on cell wall synthesis by inhibiting a different enzyme.

They also show low levels of toxicity.

10.5.4 The mechanism of action of penicillins and cephalosporins

Bacteria have to survive a large range of environmental conditions such as varying pH, temperature, and osmotic pressure. Therefore, they require a robust cell wall. Since this cell wall is not present in animal cells, it is the perfect target for antibacterial agents such as penicillins and cephalosporins.

The wall is a peptidoglycan structure (Fig. 10.59). In other words, it is made up of peptide units and sugar units. The structure of the wall consists of a parallel series of sugar backbones containing two types of sugar (N-acetylmuramic acid (NAM) and N-acetyl glucosamine (NAG)) (Fig. 10.58). Peptide chains are bound to the NAM sugars, and in the final step of cell wall biosynthesis, these peptide chains are linked together by the displacement of D-alanine from one chain by glycine in another.

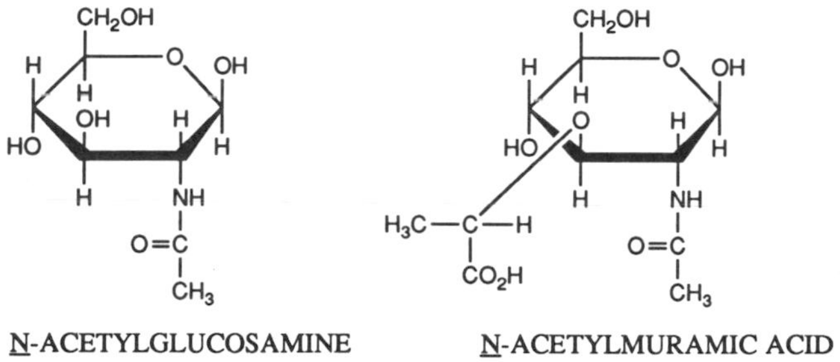

Fig. 10.58 Sugars contained in cell wall structure of bacteria.

N-ACETYLGLUCOSAMINE   N-ACETYLMURAMIC ACID

It is this final cross-linking reaction which is inhibited by penicillins and cephalosporins, such that the cell wall framework is not meshed together (Fig. 10.60). As a result, the wall becomes 'leaky'. Since the salt concentrations inside the cell are

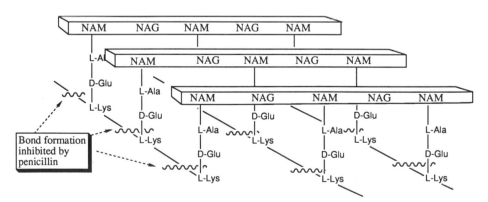

Fig. 10.59 Peptidoglycan structure.

greater than those outside the cell, water enters the cell, the cell swells, and eventually lyses (bursts).

The enzyme responsible for the cross-linking reaction is known as the transpeptidase enzyme.

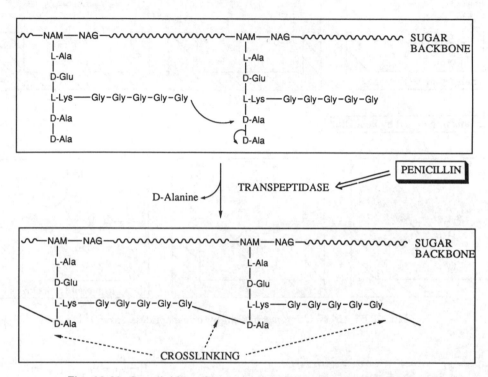

Fig. 10.60 Crosslinking of bacteria cell walls inhibited by penicillin.

It has been proposed that penicillin has a conformation which is similar to the transition-state conformation taken up by D-Ala–D-Ala—the portion of the amino acid chain involved in the cross-linking reaction (Fig. 10.61). Since this is the reaction centre for the transpeptidase enzyme, it is quite an attractive theory to postulate that the enzyme mistakes the penicillin molecule for the D-Ala–D-Ala moiety and accepts the penicillin into its active site. Once penicillin is in the active site, the normal enzymatic reaction would be carried out on the penicillin.

In the normal mechanism (Fig. 10.61), the amide bond between the two alanine units on the peptide chain is split. The terminal alanine departs the active site, leaving the peptide chain bound to the active site. The terminal glycine of the pentaglycyl chain can then enter the active site and form a peptide bond to the alanine group and thus remove it from the active site.

The enzyme can attack the β-lactam ring of penicillin and open it in the same way as it did with the amide bond. However, penicillin is cyclic and as a result the

Normal Mechanism

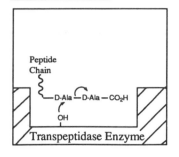
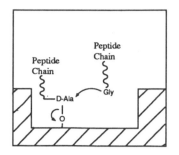
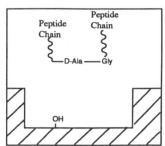

Mechanism Inhibited by Penicillin

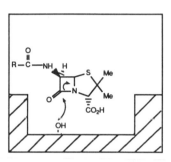
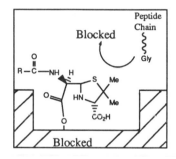
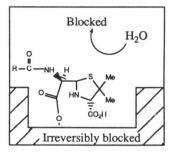

Fig. 10.61 Crosslinking mechanism by transpeptidase enzyme.

Penicillin   6-Methyl Penicillin   Acyl-D-Ala-D-Ala

Fig. 10.62

molecule is not split in two and nothing leaves the active site. Subsequent hydrolysis of the acyl group does not take place, presumably because glycine is unable to reach the site due to the bulkiness of the penicillin molecule.

However, there is some doubt over this theory since there are one or two anomalies. For example, 6-methylpenicillin (Fig. 10.62) is a closer analogue to D-Ala–D-Ala. It should fit the active site better and have higher activity. On the contrary, it is found to have lower activity.

An alternative proposition is that penicillin does not bind to the active site itself,

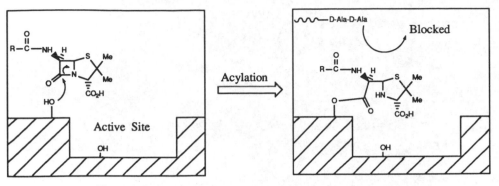

Fig. 10.63 Alternative 'umbrella' mechanism of inhibition.

but binds instead to a site nearby. By doing so, the penicillin structure overlaps the active site and prevents access to the normal reagents—the umbrella effect (see Section 5.7.2.). If a nucleophilic group (not necessarily in the active site) attacks the β-lactam ring, the penicillin becomes bound irreversibly, permanently blocking the active site (Fig. 10.63).

10.6 *Antibacterial agents which act on the plasma membrane structure*

The peptides valinomycin (Fig. 10.64) and gramicidin A (Fig. 10.67) both act as ion conducting antibiotics and allow the uncontrolled movement of ions across the cell membrane. Unfortunately, both these agents show no selective toxicity for bacterial over mammalian cells and are therefore useless as therapeutic agents. Their mechanism of action is interesting nevertheless.

Valinomycin is a cyclic structure containing three molecules of L-valine, three molecules of D-valine, three molecules of L-lactic acid, and three molecules of D-hydroxyisovalerate. These four components are linked in an ordered fashion such that there is an alternating sequence of ester and amide linking bonds around the cyclic structure. This is achieved by the presence of a lactic or hydroxyvaleric acid unit between each of the six valine units. Further ordering can be observed by noting that the L and D portions of valine alternate around the cycle, as do the lactate and hydroxyisovalerate units.

Valinomycin acts as an ion carrier and in some ways could be looked upon as an inverted detergent. Since it is cyclic, it forms a doughnut-type structure where the polar carbonyl oxygens of the ester and amide groups face inside, while the hydrophobic side-chains of the valine and hydroxyisovalerate units point outwards. This is clearly favoured since the hydrophobic side-chains can interact via van der Waals forces with the fatty lipid interior of the cell membrane, while the polar hydrophilic

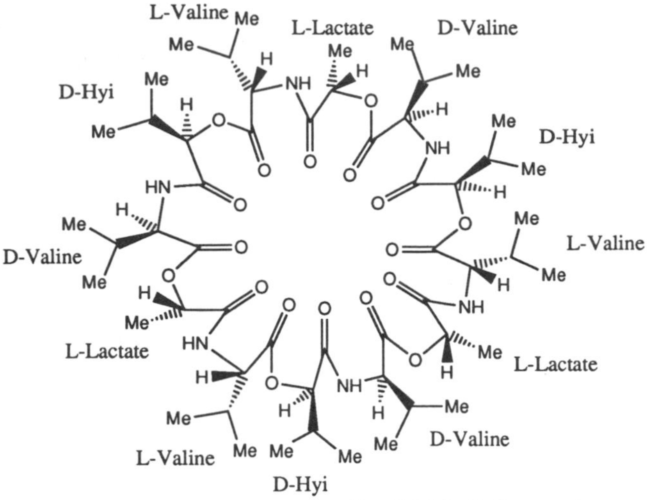

Fig. 10.64 Valinomycin.

D-Hyi =
D-Hydroxyisovaleric acid

groups are clustered together in the centre of the doughnut to produce a hydrophilic environment.

This hydrophilic centre is large enough to accommodate an ion and it is found that a 'naked' potassium ion (i.e. no surrounding water molecules) fits the space and is complexed by the amide carboxyl groups (Fig. 10.65).

Valinomycin can therefore 'collect' a potassium ion from the inner surface of the membrane, carry it across the membrane and deposit it outside the cell, thus disrupting the ionic equilibrium of the cell (Fig. 10.66). Normally, cells have a high concentration of potassium and a low concentration of sodium. The fatty cell membrane prevents passage of ions between the cell and its environment, and ions can only pass through the cell membrane aided by specialized and controlled ion transport systems. Valinomycin introduces an uncontrolled ion transport system which proves fatal.

Valinomycin is specific for potassium ions over sodium ions. One might be tempted to think that sodium ions would be too small to be properly complexed. However, the real reason is that sodium ions do not lose their surrounding water 'coat' very easily and would have to be transported as the hydrated ion. As such, they are too big for the central cavity of valinomycin.

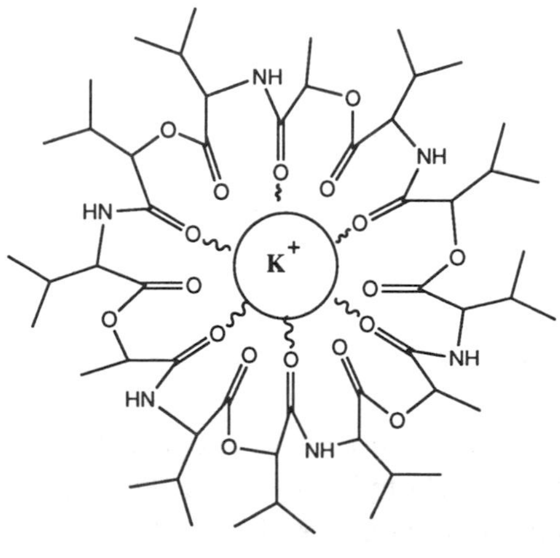

Fig. 10.65 Potassium ion in the hydrophilic centre of valinomycin.

Gramicidin A (Fig. 10.67) is a peptide containing 15 amino acids which is thought to coil into a helix such that the outside of the helix is hydrophobic and interacts with the membrane lipids, while the inside of the helix contains hydrophilic groups, thus allowing the passage of ions. Therefore, gramicidin A could be viewed as an escape tunnel through the cell membrane.

In fact, one molecule of gramicidin would not be long enough to traverse the membrane and it has been proposed that two gramicidin helices align themselves end-to-end in order to achieve the length required (Fig. 10.68).

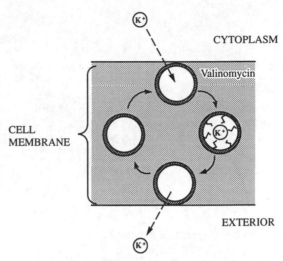

Fig. 10.66 Valinomycin disrupts the ionic equilibrium of a cell.

Val-Gly-Ala-Leu-Ala-Val-Val-Val-Trp-Leu-Trp-Leu-Trp-Leu-Trp-NH-CH$_2$-CH$_2$-OH

Fig. 10.67 Gramicidin A.

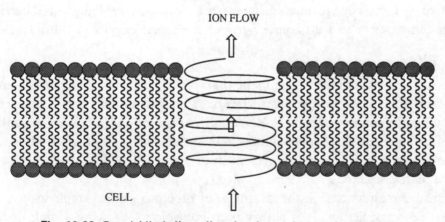

Fig. 10.68 Gramicidin helices aligned end-to-end traversing membrane.

The polypeptide antibiotic polymyxin B (Fig. 10.69) also operates within the cell membrane. It shows selective toxicity for bacterial cells over animal cells, which appears to be related to the ability of the compound to bind selectively to the different plasma membranes. The mechanism of this selectivity is not fully understood.

Polymyxin B acts like valinomycin, but it causes the leakage of small molecules

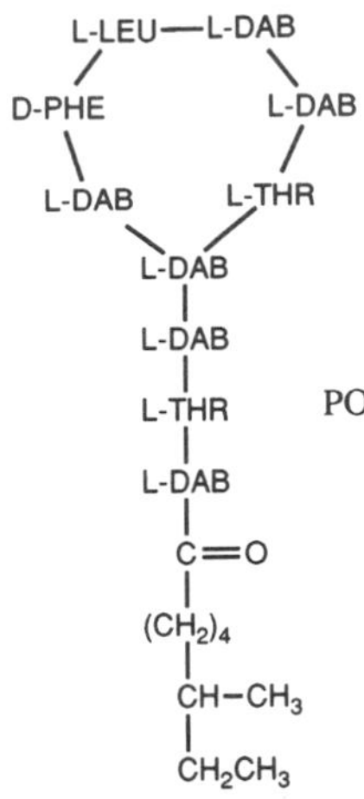

```
              L-LEU——L-DAB
             /              \
        D-PHE              L-DAB
             \              /
           L-DAB        L-THR
               \       /
                L-DAB
                 |
                L-DAB
                 |
                L-THR        POLYMYXIN B
                 |
                L-DAB
                 |
                C═O
                 |
               (CH₂)₄
                 |
               CH–CH₃
                 |
               CH₂CH₃
```

Fig. 10.69 Polypeptide antibiotic.

such as nucleosides from the cell. The drug is injected intramuscularly and is useful against *Pseudomonas* strains which are resistant to other antibacterial agents.

10.7 *Antibacterial agents which impair protein synthesis*

Examples of such agents are the rifamycins which act against RNA, and the aminoglycosides, tetracyclines, and chloramphenicol which all act against the ribosomes.

Selective toxicity is due to either different diffusion rates through the cell barriers of different cell types or to a difference between the target enzymes of different cells.

10.7.1 Rifamycins

Rifampicin (Fig. 10.70) is a semisynthetic rifamycin made from rifamycin B—an antibiotic isolated from *Streptomyces mediterranei*. It inhibits Gram-positive bacteria and works by binding non-covalently to RNA polymerase and inhibiting RNA synthesis. The DNA-dependent RNA polymerases in eukaryotic cells are unaffected, since the drug binds to a peptide chain not present in the mammalian RNA polymerase. It is therefore highly selective.

The drug is mainly used in the treatment of tuberculosis and staphylococci infections that resist penicillin. It is a very useful antibiotic, showing a high degree of selectivity against bacterial cells over mammalian cells. Unfortunately, it is also expensive, which discourages its use against a wider range of infections.

The flat naphthalene ring and several of the hydroxyl groups are essential for activity.

The selectivity of this antibiotic is interesting since both bacterial cells and mammalian cells contain the enzyme RNA polymerase. However, as we have seen, the enzyme in bacterial cells contains a peptide chain not present in mammalian RNA polymerase.

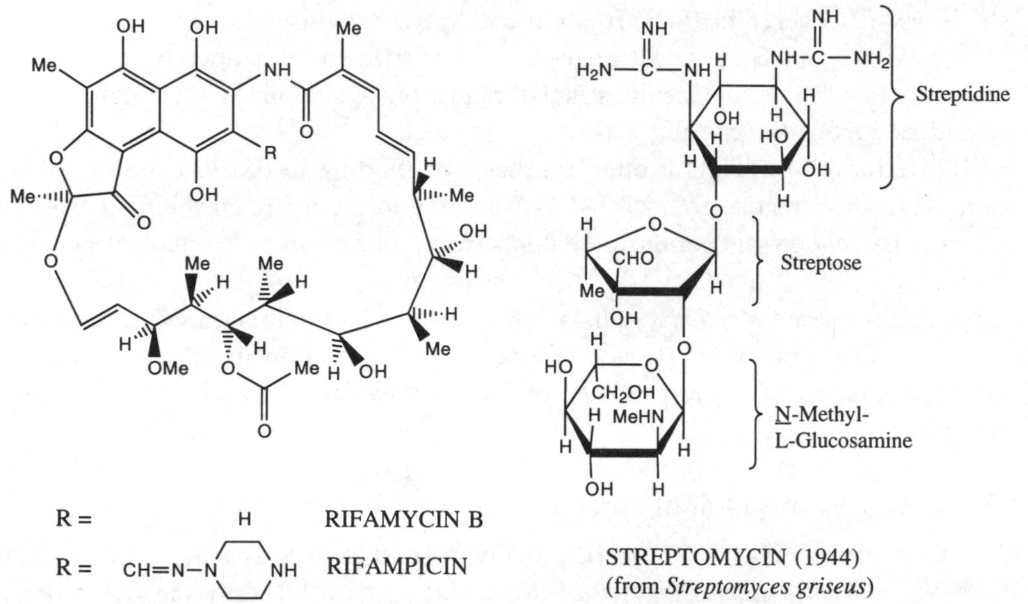

Fig. 10.70 Antibacterial agents which impair protein synthesis.

Presumably this chain was lost from the mammalian enzyme during long years of evolution.

10.7.2 Aminoglycosides

Streptomycin (Fig. 10.70) (from *Streptomyces griseus*, 1944) is an example of an important aminoglycoside. Streptomycin was the next most important antibiotic to be discovered after penicillin and proved to be the first antibiotic effective against the lethal disease tuberculous meningitis. The drug works by inhibiting protein synthesis. It binds to the 30S ribosomal subunit and prevents the growth of the protein chain as well as preventing the recognition of the triplet code on mRNA.

Aminoglycosides are fast acting, but they can also cause ear and kidney problems if the dose levels are not carefully controlled.

The aminoglycoside antibiotics used to be the only compounds effective against the particularly resistant *Pseudomonas aeruginosa* (see earlier) and it is only recently that alternative treatments have been unveiled (see above).

10.7.3 Tetracyclines

The tetracyclines as a whole have a broad spectrum of activity and are the most widely prescribed form of antibiotic after penicillins. They are also capable of attacking the malarial parasite.

One of the best known tetracyclines is chlortetracyclin (Aureomycin) (Fig. 10.71) which was discovered in 1948. It is a broad-spectrum antibiotic, active against both Gram-positive and Gram-negative bacteria. Unfortunately, it does have side-effects due to the fact that it kills the intestinal flora that make vitamin K—a vitamin which is needed as part of the clotting process.

Chlortetracyclin inhibits protein synthesis by binding to the 30S subunit of ribosomes and prevents the aminoacyl-tRNA binding to the A site on the ribosome. This prevents the codon–anticodon interaction from taking place. Protein release is also inhibited.

There is no reason why tetracyclines should not attack protein synthesis in mammalian cells as well as in bacterial cells. In fact, they can. Fortunately, bacterial cells accumulate the drug far more efficiently than mammalian cells and are therefore more susceptible

10.7.4 Chloramphenicol

Chloramphenicol (Fig. 10.72) was originally isolated from *Streptomyces venezuela*, but is now prepared synthetically. It has two chiral centres, but only the *R,R*-isomer is active.

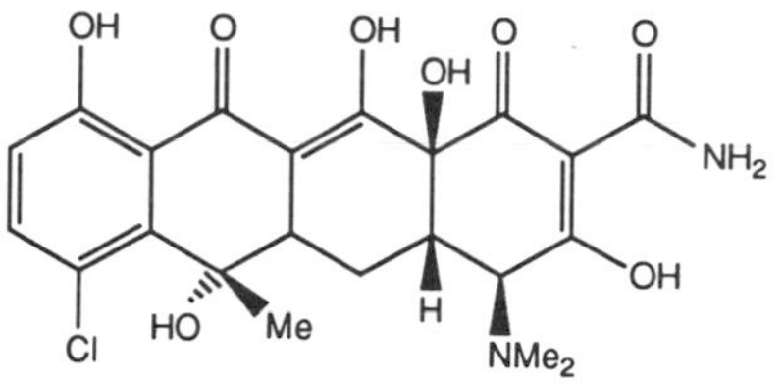

Fig. 10.71 Chlortetracyclin (aureomycin).

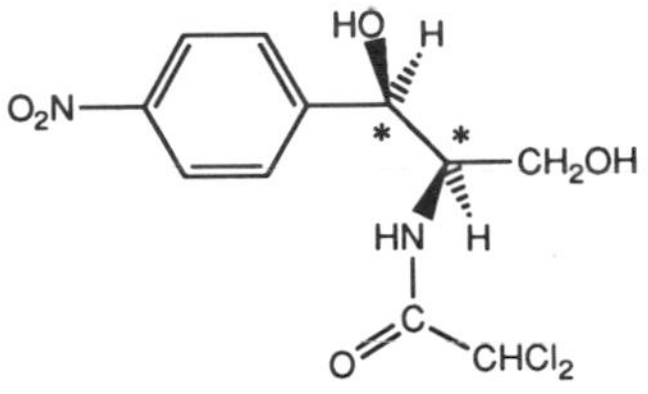

Fig. 10.72 Chloramphenicol (from *Streptomyces venezuela*).

SAR studies demonstrate that there must be a substituent on the aromatic ring which can 'resonate' with it (i.e. NO_2). The *R,R*-propanediol group is essential. The OH groups must be free and presumably are involved in hydrogen bonding. The dichloroacetamide group is important, but can be replaced by other electronegative groups.

Chloramphenicol binds to the 50S subunit of ribosomes and appears to act by inhibiting the movement of ribosomes along mRNA, probably by inhibiting the peptidyl transferase reaction by which the peptide chain is extended.

Chloramphenicol is the drug of choice against typhoid and is also used in severe bacterial infections which are insensitive to other antibacterial agents. It has also found widespread use against eye infections. However, the drug should only be used in these restricted scenarios since it is quite toxic, especially to bone marrow. The

NO_2 group is suspected to be responsible for this, although intestinal bacteria are capable of reducing this group to an amino group.

10.7.5 Macrolides

The best known example of this class of compounds is erythromycin—a metabolite produced by the microorganism *Streptomyces erythreus*. The structure (Fig. 10.73) consists of a macrocylic lactone ring with a sugar and an aminosugar attached. The sugar residues are important for activity.

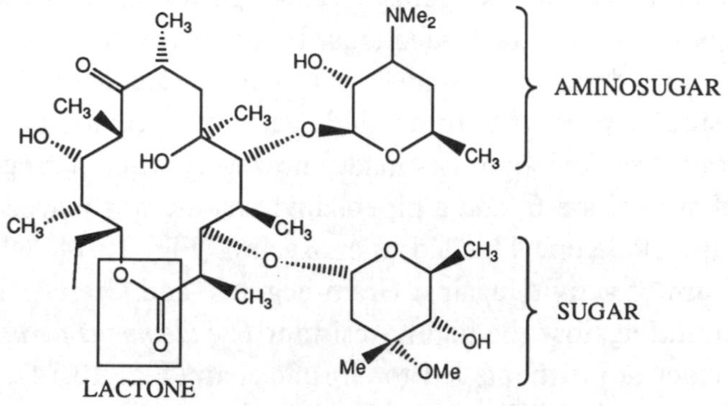

Fig. 10.73 Erythromycin.

Erythromycin acts by binding to the 50S subunit by an unknown mechanism. It works in the same way as chloramphenicol by inhibiting translocation, where the elongated peptide chain attached to tRNA is shifted back from the aminoacyl site to the peptidyl site. Erythromycin was used against penicillin-resistant staphylococci, but newer penicillins are now used for these infections. It is, however, the drug of choice against 'legionnaires disease'.

10.8 *Agents which act on nucleic acid transcription and replication*

10.8.1 Quinolones and fluoroquinolones

The quinolone and fluoroquinolone antibacterial agents are relatively late arrivals on the antibacterial scene, but are proving to be very useful therapeutic agents. They are particularly useful in the treatment of urinary tract infections and also for the treatment of infections which prove resistant to the more established antibacterial agents. In the latter case, microorganisms which have gained resistance to penicillin may have done so by mutations affecting cell wall biosynthesis. Since the quinolones and fluoroquinolones act by a different mechanism, such mutations provide no protection against these agents.

Nalidixic acid (Fig. 10.74) was the first therapeutically useful agent in this class of

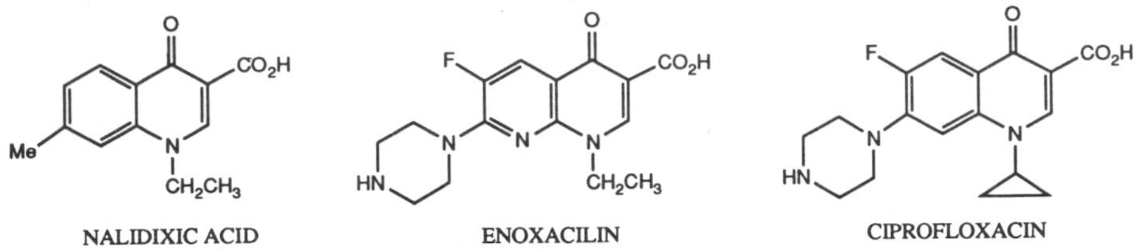

Fig. 10.74 Quinolones and fluoroquinolones.

compounds. It is active against Gram-negative bacteria and is useful in the short-term therapy of urinary tract infections. It can be taken orally, but unfortunately, bacteria can rapidly develop resistance to it. Various analogues have been synthesized which have similar properties to nalidixic acid, but provide no great advantage.

A big breakthrough was made, however, when a single fluorine atom was introduced at position 6, and a piperazinyl residue was placed at position 7 of the heteroaromatic skeleton. This led to enoxacilin (Fig. 10.74) which has a greatly increased spectrum of activity against Gram-negative and Gram-positive bacteria. Activity was also found against the highly resistant *Pseudomonas aeruginosa*.

Further adjustments led to ciprofloxacin (Fig. 10.74), now the agent of choice in treating travellers' diarrhoea. It has been used in the treatment of a large range of infections involving the urinary, respiratory, and gastrointestinal tracts as well as infections of skin, bone, and joints. It has been claimed that ciprofloxacin may be the most active broad-spectrum antibacterial agent on the market. Furthermore, bacteria are slow in acquiring resistance to ciprofloxacin, in contrast to nalidixic acid.

The quinolones and fluoroquinolones are thought to act on the bacterial enzyme deoxyribonucleic acid gyrase (DNA gyrase). This enzyme catalyses the supercoiling of chromosomal DNA into its tertiary structure. A consequence of this is that replication and transcription are inhibited and the bacterial cell's genetic code remains unread. At present, the mechanism by which these agents inhibit DNA gyrase is unclear.

10.8.2 Aminoacridines

Aminoacridines such as proflavine (Fig. 10.75) are topical antibacterial agents which were used in the Second World War for the treatment of surface wounds. Their mechanism of action is described in Chapter 6.

Fig. 10.75 Proflavine.

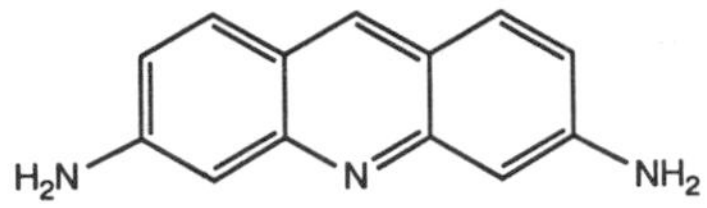

10.9 *Drug resistance*

With such a wide range of antibacterial agents available in medicine, it may seem surprising that medicinal chemists are still actively seeking new and improved antibacterial agents. The reason for this is due mainly to the worrying ability of bacteria to acquire resistance.

Drug resistance can be due to a variety of things. For example, the bacterial cell may change the structure of its cell membrane and prevent the drug from entering the cell. Alternatively, an enzyme may be produced which destroys the drug. Another possibility is that the cell counteracts the action of the drug. For example, if the drug is targeting a specific enzyme, then the bacterium may synthesize an excess of the enzyme. All these mechanisms require some form of control. In other words, the cell must have the necessary genetic information. This genetic information can be obtained by mutation or by the transfer of genes between cells.

10.9.1 Drug resistance by mutation

Bacteria multiply at such a rapid rate that there is always a chance that a mutation will render a bacterial cell resistant to a particular agent. This feature has been known for a long time and is the reason why patients should fully complete a course of antibacterial treatment even though their symptoms may have disappeared well before the end of the course.

If this rule is adhered to, the vast majority of the invading bacterial cells will be wiped out, leaving the body's own defence system to mop up any isolated survivors or resistant cells. If, however, the treatment is stopped too soon, then the body's defences struggle to cope with the survivors. Any isolated resistant cell is then given the chance to multiply, resulting in a new infection which will, of course, be completely resistant to the original drug.

These mutations occur naturally and randomly and do not require the presence of the drug. Indeed, it is likely that a drug-resistant cell is present in a bacterial population even before the drug is encountered. This was demonstrated with the identification of streptomycin-resistant cells from old cultures of a bacterium called *E. coli* which had been freeze-dried to prevent multiplication before the introduction of streptomycin into medicine.

10.9.2 Drug resistance by genetic transfer

A second way in which bacterial cells can acquire drug resistance is by gaining that resistance from another bacterial cell. This occurs because it is possible for genetic information to be passed on directly from one bacterial cell to another. There are two main methods by which this can take place—transduction and conjugation.

In transduction, small segments of genetic information known as plasmids are transferred by means of bacterial viruses (bacteriophages) leaving the resistant cell and infecting a non-resistant cell. If the plasmid brought to the infected cell contains the gene required for drug resistance, then the recipient cell will be able to use that information and gain resistance. For example, the genetic information required to synthesize β-lactamases can be passed on in this way, rendering bacteria resistant to penicillins. The problem is particularly prevalent in hospitals where currently over 90 per cent of staphylococcal infections are resistant to antibiotics such as penicillin, erythromycin, and tetracycline. It may seem odd that hospitals should be a source of drug-resistant strains of bacteria. In fact, they are the perfect breeding ground. Drugs commonly used in hospitals are present in the air in trace amounts. It has been shown that breathing in these trace amounts kills sensitive bacteria in the nose and allows the nostrils to act as a breeding ground for resistant strains.

In conjugation, bacterial cells pass genetic material directly to each other. This is a method used mainly by Gram-negative, rod-shaped bacteria in the colon, and involves two cells building a connecting bridge of sex pili through which the genetic information can pass.

10.9.3 Other factors affecting drug resistance

The more useful a drug is, the more it will be used and the greater the possibilities of resistant bacterial strains emerging. The original penicillins were used widely in human medicine, but were also commonly used in veterinary medicine. Antibacterial agents have also been used in animal feeding to increase animal weight and this, more than anything else, has resulted in drug-resistant bacterial strains. It is sobering to think that many of the original bacterial strains which were treated so dramatically with penicillin V or penicillin G are now resistant to those early penicillins. In contrast, these two drugs are still highly effective antibacterial agents in poorer, developing nations in Africa, where the use (and abuse) of the drug has been far less widespread.

The ease with which different bacteria acquire resistance varies. For example, *Staphylococcus aureus* is notorious for its ability to acquire drug resistance due to the ease with which it can undergo transduction. On the other hand, the microorganism responsible for syphilis seems incapable of acquiring resistance and is still susceptible to the original drugs used against it.

11 · The peripheral nervous system—cholinergics, anticholinergics, and anticholinesterases

In Chapter 10, we discussed the medicinal chemistry of antibacterial agents and noted the success of these agents in combating many of the diseases which have afflicted mankind over the years. This success was aided in no small way by the fact that the 'enemy' could be identified, isolated, and conquered—first in the petri dish, then in the many hiding places which it could frequent in the body. After this success, the medicinal chemist set out to tackle the many other human ailments which were not infection-based—problems such as heart disorders, depression, schizophrenia, ulcers, autoimmune disease, and cancer. In all these ailments, the body itself has ceased to function properly in some way or other. There is no 'enemy' as such.

So what can medicinal chemistry do if there is no enemy to fight, save for the human body's inefficiency? The first logical step is to understand what exactly has gone wrong.

However, the mechanisms and reaction systems of the human body can be extremely complex. A vast array of human functions proceed each day with the greatest efficiency and with the minimum of outside interference. Breathing, digestion, temperature control, excretion, posture—these are all day-to-day operations which we take for granted—until they go wrong of course! Considering the complexity of the human body, it is perhaps surprising that its workings don't go wrong more often than they do.

Even if the problem is identified, what can a mere chemical do amidst a body filled with complex enzymes and interrelated chemical reactions? If it is even possible for a single chemical to have a beneficial effect, which of the infinite number of organic compounds would we use?

The problem might be equated with finding the computer virus which has invaded your home computer software, or perhaps trying to trace where a missing letter went, or finding the reason for the country's balance of payments deficit.

However, all is not doom and gloom. There are some clues and hints to be had. The ancient herbal remedies of the past partially open the curtain to some of the body's jealously guarded secrets. Even the toxins of snakes, spiders, and plants can give important clues to the workings of the body and provide lead compounds to possible cures.

Over the last one hundred years or so, many biologically active compounds have been extracted from their natural sources, then purified and identified. Chemists subsequently rung the changes on these lead compounds until an effective drug was identified. The process depended on trial and effort, chance and serendipity, but with this effort came a better understanding of how the body works and how drugs interact with the body. Now that that has been achieved, medicinal chemistry has started to move from being a game of chance to being a science where the design of new drugs is based on logical theories.

To illustrate this, we are going to concentrate on one particular field—cholinergic and anticholinergic drugs. These are drugs which act on the peripheral nervous system, and so it is important to have some idea of how that system works before we proceed.

11.1 *The peripheral nervous system*

The peripheral nervous system (Fig. 11.1) is, as the name indicates, that part of the nervous system which is outside of the central nervous system (CNS—the brain and spinal column).

There are many divisions and subdivisions of the peripheral system which can lead to confusion. The first distinction we can make is between the following:

- sensory nerves (nerves which take messages from the body to the CNS)
- motor nerves (nerves which carry messages from the CNS to the rest of the body)

We need only concern ourselves with the latter—the motor nerves.

11.2 *Motor nerves of the peripheral nervous system*

These nerves take messages from the CNS to various parts of the body such as skeletal muscle, smooth muscle, cardiac muscle, and glands. The messages can be considered as 'electrical pulses'. However, the analogy with electricity should not be taken too far since the pulse is a result of ion flow across the membranes of nerves and not a flow of electrons (Appendix 2).

It should be evident that the workings of the human body depend crucially on an effective motor nervous system. Without it, we would not be able to operate our muscles and we would end up as flabby blobs, unable to move or breathe. We would not be able to eat, digest, or excrete our food since the smooth muscle of the gastrointestinal tract (GIT) and the urinary tract are innervated by motor nerves. We would not be able to control body temperature since the smooth muscle controlling the diameter of our peripheral blood vessels would cease to function. Finally, our heart would resemble a wobbly jelly rather than a powerful pump. In short, if the motor nerves failed

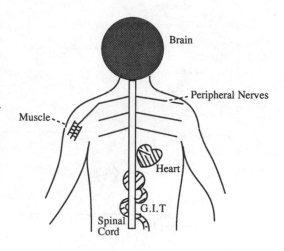

Fig. 11.1 The peripheral nervous system.

to function, we would be in a mess! Let us now look at the motor nerves in more detail.

The motor nerves of the peripheral nervous system have been divided into two subsystems (Fig. 11.2):

• the somatic motor nervous system
• the autonomic motor nervous system

11.2.1 The somatic motor nervous system

These are nerves which carry messages from the CNS to the skeletal muscles. There are no synapses (junctions) *en route* and the neurotransmitter at the neuromuscular junction is acetylcholine. The final result of such messages is contraction of skeletal muscle.

11.2.2 The autonomic motor nervous system

These nerves carry messages from the CNS to smooth muscle, cardiac muscle, and the adrenal medulla. This system can be divided into two subgroups.

Parasympathetic nerves

These leave the CNS, travel some distance, then synapse with a second nerve which then proceeds to the final synapse with smooth muscle. The neurotransmitter at both synapses is acetylcholine.

Sympathetic nerves

These leave the CNS, but almost immediately synapse with a second nerve (neuro-transmitter—acetylcholine) which then proceeds to the same target organs as the

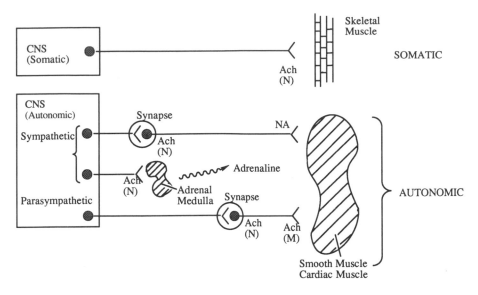

Fig. 11.2 Motor nerves of the peripheral nervous system. N, Nicotinic receptor; M, muscarinic receptor.

parasympathetic nerves. However, they synapse with different receptors on the target organs and use a different neurotransmitter—noradrenaline (for their actions, see Section 11.4).

The only exception to this are the nerves which go directly to the adrenal medulla. The neurotransmitter released here is noradrenaline and this stimulates the adrenal medulla to release the hormone adrenaline. This hormone then circulates in the blood system and interacts with noradrenaline receptors as well as other adrenaline receptors not directly 'fed' with nerves.

Note that the nerve messages are not sent along continuous 'telephone lines'. Gaps (synapses) occur between different nerves and also between nerves and their target organs (Fig. 11.3). If a nerve wishes to communicate its message to another nerve or a target organ, it can only do so by releasing a chemical. This chemical has to cross the synaptic gap and bind to receptors on the target cell in order to pass on the message. This interaction between neurotransmitter and receptor can then stimulate other processes which, in the case of a second nerve, leads to the message being continued. Since these chemicals effectively carry the message from one nerve to another, they have become known as chemical messengers or neurotransmitters. The very fact that they are chemicals and that they carry out a crucial role in nerve transmission allows the medicinal chemist to design and synthesize organic compounds which can mimic (agonists) or block (antagonists) the natural neurotransmitters.

We shall now look at the two neurotransmitters involved in the peripheral nervous system.

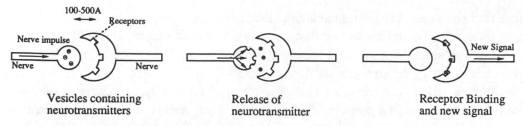

Fig. 11.3 Signal transmission at a synapse.

11.3 *The neurotransmitters*

There are a large variety of neurotransmitters in the CNS, but as far as the peripheral nervous system is concerned we need only consider two—acetylcholine and noradrenaline (Fig. 11.4).

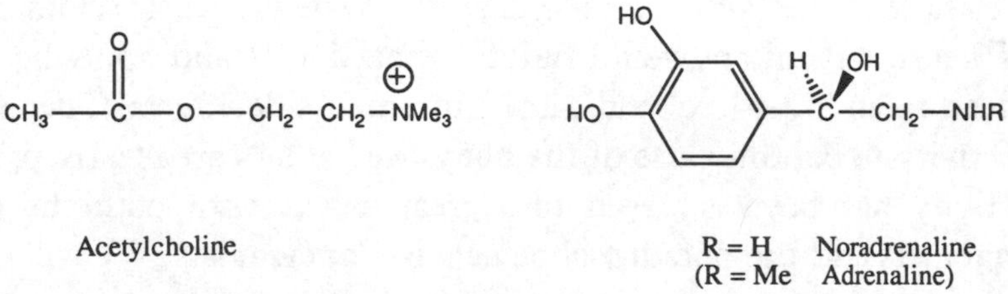

Fig. 11.4 Two major neurotransmitters of the peripheral nervous system.

11.4 *Actions of the peripheral nervous system*

Somatic

Stimulation leads to the contraction of skeletal muscle.

Autonomic

- Sympathetic.

  Noradrenaline is released at target organs and leads to the contraction of cardiac muscle and an increase in heart rate. It relaxes smooth muscle and reduces the contractions of the GIT and urinary tracts. It also reduces salivation and reduces dilation of the peripheral blood vessels.

  In general, the sympathetic nervous system promotes the 'fight or flight' response by shutting down the body's housekeeping roles (digestion, defecation, urination, etc.), and stimulating the heart. The stimulation of the adrenal medulla releases the hormone adrenaline which reinforces the action of noradrenaline.

- Parasympathetic.

  The stimulation of the parasympathetic system leads to the opposite effects from those of the sympathetic system. Acetyl choline is released at the target organs and reacts with receptors specific to it and not to noradrenaline.

Note that the sympathetic and parasympathetic nervous systems oppose each other in their actions and could be looked upon as a brake and an accelerator. The analogy is not quite apt since both systems are always operating and the overall result depends on which effect is the stronger.

Failure in either of these systems would clearly lead to a large variety of ailments involving heart, skeletal muscle, digestion, etc. Such failure might be the result of either a deficit or an excess of neurotransmitter. Therefore, treatment would involve the administration of drugs which could act as agonists or antagonists depending on the problem.

However, there is a difficulty with this approach. Usually, the problem which we wish to tackle occurs at a certain location where there might, for example, be a lack of neurotransmitter. Application of an agonist to make up for low levels of neurotransmitter at the heart, for example, might solve the problem there, but would lead to problems elsewhere in the body (e.g. the digestion system). At these other locations, the levels of neurotransmitter would be at normal levels and applying an agonist would then lead to an 'overdose' and cause unwanted side-effects. Therefore, drugs showing selectivity to certain parts of the body over others are clearly preferred.

This selectivity has been achieved to a great extent with both the cholinergic agonists/antagonists and the noradrenaline agonists/antagonists. We will concentrate on the former.

11.5 *The cholinergic system*

Let us look first at what happens at synapses involving acetylcholine as the neurotransmitter. Figure 11.5 shows the synapse between two nerves and the events involved when a message is transmitted from one nerve cell to another. The same general process takes place when a message is passed from a nerve cell to a muscle cell.

1. Biosynthesis of acetylcholine (Fig. 11.6).
   Acetylcholine is synthesized in the nerve ending of the pre-synaptic nerve from choline and acetyl coenzyme A. The reaction is catalysed by the enzyme choline acetyltransferase.
2. Acetylcholine is incorporated into membrane-bound vesicles.
3. The arrival of a nerve signal leads to the release of acetylcholine. The mechanism of this process is poorly understood. Conventionally, it is thought that vesicles containing the neurotransmitter merge with the cell membrane and in doing so release the transmitter into the synaptic gap. Other mechanisms have been proposed however.
4. Acetylcholine crosses the synaptic gap and binds to the cholinergic receptor leading to stimulation of the second nerve.

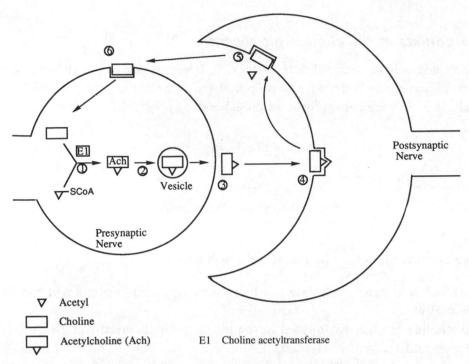

Fig. 11.5 Synapse with acetylcholine as neurotransmitter.

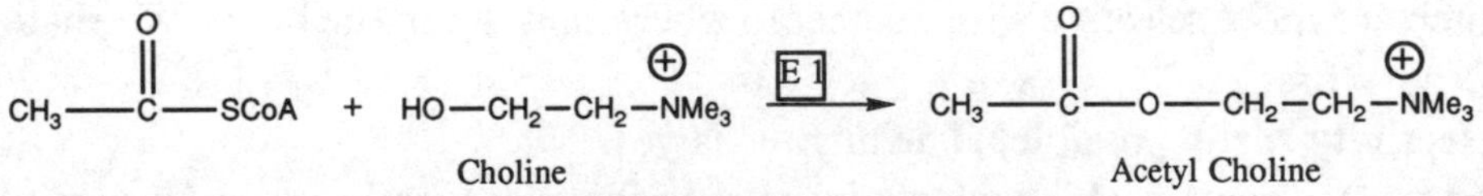

Fig. 11.6 Biosynthesis of acetylcholine.

5. Acetylcholine moves to an enzyme called acetylcholinesterase which is situated on the postsynaptic nerve and which catalyses the hydrolysis of acetylcholine to produce choline and ethanoic acid.

6. Choline binds to the choline receptor on the presynaptic nerve and is taken up into the cell by an efficient transport system to continue the cycle.

The most important thing to note about this process is that there are several stages where it is possible to use drugs to either promote or inhibit the overall process. The greatest success so far has been with drugs targeted at stages 4 and 5 (i.e. the cholinergic receptor and the acetylcholinesterase enzyme).

We will now look at these in more detail.

11.6 *Agonists at the cholinergic receptor*

One point might have occurred to the reader. If there is a lack of acetylcholine acting at a certain part of the body, why do we not just give the patient more acetylcholine? After all, it is easy enough to make in the laboratory (Fig. 11.7).

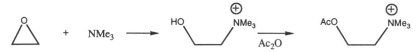

Fig. 11.7 Synthesis of acetylcholine.

There are three reasons why this is not feasible.

- Acetylcholine is easily hydrolysed in the stomach by acid catalysis and cannot be given orally.
- Acetylcholine is easily hydrolysed in the blood, both chemically and by enzymes (esterases and acetylcholinesterase).
- There is no selectivity of action. Acetylcholine will switch on all acetylcholine receptors in the body.

Therefore, we need analogues of acetylcholine which are more stable to hydrolysis and which are more selective with respect to where they act in the body. We shall look at selectivity first.

Is selectivity really possible? The answer is yes.

There are two ways in which selectivity can be achieved. Firstly, some drugs might be distributed more efficiently to one part of the body than another. Secondly, cholinergic receptors in various parts of the body might be slightly different. This difference would have to be quite subtle—not enough to affect the interaction with the natural neurotransmitter acetylcholine, but enough to distinguish between two different synthetic analogues.

We could, for example, imagine that the binding site for the cholinergic receptor is a hollow into which the acetylcholine molecule could fit (Fig. 11.8). We might then imagine that some cholinergic receptors in the body have a 'wall' bordering this hollow, while other cholinergic receptors do not.

Thus, a synthetic analogue of acetylcholine which is slightly bigger than acetylcholine itself would bind to the latter receptor, but would be unable to bind to the former receptor because of the wall.

This theory might appear to be wishful thinking, but it is now established that cholinergic receptors in different parts of the body are indeed subtly different.

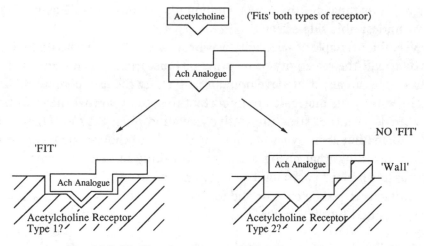

Fig. 11.8 Binding sites for two cholinergic receptors.

This is not just a peculiarity of acetylcholine receptors. Subtle differences have been observed for other types of receptors such as those for dopamine, noradrenaline, and serotonin.

To return to the acetylcholine receptor, how do we know if there are different subtypes? As is often the case, the first clues came from the action of natural compounds. It was discovered that the compounds nicotine and muscarine (Fig. 11.9) were both acetylcholine agonists, but that they had different physiological effects.

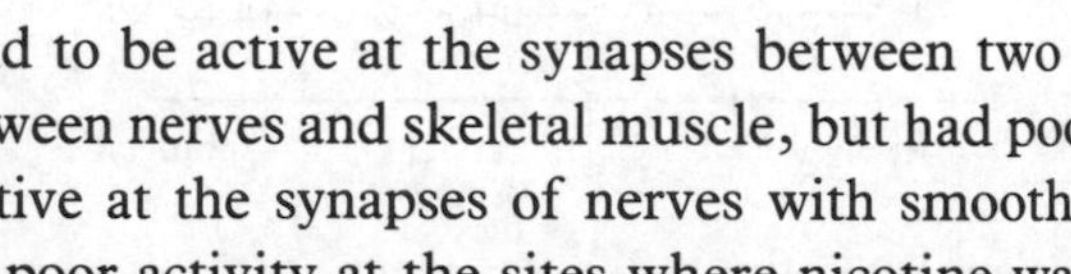

Fig. 11.9   NICOTINE   L (+) MUSCARINE

Nicotine was found to be active at the synapses between two different nerves and also the synapses between nerves and skeletal muscle, but had poor activity elsewhere.

Muscarine was active at the synapses of nerves with smooth muscle and cardiac muscle, but showed poor activity at the sites where nicotine was active.

From these results it was concluded that there was one type of acetylcholine receptor on skeletal muscles and at nerve synapses (the nicotinic receptor), and a different sort of acetylcholine receptor on smooth muscle and cardiac muscle (the muscarinic receptor).

Therefore, muscarine and nicotine were the first compounds to show receptor selectivity. Unfortunately, these two compounds are not suitable as medicines since they have undesirable side-effects.[1]

However, the principle of selectivity was proven and the race was on to design novel drugs which had the selectivity of nicotine or muscarine, but not the side-effects.[2]

The first stage in any drug development is to study the lead compound and to find out which parts of the molecule are important to activity so that they can be retained in future analogues (i.e. structure–activity relationships (SAR)). These results also provide information about what the binding site of the cholinergic receptor looks like and help in deciding what changes are worth making in new analogues.

In this case, the lead compound is acetylcholine itself.

The results described below are valid for both the nicotinic and muscarinic receptors and were obtained by the synthesis of a large range of analogues.

11.7 *Acetylcholine—structure, SAR, and receptor binding*

- The positively charged nitrogen atom is essential to activity. Replacing it with a neutral carbon atom eliminates activity.
- The distance from the nitrogen to the ester group is important.
- The ester functional group is important.
- The overall size of the molecule cannot be altered greatly. Bigger molecules have poorer activity.
- The ethylene bridge between the ester and the nitrogen atom cannot be extended.
- There must be two methyl groups on the nitrogen. A larger, third alkyl group is tolerated, but more than one large alkyl group leads to loss of activity.
- Bigger ester groups lead to a loss of activity.

Conclusions: clearly, there is a tight fit between acetylcholine and its binding site which leaves little scope for variation. The above findings fit in with a receptor site as shown in Fig. 11.11.

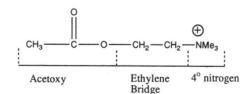

Fig. 11.10 Acetylcholine.

Acetoxy   Ethylene   4° nitrogen
                   Bridge

[1] This is due to interactions with other types of receptor, such as the receptors for dopamine or noradrenaline. In the search for a good drug, it is important to gain two types of selectivity—selectivity for one type of receptor over another (e.g. the acetylcholine receptor in preference to a noradrenaline receptor), and selectivity for receptor subtypes (e.g. the muscarinic receptor in preference to a nicotinic receptor).

[2] The search for increasingly selective drugs has led to the discovery that there are subtypes of receptors within subtypes. In other words, not every muscarinic receptor is the same throughout the body. At present, three subtypes of the muscarinic receptor have been discovered and have been labelled M1, M2, and M3. More may still be discovered.

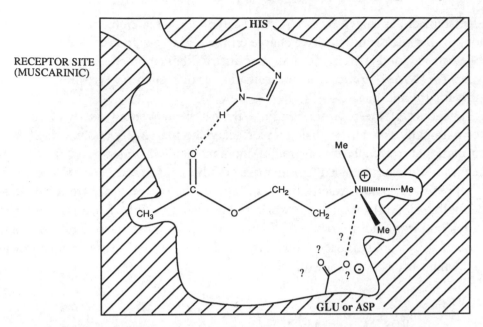

RECEPTOR SITE
(MUSCARINIC)

Fig. 11.11

It is proposed that an important hydrogen bonding interaction exists between the ester group of the acetylcholine molecule and a histidine residue. It is also thought that a small hydrophobic pocket exists which can accommodate the methyl group of the ester, but nothing larger. This interaction is thought to be more important in the muscarinic receptor than the nicotinic receptor.

Now let us look at the NMe_3^+ group. The evidence suggests two small hydrophobic pockets in the receptor, which are large enough to accommodate two of the three methyl substituents on the NMe_3^+ group. The third methyl substituent on the nitrogen is not bound and can be replaced with other groups. A strong ionic interaction has been proposed between the charged nitrogen atom and the anionic side-group of either a glutamic acid or an aspartic acid residue. The existence of this ionic interaction represents the classical view of the cholinergic receptor, but recent opinion has moved away from this position. Why is this?

First of all, the positive charge on the NMe_3^+ group is not localized on the nitrogen atom. It is also spread over the three methyl groups. Such a diffuse charge is less likely to be involved in an ionic interaction. Secondly, a suitable aspartate or glutamate residue has not been identified. In fact, there is evidence that the NMe_3^+ group is bound to a hydrophobic region of the receptor. Thirdly, model studies have shown that NMe_3^+ groups can be stabilized by binding to aromatic rings. It might seem strange that a hydrophobic group like an aromatic ring should be capable of stabilizing a positively charged group. However, it has to be remembered that aromatic rings are electron-rich, as shown by the fact they can undergo reaction with electrophiles. It is thought that the diffuse positive charge on the NMe_3^+ group is capable of distorting the pi electron cloud of aromatic rings to induce a dipole moment. Dipole interactions between the NMe_3^+ group and an aromatic residue such as tyrosine would then account for the binding.

A 3D model of the receptor binding site has been worked out with the aid of conformationally restrained analogues of acetylcholine. Acetylcholine itself has no conformational restraints. It is a straight-chain molecule in which bond rotation along the length of its chain can lead to numerous possible conformations (or shapes). Thus, it is impossible to know exactly the 3D shape of the receptor site from considering acetylcholine alone. In the past, it was assumed that a flexible neurotransmitter such as acetylcholine would interact with its receptor in its most stable conformation. In the case of acetylcholine, that would be the conformation represented by the sawhorse and Newman projections shown in Fig. 11.12.

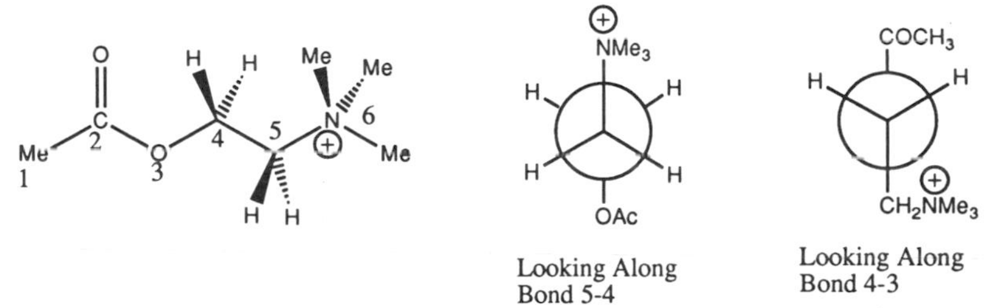

Fig. 11.12 The sawhorse and Newman projections of acetylcholine.

This assumption is invalid since there is not a great energy difference between alternative conformations such as the gauche or staggered conformations (Fig. 11.13). The energy gained from the neurotransmitter–receptor binding interaction would be more than sufficient to compensate for the difference.

In order to establish the 'active' conformation of a flexible neurotransmitter (the conformation taken up by the neurotransmitter once it is bound to the receptor), it is necessary to study structures which contain 'locked' conformations of acetylcholine within their structures. Muscarine and the analogue shown in Fig. 11.14 are known to bind to the cholinergic receptor. These molecules contain the acetylcholine skeleton,

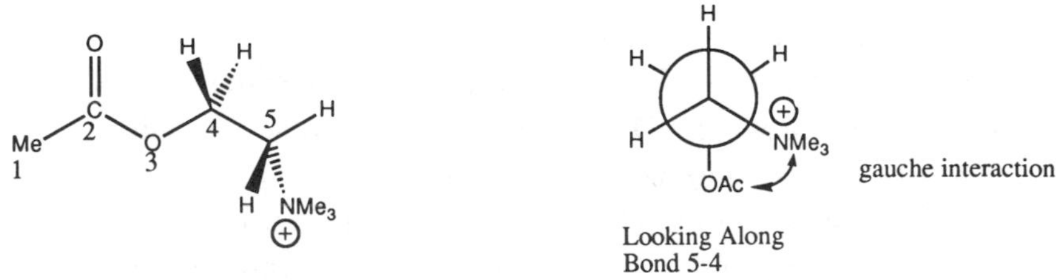

Fig. 11.13 The gauche or staggered conformation.

but since they are ring structures, the left-hand portion of the acetylcholine molecule is now restricted to one conformation. This in turn gives an accurate 3D representation of the receptor binding site interacting with that part of the molecule.

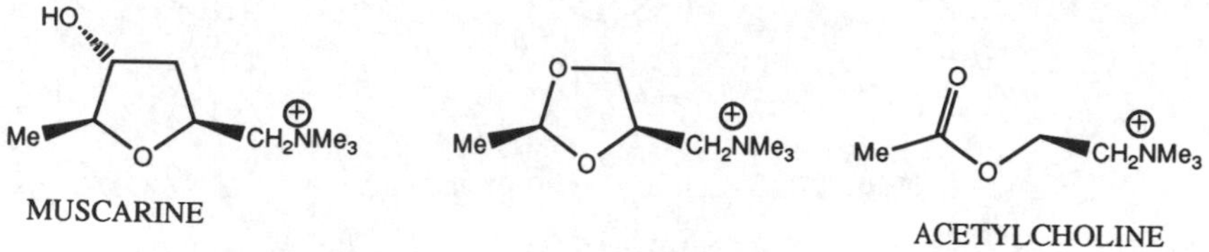

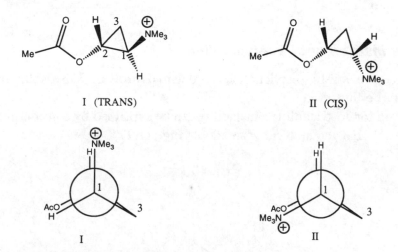

MUSCARINE   ACETYLCHOLINE

Fig. 11.14 Muscarine and the analogue.

In both molecules shown, rotation is still possible round the ring–CH_2NMe_3 bond, which means that the relative position of the nitrogen atom with respect to the ester remains uncertain. However, a third conformationally restrained molecule (structure I in Fig. 11.15) is known to bind to the muscarinic receptor site (but not the nicotinic receptor). In this molecule, the right-hand portion of the molecule is locked in one conformation—represented by the Newman projection shown in Fig. 11.15. This demonstrates that the ester and ammonium groups are staggered with respect to each other.

I (TRANS)   II (CIS)

I   II

Fig. 11.15 Conformationally restrained analogues of acetylcholine.

Since structure I (Fig. 11.15)) is found to bind to the muscarinic receptor as efficiently as acetylcholine, it suggests that the 'active' conformation of acetylcholine is the *gauche* conformation shown in Newman projection (**1**) (Fig. 11.16) rather than the *anti* or eclipsed conformations (**2**) and (**3**).

Further evidence is provided by the cyclic structure II (Fig. 11.15) which has the ester and ammonium groups *cis* to each other and therefore fully eclipsed. This shows

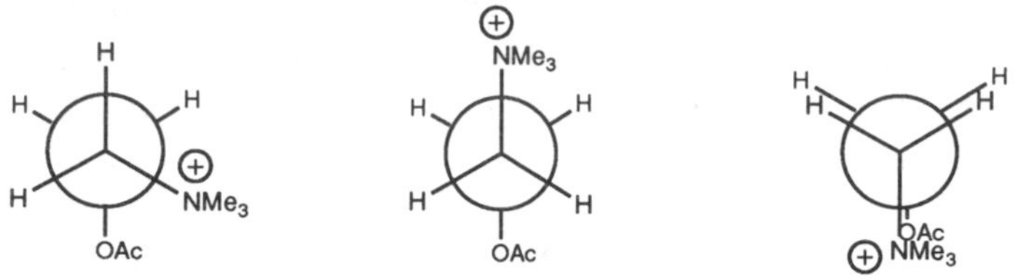

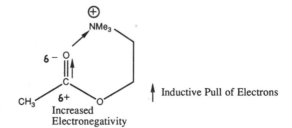

Fig. 11.16 Conformations of acetylcholine.

virtually no activity and suggests that the eclipsed conformation of acetylcholine (**3**) (Fig. 11.16) is not an 'active' conformation.

From these considerations, it can be shown that the distance between the quaternary nitrogen and the ester group is 5–7 Å.

The results from these experiments can be used to give an overall 3D shape for the receptor site as well as showing the active conformation of acetylcholine. Naturally, once the 3D shape of the receptor binding site is known, the design of novel cholinergic agents becomes much simpler. Any new agent capable of adopting a conformation whereby the important bonding groups are properly positioned is worthy of study. With this knowledge, acetylcholine analogues can be designed with improved stability.

11.8 *The instability of acetylcholine*

As described previously, acetylcholine is prone to hydrolysis. Why is this and how can the stability be improved?

The reason for acetylcholine's instability can be explained by considering one of the conformations that the molecule can adopt (Fig. 11.17).

Fig. 11.17 Neighbouring group participation.

In this conformation, the positively charged nitrogen interacts with the carbonyl oxygen and has an electron withdrawing effect. To compensate for this, the oxygen atom pulls electrons towards it from the neighbouring carbon atom and as a result makes that carbon atom electron deficient and more prone to nucleophilic attack.

Water is a poor nucleophile, but since the carbonyl group is now more electrophilic, hydrolysis takes place relatively easily.

This influence of the nitrogen ion is known as neighbouring group participation or anchimeric assistance.

We shall now look at how the problem of hydrolysis was overcome, but it should be appreciated that we are doing so with the benefit of hindsight. At the time the problem was tackled, the SAR studies were incomplete and the exact shape of the cholinergic receptor binding site was unknown. In fact, it was the very analogues which were made to try and solve the problem of hydrolysis that led to a better understanding of the receptor binding site.

11.9 *Design of acetylcholine analogues*

In order to tackle the inherent instability of acetylcholine, two approaches are possible:

- steric hindrance
- electronic stabilization

11.9.1 Steric hindrance

The principle involved here can be demonstrated with methacholine (Fig. 11.18).

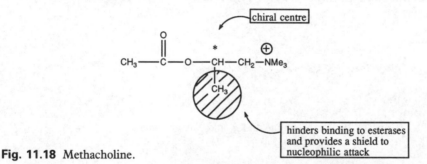

Fig. 11.18 Methacholine.

This analogue of acetylcholine contains an extra methyl group on the ethylene bridge. The reasons for putting it there are twofold. Firstly, it is to try and build in a shield for the carbonyl group. The bulky methyl group should hinder the approach of any potential nucleophile and slow down the rate of hydrolysis. It should also hinder binding to the esterase enzymes, thus slowing down enzymatic hydrolysis.

The results were encouraging, with methacholine proving three times more stable to hydrolysis than acetylcholine.

The obvious question to ask now is, Why not put on a bigger alkyl group like an ethyl group or a propyl group? Alternatively, why not put a bulky group on the acyl half of the molecule, since this would be closer to the carbonyl centre and have a greater shielding effect?

In fact, these approaches were tried, but failed. We should already know why—the fit between acetylcholine and its receptor is so tight that there is little scope for enlarging the molecule. The extra methyl group is as much as we can get away with. Larger substituents certainly cut down the chemical and enzymatic hydrolysis, but they also prevent the molecule binding to the cholinergic receptor.

In conclusion, attempts to increase the steric shield beyond the methyl group certainly increase the stability of the molecule, but decrease its activity since it cannot fit the cholinergic receptor.

One other very useful result was obtained from methacholine. It was discovered that the introduction of the methyl group led to significant muscarinic activity and very little nicotinic activity. Therefore, methacholine showed a good selective action for the muscarinic receptor. This result is perhaps more important than the gain in stability.

The good binding to the muscarinic receptor can be explained if we compare the active conformation of methacholine with muscarine (Fig. 11.19). The methyl group of methacholine can occupy the same position as a methylene group in muscarine.

Note, however, that methacholine can exist as two enantiomers (R and S) and only the S-enantiomer matches the structure of muscarine. The two enantiomers of methacholine have been isolated and the S-enantiomer is the more active enantiomer, as expected. It is not used therapeutically, however.

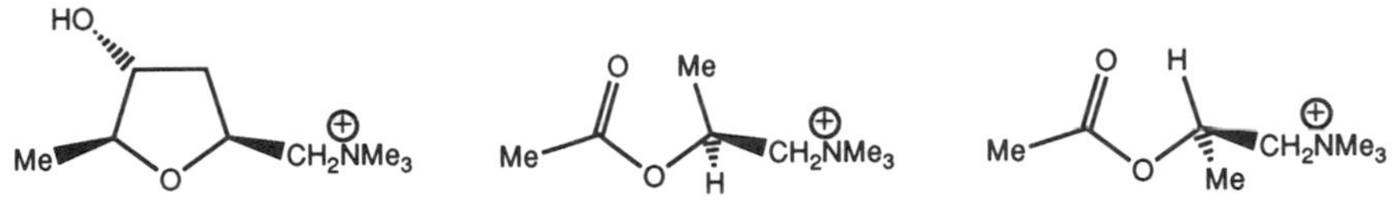

Fig. 11.19 Methacholine and R and S enantiomers.

11.9.2 Electronic effects

The best example of this approach is provided by carbachol (Fig. 11.20), a long acting cholinergic agent which is resistant to hydrolysis. In carbachol, the acyl methyl group has been replaced by an NH_2 group which is of comparable size and can therefore fit the receptor.

The resistance to hydrolysis is due to the electronic effect of the carbamate group. The resonance structures shown in Fig. 11.21 demonstrate how the lone pair from the nitrogen atom is fed into the carbonyl group such that the group's electrophilic

Fig. 11.20 Carbachol.

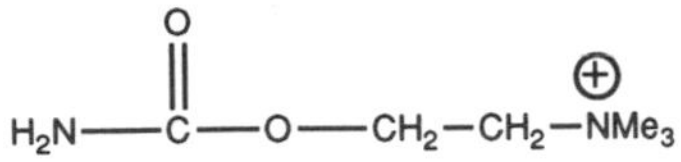

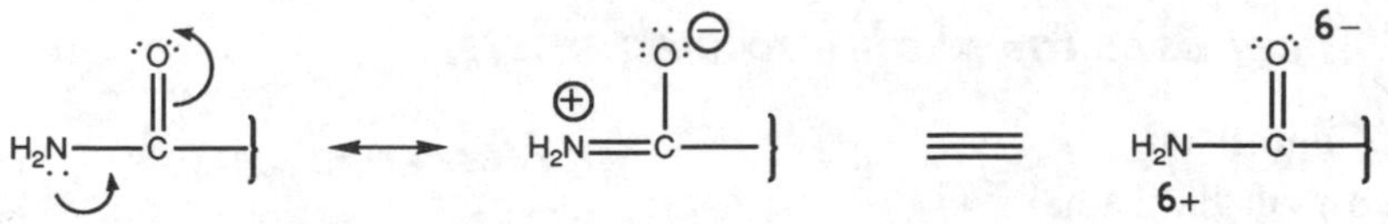

Fig. 11.21 Resonance structures of carbachol.

character is eliminated. As a result, the carbonyl is no longer susceptible to nucleophilic attack.

Carbachol is certainly stable to hydrolysis and is the right size to fit the cholinergic receptor, but it is by no means a foregone conclusion that it will be active. After all, a hydrophobic methyl group has been replaced with a polar NH_2 group and this implies that a polar group has to fit into a hydrophobic pocket in the receptor.

Fortunately, carbachol does fit and is active. Since the methyl group of acetylcholine has been replaced with an amino group without affecting the biological activity, we can call the amino group a 'bioisostere' of the methyl group.

It is worth emphasizing that a bioisostere is a group which can replace another group without affecting the pharmacological activity of interest. Thus, the amino group is a bioisostere of the methyl group as far as the cholinergic receptor is concerned, but not as far as the esterase enzymes are concerned.

Therefore, the inclusion of an electron donating group such as the amino group has greatly increased the chemical and enzymatic stability of our cholinergic agonist. Unfortunately, it is found that carbachol shows very little selectivity between the muscarinic and nicotinic sites.

Carbachol is used clinically for the treatment of glaucoma—an eye problem. The drug is applied locally and so selectivity is not a great problem.

11.9.3 Combining steric and electronic effects

We have already seen that a β-methyl group slightly increases the stability of acetyl-choline analogues through steric effects and also has the advantage of introducing some selectivity.

Clearly, it would be interesting to add a β-methyl group to carbachol. The compound obtained is bethanechol (Fig. 11.22) which, as expected, is both stable to hydrolysis and selective in its action. It is used therapeutically in stimulating the gastrointestinal tract and urinary bladder after surgery. (Both these organs are 'shut down' with drugs during surgery.)

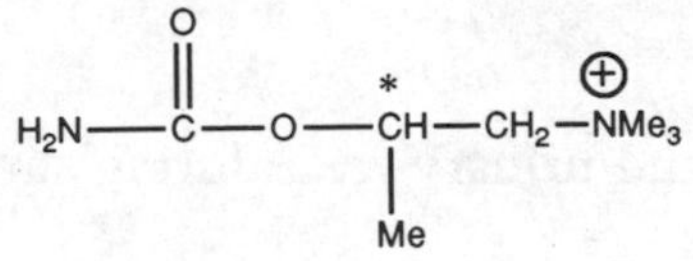

Fig. 11.22 Bethanechol.

11.10 *Clinical uses for cholinergic agonists*

Muscarinic agonists:
- Treatment of glaucoma.
- 'Switching on' the GIT and urinary tract after surgery.
- Treatment of certain heart defects by decreasing heart muscle activity and heart rate.

Nicotinic agonists:
- Treatment of myasthenia gravis, an autoimmune disease where the body has produced antibodies against its own acetylcholine receptors. This leads to a reduction in the number of available receptors and so fewer messages reach the muscle cells. This in turn leads to severe muscle weakness and fatigue. Administering an agonist increases the chance of activating what few receptors remain.

An example of a selective nicotinic agonist is shown in Fig. 11.23.

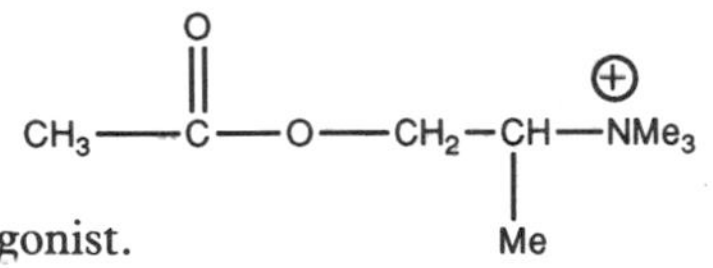

$$CH_3-\overset{\displaystyle O}{\overset{\displaystyle \|}{C}}-O-CH_2-\underset{\displaystyle Me}{CH}-\overset{\displaystyle \oplus}{N}Me_3$$

Fig. 11.23 A selective nicotinic agonist.

11.11 *Antagonists of the muscarinic cholinergic receptor*

11.11.1 Actions and uses of muscarinic antagonists

Antagonists of the cholinergic receptor are drugs which bind to the receptor but do not 'switch it on'. By binding to the receptor, an antagonist acts like a plug at the receptor site and prevents the normal neurotransmitter (i.e. acetylcholine) from binding (Fig. 11.24). Since acetylcholine cannot 'switch on' its receptor, the overall effect on the body is the same as if there was a lack of acetylcholine. Therefore, antagonists have the opposite clinical effect from agonists.

The following antagonists act only at the muscarinic receptor and therefore affect nerve transmissions to the smooth muscle of the gastrointestinal tract, urinary tract, and glands. The clinical effects and uses of these antagonists reflect this fact.

Clinical effects:
- Reduction of saliva and gastric secretions.
- Reduction of the motility of the GIT and the urinary tracts by relaxing smooth muscle.
- Dilation of eye pupils.

Clinical uses:
- Shutting down the GIT and urinary tract during surgery.
- Ophthalmic examinations.

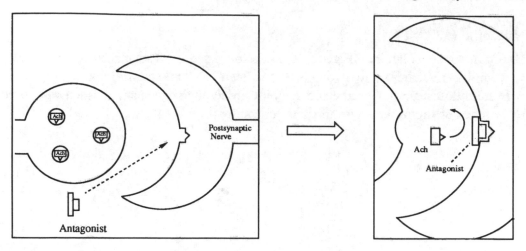

Fig. 11.24 Action of an antagonist.

- Relief of peptic ulcers.
- Treatment of Parkinson's disease.

11.11.2 Muscarinic antagonists

The first antagonists were natural products and in particular alkaloids (nitrogen-containing compounds derived from plants).

Atropine

Atropine (Fig. 11.25) has a chiral centre (★) and therefore two enantiomers are possible. Usually, natural products exist exclusively as one enantiomer. This is also true for atropine which is present in the plant species *Solanaceae* as a single enantiomer called hyoscyamine. However, as soon as the natural product is extracted into solution, the chiral centre racemizes such that atropine is obtained as a racemic mixture and not as a single enantiomer.

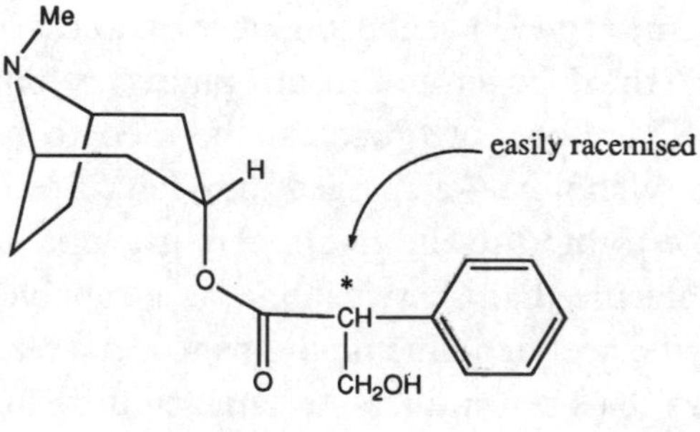

Fig. 11.25 Atropine.

The chiral centre in atropine is easily racemized since it is next to a carbonyl group. The proton attached to the chiral centre is acidic and as a result is easily replaced.

Atropine was obtained from the roots of belladonna (deadly nightshade) in 1831. It was once used by Italian women to dilate the pupils of the eye in order to appear more beautiful (hence the name belladonna).

Hyoscine (1879–84)

Hyoscine (or scopolamine) (Fig. 11.26) is also obtained from solanaceous plants and is very similar in structure to atropine. It has been used as a truth drug.

In high doses, both hyoscine and atropine are hallucinogens and as such were used by witches of the middle ages in their concoctions.

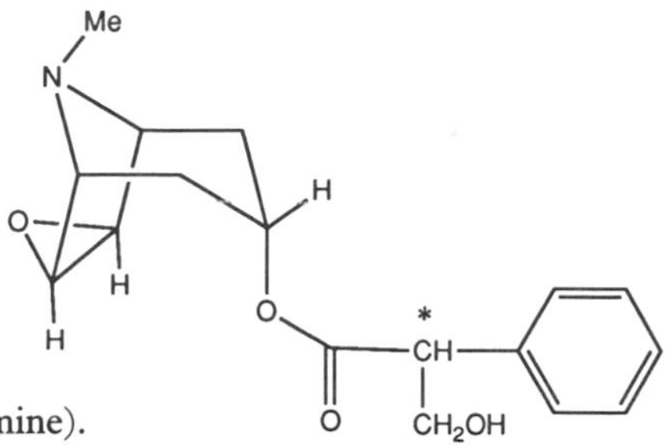

Fig. 11.26 Hyoscine (scopolamine).

These two compounds can bind to and block the cholinergic receptor. But why should they? At first sight, they do not look anything like acetylcholine.

If we look more closely though, we can see that a basic nitrogen and an ester group are present, and if we superimpose the acetylcholine skeleton on to the atropine skeleton, the distance between the ester and the nitrogen groups are similar in both molecules (Fig. 11.27).

There is, of course, the problem that the nitrogen in atropine is uncharged, whereas the nitrogen in acetylcholine is quaternary and has a full positive charge. This implies that the nitrogen atom in atropine is protonated when it binds to the cholinergic receptor.

Therefore, atropine can be seen to have the two important binding features of acetylcholine—a charged nitrogen (if protonated) and an ester group. It is, therefore, able to bind to the receptor, but is unable to 'switch it on'. Since atropine is a larger molecule than acetylcholine, it is capable of binding to other binding groups outside of the acetylcholine binding site. As a result, it interacts differently with the receptor, and does not induce the same conformational changes as acetylcholine.

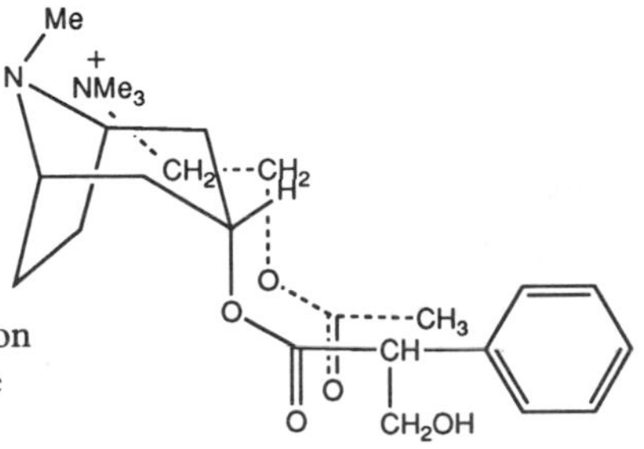

Fig. 11.27 Acetylcholine skeleton superimposed on to the atropine skeleton.

Structural analogues based on atropine

Analogues of atropine were synthesized to 'slim down' the structure to the essentials. This resulted in a large variety of active antagonists (e.g. tridihexethyl bromide and propantheline chloride) (Fig. 11.28).

TRIDIHEXETHYL BROMIDE   PROPANTHELINE CHLORIDE

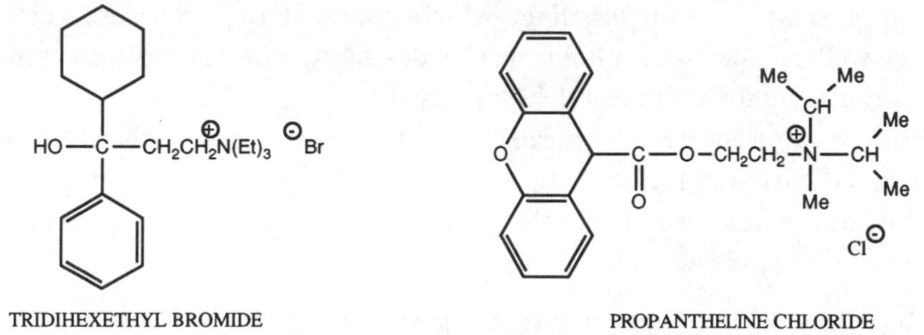

Fig. 11.28 Two analogues of atropine.

SAR studies have come up with the following generalizations (Fig. 11.29):

R' = Aromatic or Heteroaromatic

Fig. 11.29

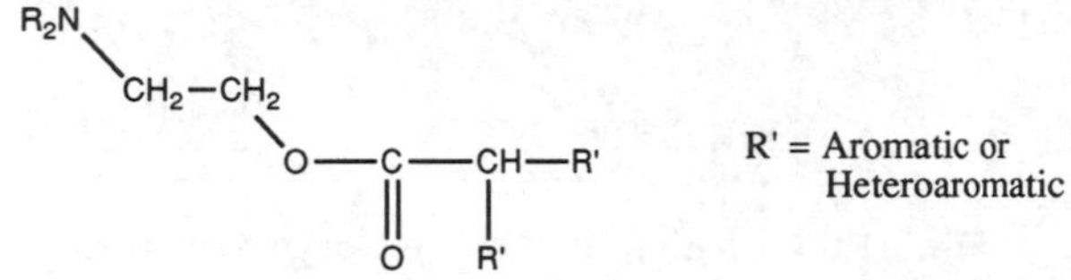

- The alkyl groups on nitrogen (R) can be larger than methyl (in contrast to agonists).
- The nitrogen can be tertiary or quaternary, whereas agonists must have a quaternary nitrogen (note, however, that the tertiary nitrogen is probably charged when it interacts with the receptor).
- Very large acyl groups are allowed (R' = aromatic or heteroaromatic rings). This is in contrast with agonists where only the acetyl group is permitted.

It is the last point which appears to be the most crucial in determining whether a compound will act as an antagonist or not. The acyl group has to be bulky, but it also has to have that bulk arranged in a certain manner (i.e. there must be some sort of branching in the acyl group). For example, the molecule shown in Fig. 11.30 has a large unbranched acyl group but is not an antagonist.

Fig. 11.30 Analogue with no branching in the acyl group.

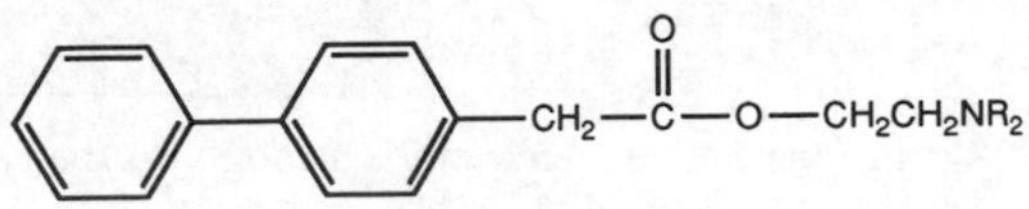

The conclusion which can be drawn from these results is that there must be another binding site on the receptor surface next to the normal acetylcholine binding site. This area must be hydrophobic since most antagonists have aromatic rings. The overall shape of the acetylcholine binding site plus the extra binding site would have to be T- or Y-shaped in order to explain the importance of branching in antagonists (Fig. 11.31).

A structure such as propantheline, which contains the complete acetylcholine skeleton as well as the hydrophobic acyl side-chain, not surprisingly binds more strongly to the receptor than acetylcholine itself (Fig. 11.32).

The extra bonding interaction means that the conformational changes induced in the receptor (if any are induced at all) will be different from those induced by acetylcholine and will fail to induce the secondary biological response. As long as the antagonist is bound, acetylcholine is unable to bind and pass on its message.

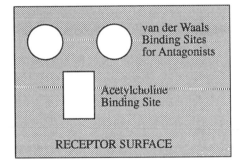

Fig. 11.31 Binding sites on the receptor surface.

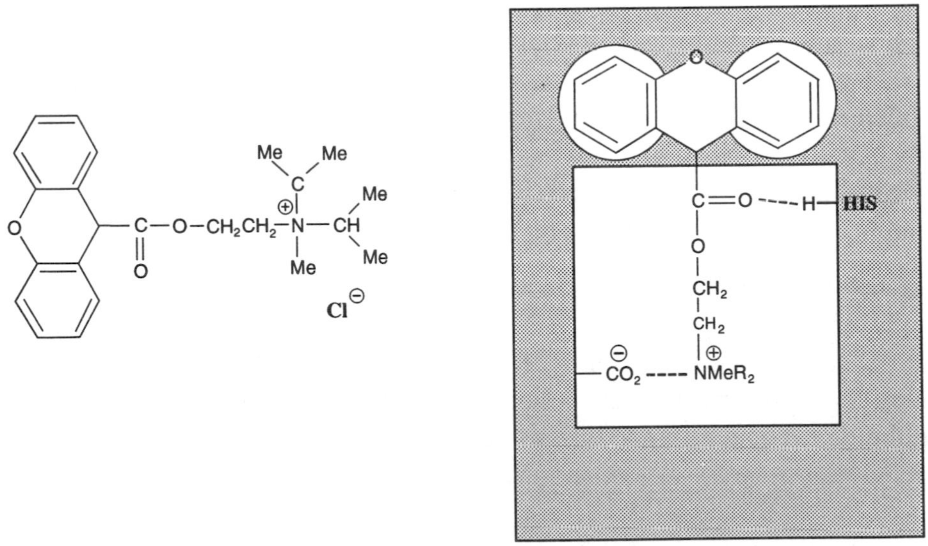

Fig. 11.32 Propantheline which binds strongly to the receptor.

A large variety of antagonists have proved to be useful medicines, with many showing selectivity for specific organs. For example, some act at the intestine to relieve spasm, some act selectively to decrease gastric secretions, while others are useful in ulcer therapy. This selectivity of action owes more to the distribution properties of the drug than to receptor selectivity (i.e. the compounds can reach one part of the body more easily than another).

However, the antagonist pirenzepine (Fig. 11.33), which is used in the treatment of peptic ulcers, is a selective M1 antagonist with no activity against M2 receptors.[2]

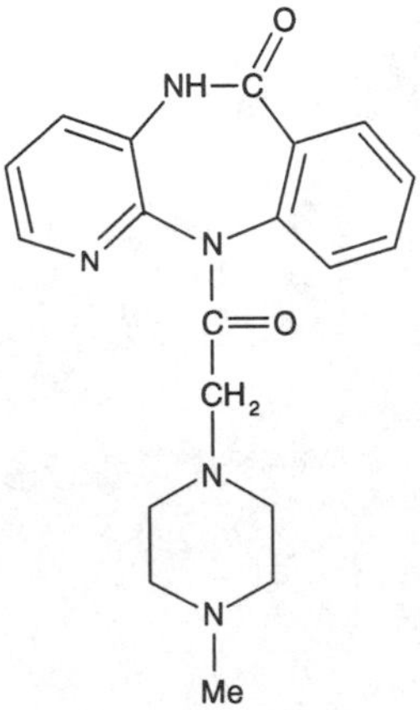

Fig. 11.33 Pirenzepine.

Since antagonists bind more strongly than agonists, they are better compounds to use for the labelling and identification of receptors on tissue preparations.

The antagonist, labelled with a radioactive isotope of H or C, binds strongly to the receptor and the radioactivity reveals where the receptor is located.

Ideally, we would want the antagonist to bind irreversibly in this situation. Such binding would be possible if the antagonist could form a covalent bond to the receptor. One useful tactic is to take an established antagonist and to incorporate a reactive chemical centre into the molecule. This reactive centre is usually electrophilic so that it will react with any suitably placed nucleophile close to the binding site (for example, the OH of a serine residue or the SH of a cysteine residue). In theory, the antagonist should bind to the receptor in the usual way and the electrophilic group will react with any nucleophilic amino acid within range. The resulting alkylation irreversibly binds the antagonist to the receptor through a covalent bond.

---

[2] The search for increasingly selective drugs has led to the discovery that there are subtypes of receptors within subtypes. In other words, not every muscarinic receptor is the same throughout the body. At present, three subtypes of the muscarinic receptor have been discovered and have been labelled M1, M2, and M3. More may still be discovered.

In practice, the procedure is not always as simple as this, since the highly reactive electrophilic centre might react with another nucleophilic group before it reaches the receptor binding site. One way to avoid this problem is to include a latent reactive centre which can only be activated once the antagonist has bound to the receptor binding site. One favourite method is photoaffinity labelling, where the reactive centre is activated by light. Chemical groups such as diazoketones or azides can be converted to highly reactive carbenes and nitrenes respectively, when irradiated (Fig. 11.34).

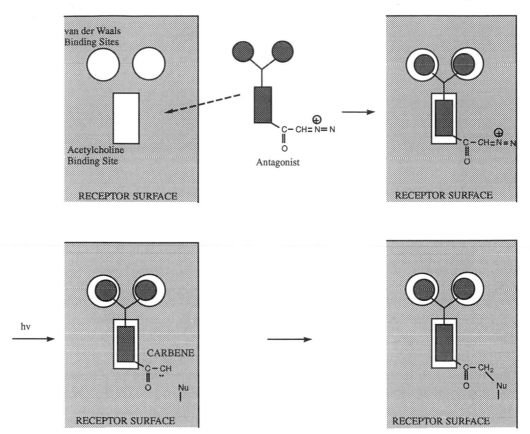

Fig. 11.34 Photoaffinity labelling.

11.12 *Antagonists of the nicotinic cholinergic receptor*

11.12.1 Applications of nicotinic antagonists

Nicotinic receptors are present in nerve synapses at ganglia, as well as at the neuro-muscular synapse. However, drugs are able to show a level of selectivity between these two sites, mainly because of the distinctive routes which have to be taken to reach them.

Antagonists of ganglionic nicotinic receptor sites are not therapeutically useful since they cannot distinguish between the ganglia of the sympathetic nervous system and the ganglia of the parasympathetic nervous system (both use nicotinic receptors). Consequently, they have many side-effects.

However, antagonists of the neuromuscular junction are therapeutically useful and are known as neuromuscular blocking agents.

11.12.2 Nicotinic antagonists

Curare (1516) and tubocurarine

Curare was first identified when Spanish soldiers in South America found themselves the unwilling victims of poisoned arrows. It was discovered that the Indians were putting a poison on to the tips of their arrows. This poison was a crude, dried extract from a plant called *Chondrodendron tomentosum* and caused paralysis as well as stopping the heart. We now know that curare is a mixture of compounds. The active principle, however, is an antagonist of acetylcholine which blocks nerve transmissions from nerve to muscle.

It might seem strange to consider such a compound for medicinal use, but at the right dose levels and under proper control, there are very useful applications for this sort of action. The main application is in the relaxation of abdominal muscles in preparation for surgery. This allows the surgeon to use lower levels of general anaesthetic than would otherwise be required and therefore increase the safety margin for operations.

Curare, as mentioned above, is actually a mixture of compounds, and it was not until 1935 that the active principle (Tubocurarine) was isolated. The determination of the structure took even longer and was not established until 1970 (Fig. 11.35).

The structure of tubocurarine presents a problem to our theory of receptor binding, since, although it has a couple of charged nitrogen centres, there is no ester present to interact with the acetyl binding site. Studies on the compounds discussed so far show that the positively charged nitrogen on its own is not sufficient for good binding, so why should tubocurarine bind to and block the cholinergic receptor?

The answer lies in the fact that the molecule has *two* positively charged nitrogen atoms (one tertiary which is protonated, and one quaternary). Originally, it was believed that the distance between the two centres (1.4 nm) might be equivalent to the distance between two separate cholinergic receptors and that the large tubocurarine molecule could act as a bridge between the two receptor sites, thus spreading a blanket over the two receptors and blocking access to acetylcholine. However

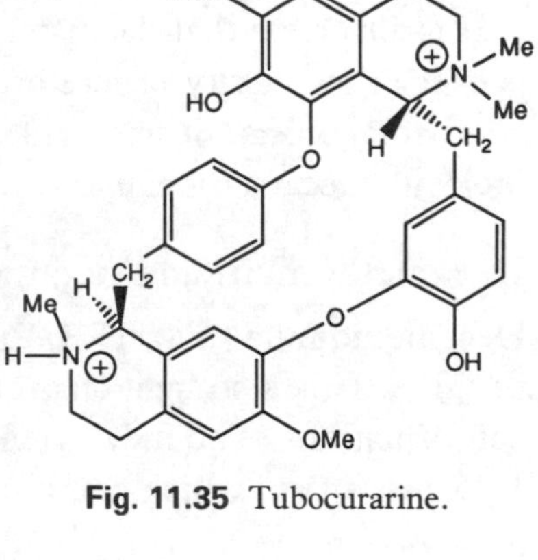

Fig. 11.35 Tubocurarine.

pleasing that theory may be, the dimensions of the nicotinic receptor make this unlikely. The receptor, as we shall see later, is a protein dimer made up of two identical protein complexes separated by 9–10 nm—far too large for the tubocurarine molecule to bridge (Fig. 11.36(a)). Another possibility is that the tubocurarine molecule bridges two acetylcholine binding sites within the one protein complex. Since there are two such sites within the complex, this appears an attractive alternative theory. However, the two sites are further apart than 1.4 nm and so this too seems unlikely. It has now been proposed that one of the positively charged nitrogens on tubocurarine binds to the anionic binding site of the acetylcholine receptor in the protein complex, while the other nitrogen binds to a nearby cysteine residue 0.9–1.2 nm away (Fig. 11.36(b)).

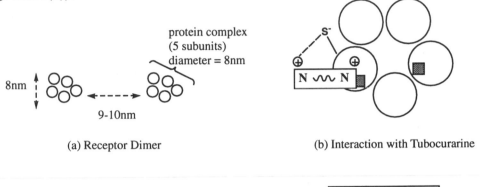

(a) Receptor Dimer   (b) Interaction with Tubocurarine

Fig. 11.36 Tubocurarine binding to and blocking the cholinergic receptor.

Despite the uncertainty surrounding the bonding interactions of tubocurarine, it seems highly probable that two ionic bonding sites are involved. Such an interaction is extremely strong and would more than make up for the lack of the ester binding interaction.

It is also clear that the distance between the two positively charged nitrogen atoms is crucial to activity. Therefore, analogues which retain this distance should also be good antagonists. Strong evidence that this is so comes from the fact that the simple molecule decamethonium is a good antagonist.

Decamethonium and suxamethonium

Decamethonium (Fig. 11.37) is as simple an analogue of tubocurarine as one could imagine. It is a straight-chain molecule and as such is capable of a large number of conformations. The fully extended conformation would position the nitrogen centres 1.35 nm apart, which compares well with the equivalent distance in tubocurarine

Fig. 11.37 Decamethonium. $Me_3\overset{\oplus}{N}(CH_2)_{10}\overset{\oplus}{N}Me_3$

(1.4 nm). The drug binds strongly to cholinergic receptors and has proved a useful clinical agent. However, it suffers from several disadvantages. For example, when it binds initially to the acetylcholine receptor, it acts as an agonist rather than an antagonist. In other words, it switches on the receptor and this leads to a brief contraction of the muscle. Once this effect has passed, the drug remains bound to the receptor—blocking access to acetylcholine—and thus acts as an antagonist. (A theory on how such an effect might take place is described in Chapter 5.) Unfortunately, it binds too strongly and as a result patients take a long time to recover from its effects. It is also not completely selective for the neuromuscular junction and has an effect on acetylcholine receptors at the heart. This leads to an increased heart rate and a fall in blood pressure.

The problem we now face in designing a better drug is the opposite problem from the one we faced when trying to design acetylcholine agonists. Instead of stabilizing a molecule, we now want to introduce some sort of instability—a sort of timer control whereby the molecule can be switched off quickly and become inactive. Success was first achieved by introducing ester groups into the chain while retaining the distance between the two charged nitrogens to give suxamethonium (Fig. 11.38).

Fig. 11.38 Suxamethonium. $Me_3\overset{\oplus}{N}CH_2CH_2-O-\overset{O}{\overset{\|}{C}}-CH_2CH_2-\overset{O}{\overset{\|}{C}}-O-CH_2CH_2\overset{\oplus}{N}Me_3$

The ester groups are susceptible to chemical and enzymatic hydrolysis. Once hydrolysis occurs, the molecule can no longer bridge the two receptor sites and becomes inactive. Suxamethonium has a duration of action of five minutes, but suffers from other side-effects.[3] Furthermore, about one person in every two thousand lacks the enzyme which hydrolyses suxamethonium.

Pancuronium and vecuronium

Pancuronium and vecuronium (Fig. 11.39) were designed to act like tubocurarine, but with a steroid nucleus acting as the 'spacer'. The distance between the quaternary nitrogens is 1.1 nm as compared to 1.4 nm in tubocurarine. Acyl groups were also added to introduce two acetylcholine skeletons into the molecule in order to improve affinity for the receptor sites. These compounds have a rapid onset of action and do not affect blood pressure. However, they are not as rapid in onset as suxamethonium and also last too long (45 minutes).

---

[3] Both decamethonium and suxamethonium have effects on the autonomic ganglia which explains some of their side-effects.

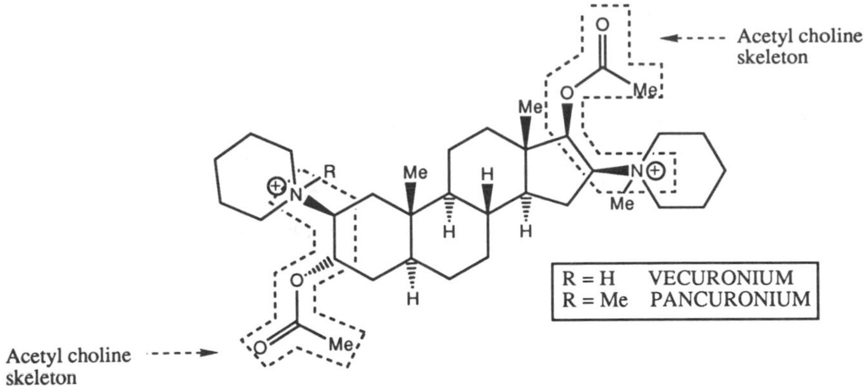

Fig. 11.39 Pancuronium and vecuronium.

Atracurium

The design of atracurium (Fig. 11.40) was based on the structures of tubocurarine and suxamethonium. It is superior to both since it lacks cardiac side-effects and is rapidly broken down in blood. This rapid breakdown allows the drug to be administered as an intravenous drip.

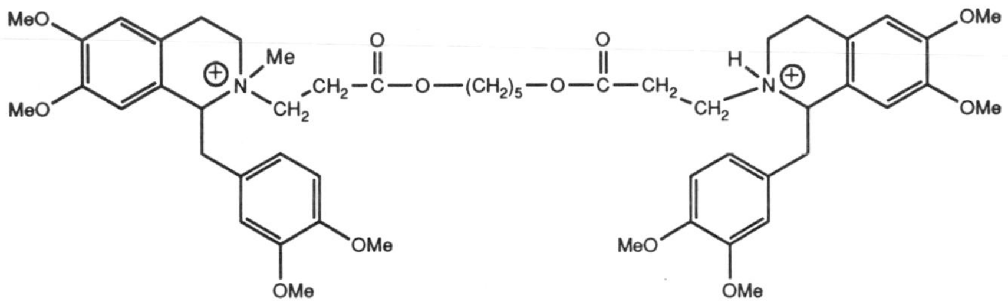

Fig. 11.40 Atracurium.

The rapid breakdown was designed into the molecule by incorporating a self-destruct mechanism. At blood pH (slightly alkaline at 7.4), the molecule can undergo a Hofmann elimination (Fig. 11.41). Once this happens, the compound is inactivated since the positive charge on the nitrogen is lost. It is a particularly clever example of

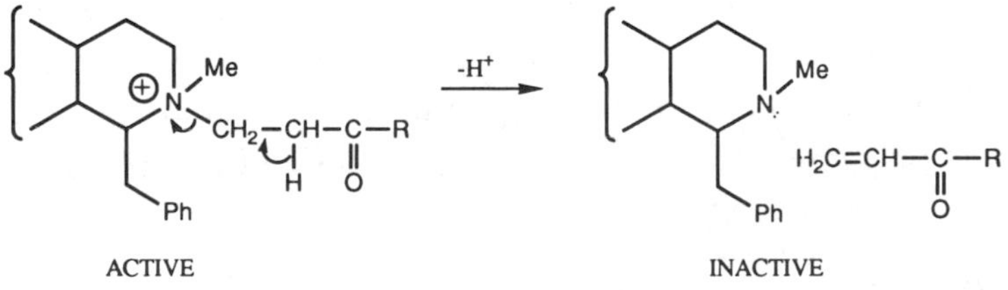

Fig. 11.41 Hofmann elimination of atracurium.

drug design in that the very element responsible for the molecule's biological activity promotes its deactivation.

The important features of atracurium are:

- The spacer.
  This is the 13-atom connecting chain which connects the two quaternary centres and separates the two centres.
- The blocking units.
  The cyclic structures at either end of the molecule block the receptor site from acetylcholine.
- The quaternary centres.
  These are essential for receptor binding. If one is lost through Hofmann elimination, the binding interaction is too weak and the antagonist leaves the binding site.
- The Hofmann elimination.
  The ester groups within the spacer chain are crucial to the rapid deactivation process. Hofmann eliminations normally require strong alkaline conditions and high temperatures—hardly normal physiological conditions. However, if a good electron withdrawing group is present on the carbon, beta to the quaternary nitrogen centre, it allows the reaction to proceed under much milder conditions. The electron withdrawing group increases the acidity of the hydrogens on the beta-carbon such that they are easily lost.

  The Hofmann elimination does not occur at acid pH, and so the drug is stable in solution at a pH of 3–4 and can be stored safely in a refrigerator.

Since the drug only acts very briefly, it has to be added intravenously for as long as it is needed. As soon as surgery is over, the intravenous drip is stopped and antagonism ceases almost instantaneously.

Another major advantage of a drug which is deactivated by a chemical mechanism rather than by an enzymatic mechanism is that deactivation occurs at a constant rate between patients. With previous neuromuscular blockers, deactivation depended on metabolic mechanisms involving enzymic deactivation and/or excretion. The efficiency of these processes varies from patient to patient and is particularly poor for patients with kidney failure or with low levels of plasma esterases.

11.13 *Other cholinergic antagonists*

Local anaesthetics and barbiturates appear to prevent the changes in ion permeability which would normally result from the interaction of acetylcholine with its receptor. They do not, however, bind to the acetylcholine binding site. It is believed that they bind instead to the part of the receptor which is on the inside of the cell membrane, perhaps binding to the ion channel itself and blocking it.

Certain snake toxins have been found to bind irreversibly to the acetylcholine receptor, thus blocking cholinergic transmissions. These include toxins such as alpha-bungarotoxin from the Indian cobra. The toxin is a polypeptide containing 70 amino acids which cross-links the alpha and beta subunits of the cholinergic receptor (see Section 11.14.).

11.14 *The nicotinic receptor—structure*

The nicotinic receptor has been successfully isolated from the electric ray (*Torpedo marmorata*) found in the Atlantic Ocean and Mediterranean sea, allowing the receptor to be carefully studied. As a result, a great deal is known about its structure and operation.

It is a protein complex made up of five subunits, two of which are the same. The five subunits (two alpha, one beta, gamma, and delta) form a cylindrical or barrel shape which traverses the cell membrane as shown in Fig. 11.42.

The centre of the cylinder can therefore act as an ion channel for sodium. A gating or lock system is controlled by the interaction of the receptor with acetylcholine. When acetylcholine is unbound the gate is shut. When acetylcholine binds the gate is opened.

The amino acid sequence for each subunit has been established and it is known that there is extensive secondary structure.

The binding site for acetylcholine is situated on the alpha subunit and therefore there are two binding sites per receptor protein.

It is usually found that the nicotinic receptors occur in pairs linked together by a disulfide bridge between the delta subunits (Fig. 11.43).

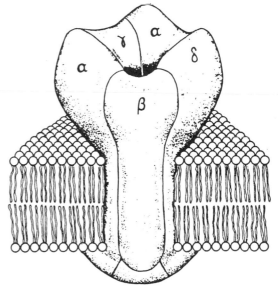

Fig. 11.42 Schematic diagram of the nicotinic receptor. Taken from C. M. Smith and A. M. Reynard, *Textbook of pharmacology*, W. B. Saunders and Co. (1992).

11.15 *The muscarinic receptor—structure*

The muscarinic receptor has not been studied in as much detail as the nicotinic receptor, since it is more difficult to isolate. However, it is now known that there is a subtle difference between muscarinic receptors in different parts of the body. Muscar-

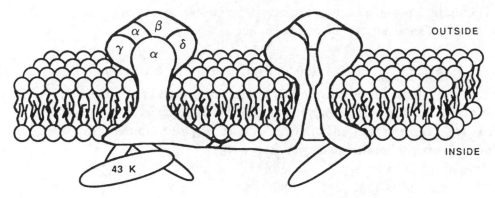

Fig. 11.43 Nicotinic receptor pair. Taken from T. Nogrady, *Medicinal chemistry, a biochemical approach*, 2nd edn, Oxford University Press (1988).

inic receptors have therefore been subdivided into three subgroups M1, M2, and M3. More subgroups are suspected.

The M2 receptor is located in heart muscle and parts of the brain. Unlike the nicotinic receptor, it appears to act by controlling the enzyme synthesis of secondary messengers (see Appendix 3) rather than by directly controlling an ion channel.

11.16 *Anticholinesterases and acetylcholinesterase*

11.16.1 Effect of anticholinesterases

Anticholinesterases are antagonists of the enzyme acetylcholinesterase—the enzyme which hydrolyses acetylcholine. If acetylcholine is not destroyed, it can return to activate the cholinergic receptor again and so the effect of an anticholinesterase is to increase levels of acetylcholine and to increase cholinergic effects (Fig. 11.44).

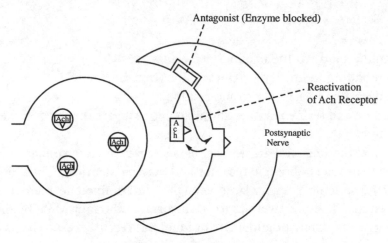

Fig. 11.44 Effect of anticholinesterases.

Therefore, an antagonist at the acetylcholinesterase enzyme will have the same biological effect as an agonist at the cholinergic receptor.

11.16.2 Structure of the acetylcholinesterase enzyme

The acetylcholinesterase enzyme has a fascinating tree-like structure (Fig. 11.45). The trunk of the tree is a collagen molecule which is anchored to the cell membrane. There are three branches (disulfide bridges) leading off from the trunk, each of which hold the acetylcholinesterase enzyme above the surface of the membrane. The enzyme itself is made up of four protein subunits, each of which has an active site. Therefore, each enzyme tree has twelve active sites. The trees are rooted immediately next to the acetylcholine receptors so that they will efficiently capture acetylcholine molecules as they depart the recaptor. In fact, the acetylcholinesterase enzyme is one of the most efficient enzymes known.

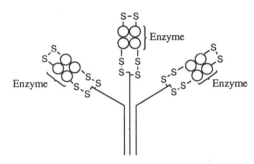

Fig. 11.45 The acetylcholinesterase enzyme.

11.16.3 The active site of acetylcholinesterase

The design of anticholinesterases depends on the shape of the enzyme active site, the binding interactions involved with acetylcholine, and the mechanism of hydrolysis.

Binding interactions at the active site

There are two important areas to be considered—the anionic binding site and the ester binding site (Fig. 11.46).

Note that:
- Acetylcholine binds to the cholinesterase enzyme by
  - (a) ionic bonding to an Asp or Glu residue (but see below),
  - (b) hydrogen bonding to a tyrosine residue.
- The histidine and serine residues at the catalytic site are involved in the mechanism of hydrolysis.
- The anionic binding site in acetylcholinesterase is very similar to the anionic binding site in the cholinergic receptor and may be identical. There are thought to be two hydrophobic pockets large enough to accommodate methyl residues but nothing larger. The positively charged nitrogen is thought to be bound to a negatively charged aspartate or glutamate residue, but recently research has placed some doubt on this assumption (see Section 11.7.).

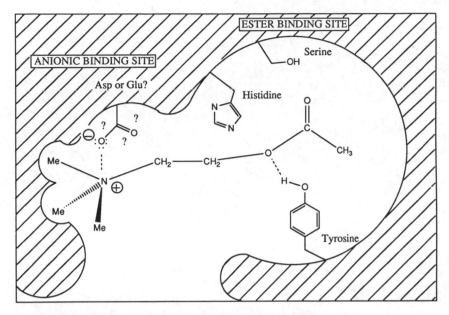

Fig. 11.46 Binding interactions at the active site.

Mechanism of hydrolysis

The histidine residue acts as an acid/base catalyst throughout the mechanism, while serine plays the part of a nucleophile. This is not a particularly good role for serine since an aliphatic alcohol is a poor nucleophile. In fact, serine by itself is unable to hydrolyse an ester. However, the fact that histidine is close by to provide acid/base catalysis overcomes that disadvantage. There are several stages to the mechanism.

Stage 1. Acetylcholine approaches and binds to the acetylcholinesterase enzyme. The histidine residue acts as a base to remove a proton from the serine hydroxyl group, thus making it strongly nucleophilic. Nucleophilic addition to the ester takes place and opens up the carbonyl group.

Stage 2. The carbonyl group reforms and expels the alcohol portion of the ester (i.e. choline). This process is aided by histidine which now acts as an acid catalyst by donating a proton to the departing alcohol.

Stage 3. The acyl portion of acetyl choline is now covalently bound to the receptor site. Choline leaves the active site and is replaced by water.

Stage 4. Water is normally a poor nucleophile, but once again histidine acts as a basic catalyst and nucleophilic addition takes place, once more opening up the carbonyl group.

Stage 5. The carbonyl group is reformed and the serine residue is released with the aid of acid catalysis from histidine.

Stage 6. Ethanoic acid leaves the active site and the cycle can be repeated.

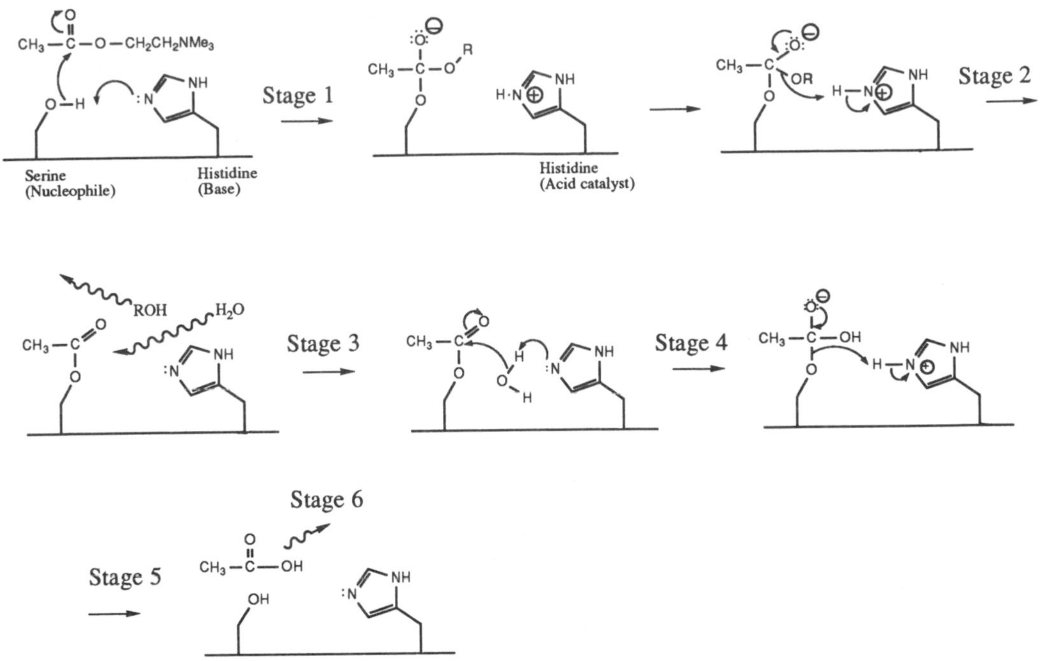

Fig. 11.47 Mechanism of hydrolysis.

The enzymatic process is remarkably efficient due to the close proximity of the serine nucleophile and the histidine acid/base catalyst. As a result, enzymatic hydrolysis by cholinesterase is one hundred million times faster than chemical hydrolysis. The process is so efficient that acetylcholine is hydrolysed within a hundred microseconds of reaching the enzyme.

The story of how this mechanism was worked out and how the structure of the active site was derived makes interesting reading but is not included here.

11.17 *Anticholinesterase drugs*

Obviously, by the very nature of their being, anticholinesterase drugs must be antagonists; that is, they stop the enzyme from hydrolysing acetylcholine. This antagonism can be either reversible or irreversible depending on how the drug reacts with the active site.

There are two main groups of acetylcholinesterases which we shall consider—carbamates and organophosphorus agents.

11.17.1 The carbamates

Physostigmine

Once again, it was a natural product which provided the lead to this group of compounds. The natural product was physostigmine (also called eserine) which was

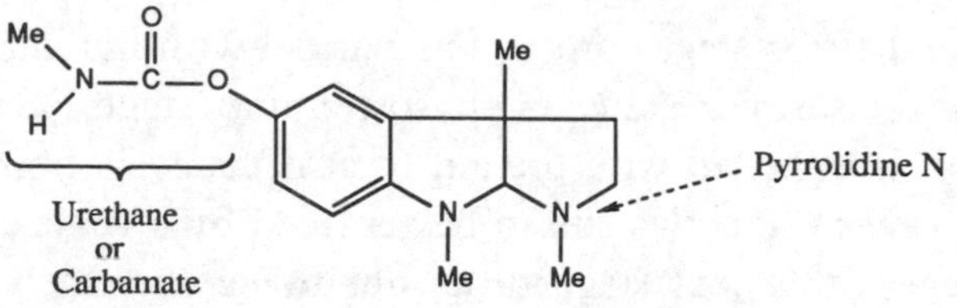

Fig. 11.48 Physostigmine.

discovered in 1864 as a product of the poisonous calabar beans from West Africa. The structure was established in 1925 (Fig. 11.48).

Structure–activity relationships:
• The carbamate group is essential to activity.
• The benzene ring is important.
• The pyrrolidine nitrogen (which is ionized at blood pH) is important.

Working backwards, the positively charged pyrrolidine nitrogen is clearly important since it must bind to the anionic receptor site of the enzyme.

The benzene ring may be involved in some extra hydrophobic bonding with the receptor site or, alternatively, it may be important in the mechanism of inhibition since it provides a good leaving group.

The carbamate group is the crucial group responsible for physostigmine's antagonistic properties, and to understand why, we have to look again at the mechanism of hydrolysis at the active site (Fig. 11.49). This time we shall see what happens when physostigmine and not acetylcholine is the substrate for the reaction.

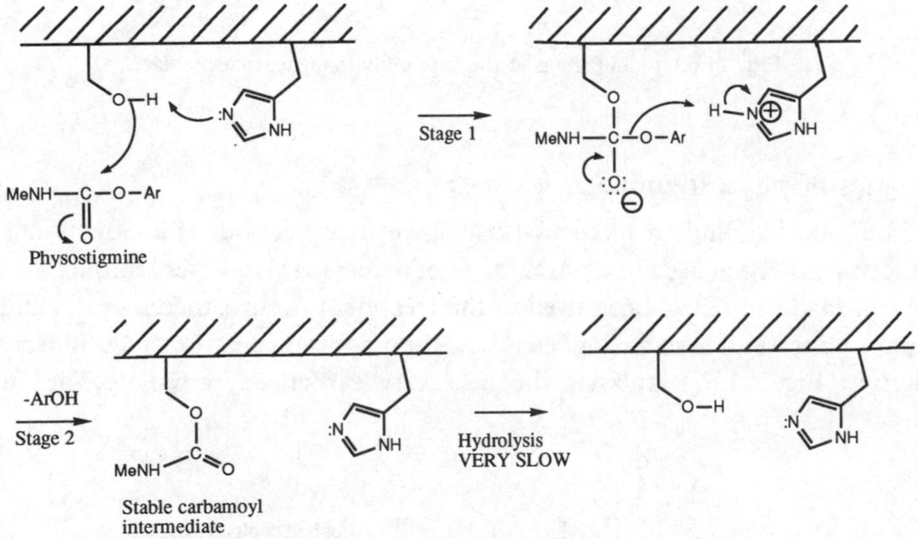

Fig. 11.49 Mechanism of inhibition.

The first three stages proceed as normal with histidine catalysing the nucleophilic attack of the serine residue on physostigmine (stage 1). The alcohol portion (this time a phenol) is expelled with the aid of acid catalysis from histidine (stage 2), and the phenol leaves the active site to be replaced by a water molecule.

However, the next stage turns out to be extremely slow. Despite the fact that histidine can still act as a basic catalyst, water finds it difficult to attack the carbamoyl intermediate. This step becomes the rate determining step for the whole process and the overall rate of hydrolysis of physostigmine compared to acetylcholine is forty million times slower. As a result, the cholinesterase active site becomes 'bunged up' and is unable to react with acetylcholine.

Why is this final stage so slow?

The carbamoyl/enzyme intermediate is stabilized because the nitrogen can feed a lone pair of electrons into the carbonyl group. This drastically reduces the electrophilic character and reactivity of the carbonyl group (Fig. 11.50).

This is the same electronic influence which stabilizes carbachol and makes it resistant towards hydrolysis (Section 11.9.2.).

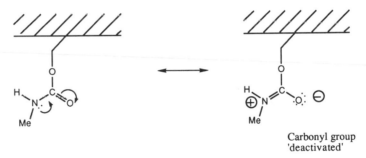

Fig. 11.50 Stabilization of the carbamoyl/enzyme intermediate.

Analogues of physostigmine

Physostigmine has limited medicinal use since it has serious side-effects, and as a result it has only been used in the treatment of glaucoma. However, simpler analogues have been made and have been used in the treatment of myasthenia gravis and as an antidote to curare. These analogues retain the important features mentioned above.

Miotine (Fig. 11.51) still has the necessary carbamate, aromatic, and tertiary

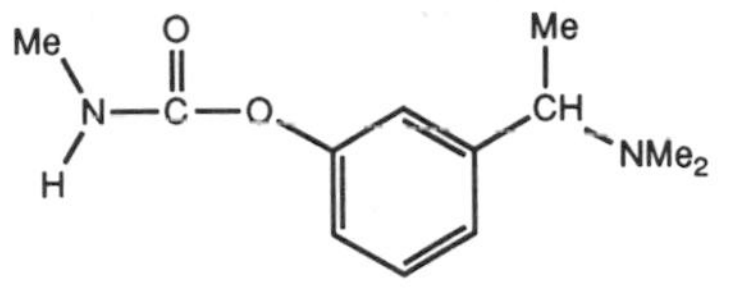

Fig. 11.51 Miotine.

aliphatic nitrogen groups. It is active as an antagonist but suffers from the following disadvantages:

- It is susceptible to chemical hydrolysis.
- It can cross the blood–brain barrier as the free base. This results in side-effects due to its action in the CNS.

Neostigmine (Fig. 11.52) was designed to deal with both the problems described above. First of all, a quaternary nitrogen atom is present so that there is no chance of the free base being formed. Since the molecule is permanently charged, it cannot cross the blood–brain barrier and cause CNS side-effects.[4]

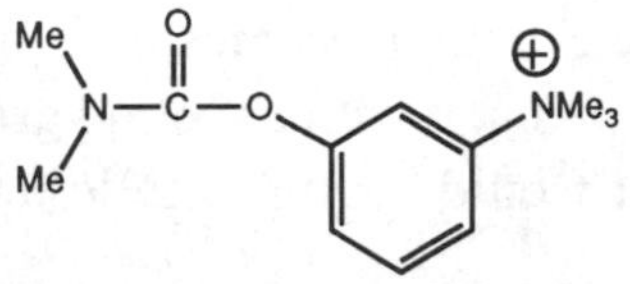

Fig. 11.52 Neostigmine.

Increased stability to hydrolysis is achieved by using a dimethylcarbamate group rather than a methylcarbamate group. There are two possible explanations for this, based on two possible hydrolysis mechanisms.

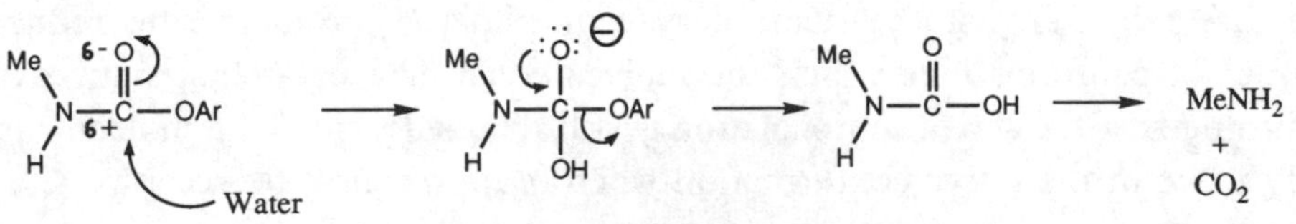

Fig. 11.53 Mechanism 1.

Mechanism 1 (Fig. 11.53) involves nucleophilic substitution by a water molecule. The rate of the reaction depends on the electrophilic character of the carbonyl group and if this is reduced, the rate of hydrolysis is reduced.

We have already seen how the lone pair of the neighbouring nitrogen can reduce the electrophilic character of the carbonyl group. The presence of a second methyl group on the nitrogen has an inductive 'pushing' effect which increases electron density on the nitrogen and further encourages the nitrogen lone pair to interact with the carbonyl group.

---

[4] The blood–brain barrier is a series of lipophilic membranes which coat the blood vessels feeding the brain and which prevent polar molecules from entering the CNS. The fact that it exists can be useful in a case like this, since polar molecules can be designed which are unable to cross it. However, its presence can be disadvantageous when trying to design drugs to act in the CNS itself.

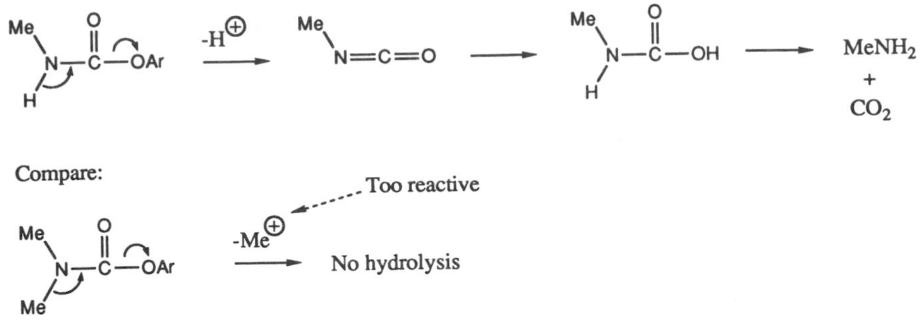

Fig. 11.54 Mechanism 2.

Mechanism 2 (Fig. 11.54) is a fragmentation whereby the phenolic group is lost before the nucleophile is added. This mechanism requires the loss of a proton from the nitrogen. Replacing this hydrogen with a methyl group would severely inhibit the reaction since the mechanism would require the loss of a methyl cation—a highly disfavoured process.

Whichever mechanism is involved, the presence of the second methyl group acts to discourage the process.

Two further points to note about neostigmine are the following:

- The quaternary nitrogen is 4.7 Å away from the ester group which matches well the equivalent distance in acetylcholine.
- The direct bonding of the quaternary centre to the aromatic ring reduces the number of conformations which the molecule can take up. This is an advantage (assuming that the active conformation is still retained), since the molecule is more likely to be in the active conformation when it approaches the receptor site.

Neostigmine has proved a useful agent and is still in use today.

11.17.2 Organophosphorus Compounds

Organophosphorus agents were designed as nerve gases during the Second World War, but were fortunately never used. In peacetime, organophosphate agents have been used as insecticides. We shall deal with the nerve gases first.

Nerve gases

The nerve gases dyflos and sarin (Fig. 11.55) were discovered and perfected long before their mode of action was known. Dyflos, which has an ID_{50} of 0.01 mg kg^{-1},

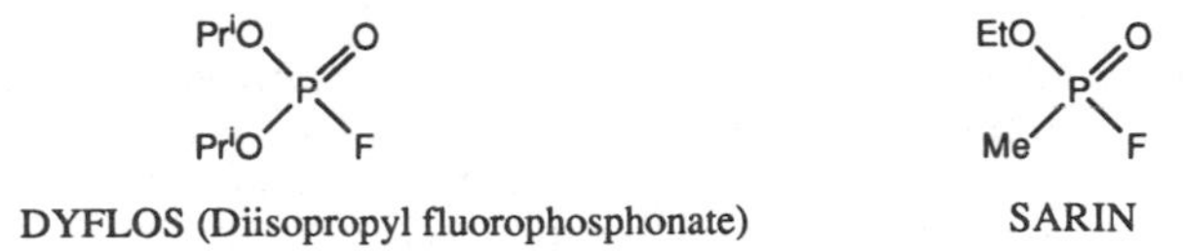

Fig. 11.55 Nerve gases.   DYFLOS (Diisopropyl fluorophosphonate)   SARIN

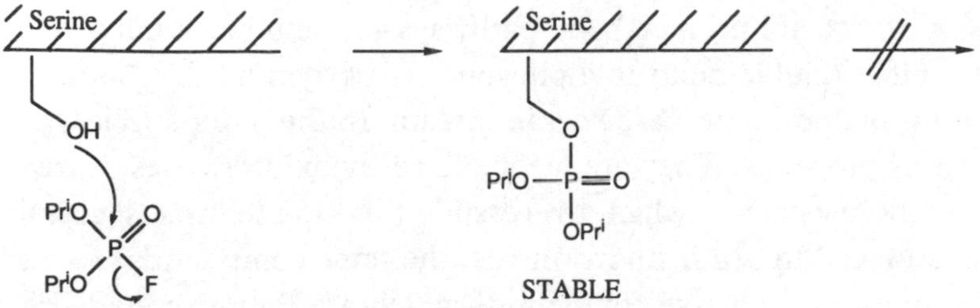

Fig. 11.56 Action of dyflos.

was developed as a nerve gas in the Second World War. It inhibits acetylcholinesterase by irreversibly phosphorylating the serine residue at the active site (Fig. 11.56).

The mechanism is the same as before, but the phosphorylated adduct which is formed after the first three stages is extremely resistant to hydrolysis. Consequently, the enzyme is permanently inactivated. Acetylcholine cannot be hydrolysed and as a result the cholinergic system is continually stimulated. This results in permanent contraction of skeletal muscle, leading to death.

As mentioned earlier, these agents were discovered before their mechanism of action was known. Once it was known that they acted on the cholinesterase enzyme, compounds such as ecothiopate (Fig. 11.57) were designed to fit the active site more accurately.

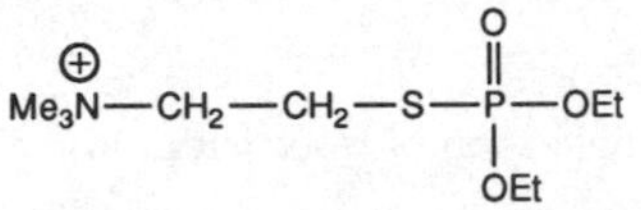

Fig. 11.57 Ecothiopate.

Ecothiopate is used medicinally in the treatment of glaucoma.

Insecticides

The insecticides parathion and malathion (Fig. 11.58) are good examples of how a detailed knowledge of biosynthetic pathways can be put to good use. Parathion and malathion are, in fact, non-toxic. The phosphorus/sulfur double bond prevents these molecules from antagonizing the active site on the cholinesterase enzyme. The equivalent compounds containing a phosphorus/oxygen double bond are, on the other hand, lethal compounds.

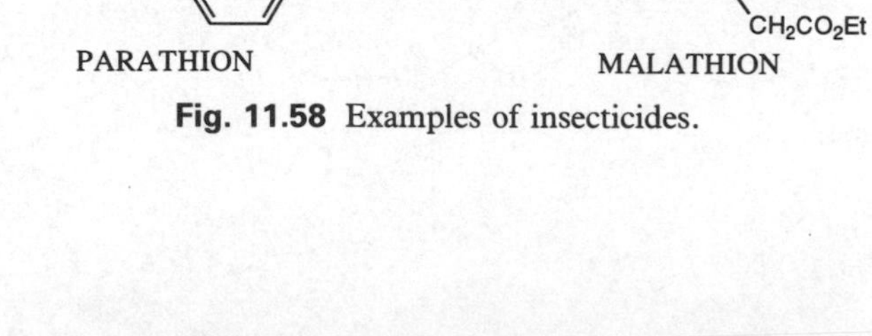

PARATHION   MALATHION

Fig. 11.58 Examples of insecticides.

Fortunately, there are no metabolic pathways in mammals which can convert the phosphorus/sulfur double bond to a phosphorus/oxygen double bond.

Such a pathway does, however, exist in insects. In the latter species, parathion and malathion act as prodrugs. They are metabolized by oxidative desulfurization to give the active anticholinesterases which irreversibly bind to the insects' acetylcholinesterase enzymes and lead to death. In mammals, the same compounds are metabolized in a different way to give inactive compounds which are then excreted (Fig. 11.59).

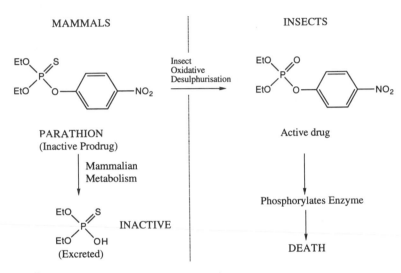

Fig. 11.59 Metabolization of insecticides in mammals and insects.

11.18 *Pralidoxime—an organophosphate antidote*

Pralidoxime (Fig. 11.60) represents one of the best examples of rational drug design. It is an antidote to organophosphate poisoning and was designed as such.

The problem faced in designing an antidote to organophosphate poisoning is to find a drug which will displace the organophosphate molecule from serine. This requires hydrolysis of the phosphate–serine bond, but this is a strong bond and not easily broken. Therefore, a stronger nucleophile than water is required.

The literature revealed that phosphates can be hydrolysed with hydroxylamine (Fig. 11.61). This proved too toxic a compound to be used on humans, so the next

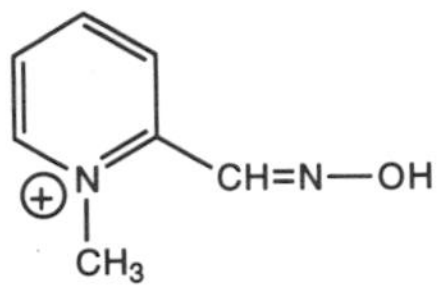

Fig. 11.60 Pralidoxime.

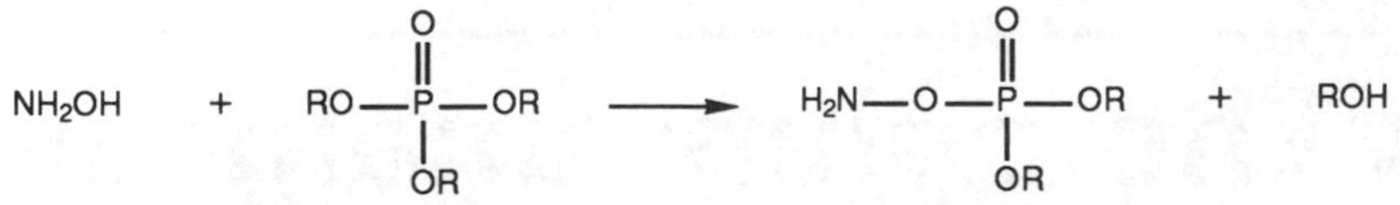

$$NH_2OH \quad + \quad RO-\overset{\overset{\displaystyle O}{\|}}{\underset{\underset{\displaystyle OR}{|}}{P}}-OR \quad \longrightarrow \quad H_2N-O-\overset{\overset{\displaystyle O}{\|}}{\underset{\underset{\displaystyle OR}{|}}{P}}-OR \quad + \quad ROH$$

Fig. 11.61 Hydrolysis of phosphoric acid esters.

stage was to design an equally reactive nucleophilic group which would specifically target the acetylcholinesterase enzyme. If such a compound could be designed, then there was less chance of the antidote taking part in toxic side-reactions.

The designers' job was made easier by the knowledge that the organophosphate group does not fill the active site and that the anionic binding site is vacant. The obvious thing to do was to find a suitable group to bind to this anionic centre and attach a hydroxylamine moiety to it. Once positioned in the active site, the hydroxylamine group could react with the phosphate ester (Fig. 11.62).

Pralidoxime was the result. The positive charge is provided by a methylated pyridine ring and the nucleophilic side-group is attached to the *ortho* position, since it was calculated that this would place the nucleophilic hydroxyl group in exactly the correct position to react with the phosphate ester. The results were spectacular, with pralidoxime showing a potency as an antidote one million times greater than hydroxylamine.

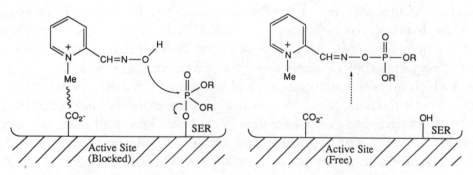

Fig. 11.62 Hydroxylamine group reaction with the phosphate ester.

12 ▪ The opium analgesics

12.1 *Introduction*

We are now going to look in detail at one of the oldest fields in medicinal chemistry, yet one where true success has proved illusive—the search for a safe, orally active, and non-addictive analgesic based on the opiate structure.

It is important to appreciate that the opiates are not the only compounds which are of use in the relief of pain and that there are several other classes of compounds, including aspirin, which combat pain. These compounds, however, operate by different mechanisms from those employed by the opiates, and therefore relieve a different, 'sharper' kind of pain. The opiates have proved ideal for the treatment of 'deep' chronic pain and work in the central nervous system (CNS).

The term 'opium alkaloids' has been used rather loosely to cover all narcotic analgesics, whether they be synthetic compounds, partially synthetic, or extracted from plant material. To be precise, we should really only use the term for those natural compounds which have been extracted from opium—the sticky exudate obtained from the poppy (*Papaver somniferum*). The term alkaloid refers to a natural product which contains a nitrogen atom and is therefore basic in character. There are, in fact, several thousand alkaloids which have been extracted and identified from various plant sources and examples of some of the better known alkaloids are shown in Fig. 12.1. These compounds provide a vast 'library' of biologically active compounds which can be used as lead compounds into many possible fields of medicinal chemistry. However, we are only interested at present in the alkaloids derived from opium.

The opiates are perhaps the oldest drugs known to man. The use of opium was recorded in China over two thousand years ago and was known in Mesopotamia before that. Over the centuries the crude extract derived from poppies has been widely used as a sedative. A tincture of opium called laudanum was introduced to England and was considered indispensable to medicine. It is ironic that a compound renowned for its sedative effects should have led to at least one war. In the nineteenth century, the Chinese authorities became so alarmed about the addictive properties of opium that they tried to ban all production of it. This was contrary to the interests of the British traders dealing in opium and as a result the British sent in the gunboats to reverse the

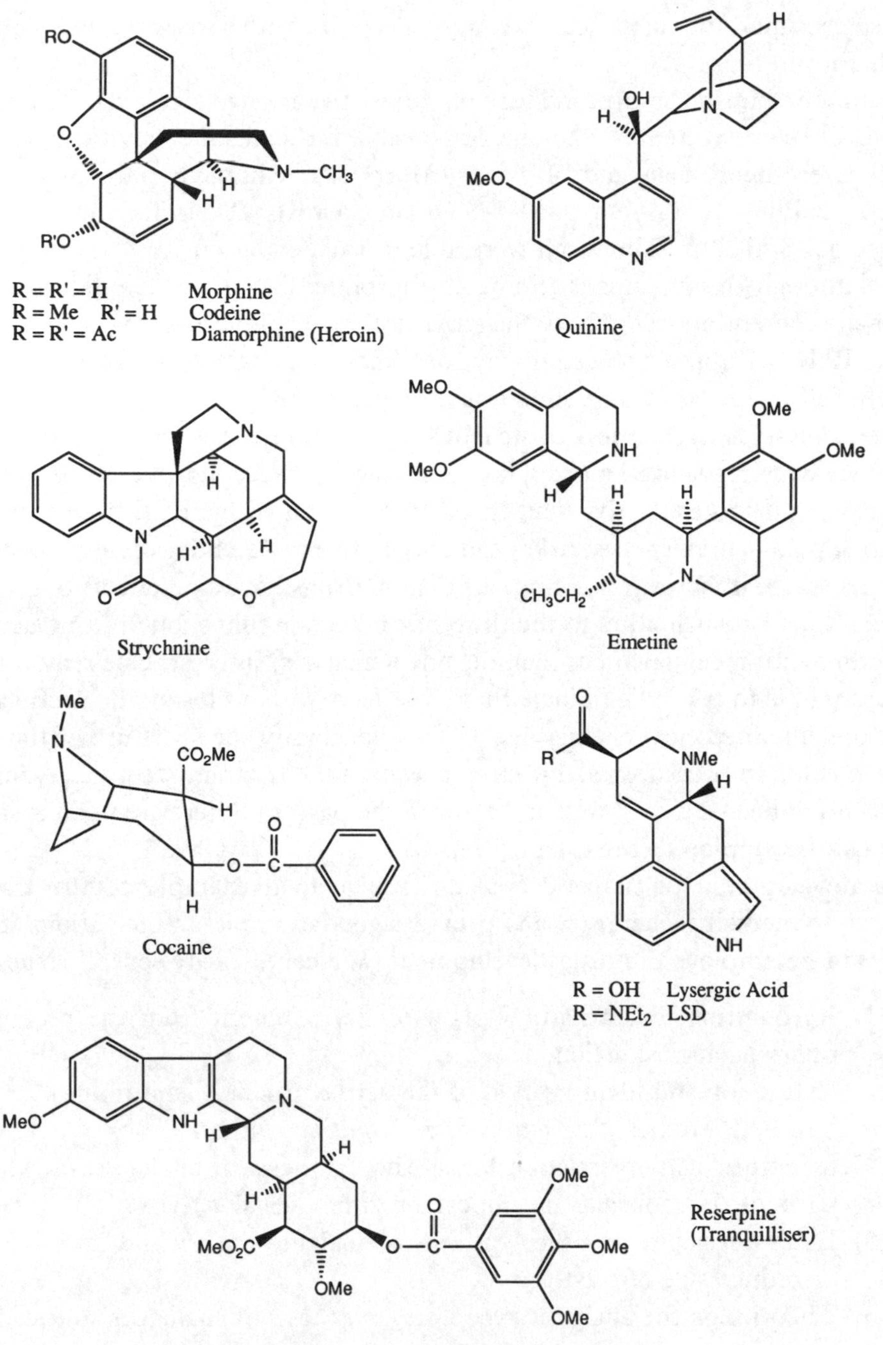

R = R' = H   Morphine
R = Me   R' = H   Codeine
R = R' = Ac   Diamorphine (Heroin)

Quinine

Strychnine

Emetine

Cocaine

R = OH   Lysergic Acid
R = NEt₂ LSD

Reserpine
(Tranquilliser)

Fig. 12.1 Examples of well-known alkaloids.

Chinese decision—one of the least savoury aspects of an otherwise relatively benevolent British Empire.

Opium contains a complex mixture of almost twenty-five alkaloids. The principle alkaloid in the mixture, and the one responsible for analgesic activity, is morphine, named after the Roman god of sleep—Morpheus. Although pure morphine was isolated in 1803, it was not until 1833 that chemists at Macfarlane & Co. (now Macfarlane–Smith) in Edinburgh were able to isolate and purify it on a commercial scale. Although the functional groups on morphine had been identified by 1881, it took many more years to establish the structure of morphine and it was not until 1925 that Sir Robert Robinson solved the puzzle. Another twenty-seven years were to pass before a full synthesis of morphine was achieved in 1952.

Nevertheless, long before the structure of morphine was realized, its analgesic properties were recognized and applied to medicine. Since morphine was in the pure form, it was far more effective than crude opium as an analgesic. But there was also a price to be paid—the increased risks of addiction, tolerance, and respiratory depression.

At this stage, it is worth pointing out that all drugs have side-effects of one sort or another. This is usually due to the drug not being specific enough in its action and interacting with receptors other than the one which is of interest. One reason for drug development is to try and eliminate the side-effects without losing the useful activity. Therefore, the medicinal chemist has to try and modify the structure of the original drug molecule in order to make it more specific for the target receptor. Admittedly, this has often been a case of trial and error in the past, but there are various strategies which can be employed (see Chapter 7).

The development of narcotic analgesics is a good example of the traditional approach to medicinal chemistry and provides good examples of the various strategies which can be employed in drug development. We can identify several stages:

Stage 1. Recognition that a natural plant or herb (opium from the poppy) has a pharmacological action.
Stage 2. Extraction and identification of the active principle (morphine).
Stage 3. Synthetic studies (full synthesis and partial synthesis).
Stage 4. Structure–activity relationships—the synthesis of analogues to see which parts of the molecule are important to biological activity.
Stage 5. Drug development—the synthesis of analogues to try and improve activity or reduce side-effects.
Stage 6. Theories on the analgesic receptors. Synthesis of analogues to test theories.

Stages 5 and 6 are the most challenging and rewarding parts of the procedure as far as the medicinal chemist is concerned, since the possibility exists of improving on what Nature has provided. In this way, the chemist hopes to gain a better understanding of the biological process involved, which in turn suggests further possibilities for new drugs.

12.2 *Morphine*

12.2.1 Structure and properties

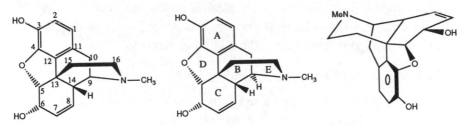

Fig. 12.2 Structure of morphine.

The active principle of opium is morphine (Fig. 12.2) and this compound is still one of the most effective painkillers available to medicine. It is especially good for treating dull, constant pain rather than sharp, periodic pain. It acts in the brain and appears to work by elevating the pain threshold, thus decreasing the brain's awareness of pain. Unfortunately, it has a large number of side-effects which include the following:

- depression of the respiratory centre
- constipation
- excitation
- euphoria
- nausea
- pupil constriction
- tolerance
- dependence

Some side-effects are not particularly serious. Some, in fact, can be advantageous. Euphoria, for example, is a useful side-effect when treating pain in terminally ill patients. Other side-effects, such as constipation, are uncomfortable but can give clues as to other possible uses for opiate-like structures. For example, opiate structures are widely used in cough medicines and the treatment of diarrhoea.

The dangerous side-effects of morphine are those of tolerance and dependence, allied with the effects morphine can have on breathing. In fact, the most common cause of death from a morphine overdose is by suffocation. Tolerance and dependence in the one drug are particularly dangerous and lead to severe withdrawal symptoms when the drug is no longer taken.

Withdrawal symptoms associated with morphine include anorexia, weight loss, pupil dilation, chills, excessive sweating, abdominal cramps, muscle spasms, hyper-irritability, lacrimation, tremor, increased heart rate, and increased blood pressure. No wonder addicts find it hard to kick the habit!

The isolation and structural identification of morphine mark the first two stages of our story and have already been described. The molecule contains five rings labelled A–E and has a pronounced T shape. It is basic because of the tertiary amino group, but it also contains a phenolic group, an alcohol group, an aromatic ring, an ether bridge, and a double bond. The next stage in the procedure is to find out which of these functional groups is essential to the analgesic activity.

12.2.2 Structure–activity relationships

The story of how morphine's secrets were uncovered is presented here in a logical step-by-step fashion. However, in reality this was not how the problem was tackled at the time. Different compounds were made in a random fashion depending on the ease of synthesis, and the logical pattern followed on from the results obtained. By presenting the development of morphine in the following manner, we are distorting history, but we do get a better idea of the general strategies and the logical approach to drug development as a whole.

The first and easiest morphine analogues which can be made are those involving peripheral modifications of the molecule (that is, changes which do not affect the basic skeleton of the molecule). In this approach, we are looking at the different functional groups and discovering whether they are needed or not.

We now look at each of these functional groups in turn.

The phenolic OH

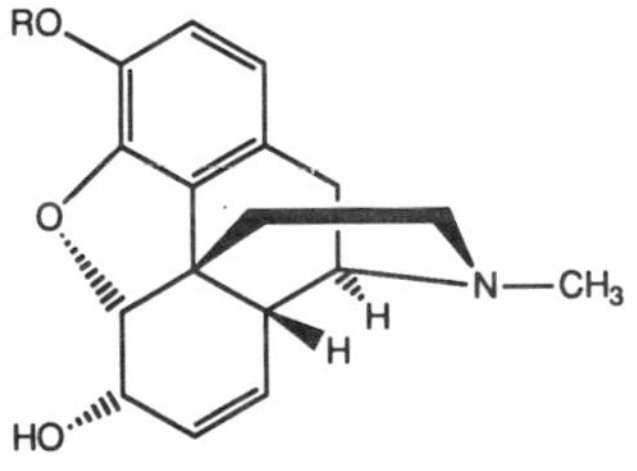

R = Me	CODEINE	Analgesic
R = Et	ETHYLMORPHINE	Activity
R = Acetyl	3-ACETYLMORPHINE	

Fig. 12.3

Codeine (Fig. 12.3) is the methyl ether of morphine and is also present in opium. It is used for treating moderate pain, coughs, and diarrhoea.

By methylating the phenolic OH, the analgesic activity drops drastically and codeine is only 0.1 per cent as active as morphine. This drop in activity is observed in other analogues containing a masked phenolic group. Clearly, a free phenolic group is crucial for analgesic activity.

However, the above result refers to isolated receptors in laboratory experiments. If codeine is administered to patients, its analgesic effect is 20 per cent that of morphine—much better than expected. Why is this so?

The answer lies in the fact that codeine can be metabolized in the liver to give morphine. The methyl ether is removed to give the free phenolic group. Thus, codeine can be viewed as a prodrug for morphine. Further evidence supporting this is provided by the fact that codeine has no analgesic effect at all if it is injected directly into the brain. By doing this, codeine is injected directly into the CNS and does not pass through the liver. As a result, demethylation does not take place.

This example shows the problems that the medicinal chemist can face in testing drugs. The manner in which the drugs are tested can be just as important as making the drug in the first place.

In all the following examples, the test procedures were carried out on animals or humans and so it must be remembered that there are several possible ways in which a change of activity could have resulted.

The 6-alcohol

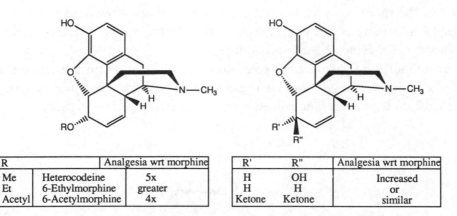

R		Analgesia wrt morphine
Me	Heterocodeine	5x
Et	6-Ethylmorphine	greater
Acetyl	6-Acetylmorphine	4x

R'	R"	Analgesia wrt morphine
H	OH	Increased
H	H	or
Ketone	Ketone	similar

Fig. 12.4 Effect of loss of alcohol group on analgesic activity.

The results in Fig. 12.4 show that masking or the complete loss of the alcohol group does not decrease analgesic activity and, in fact, often has the opposite effect. Again, it has to be emphasized that the testing of analgesics has generally been done *in vivo* and that there are many ways in which improved activity can be achieved.

In these examples, the improvement in activity is due to the pharmacodynamic properties of these drugs rather than their affinity for the analgesic receptor. In other words, it reflects how much of the drug can reach the receptor rather than how well it binds to it.

There are a number of factors which can be responsible for affecting how much of a drug reaches its target. For example, the active compound might be metabolized to an inactive compound before it reaches the receptor. Alternatively, it might be distributed more efficiently to one part of the body than another.

In this case, the morphine analogues shown are able to reach the analgesic receptor far more efficiently than morphine itself. This is because the analgesic receptors are located in the brain and in order to reach the brain, the drugs have to cross a barrier called the blood–brain barrier. The capillaries which supply the brain are lined by a series of fatty membranes which overlap more closely than in any other part of the body. In order to enter the brain, drugs have to negotiate this barrier. Since the barrier is fatty, highly polar compounds are prevented from crossing. Thus, the more polar groups a molecule has, the more difficulty it has in reaching the brain. Morphine has three polar groups (phenol, alcohol, and an amine), whereas the analogues above have either lost the polar alcohol group or have it masked by an alkyl or acyl group. They therefore enter the brain more easily and accumulate at the receptor sites in greater concentrations; hence, the better analgesic activity.

It is interesting to compare the activities of morphine, 6-acetylmorphine, and diamorphine (heroin) (Fig. 12.5). The most active (and the most dangerous) compound of the three is 6-acetylmorphine. It is four times more active than morphine. Heroin is also more active than morphine by a factor of two, but less active than 6-acetylmorphine. How do we explain this?

6-Acetylmorphine, as we have seen already, is less polar than morphine and will enter the brain more quickly and in greater concentrations. The phenolic group is free and therefore it will interact immediately with the analgesic receptors.

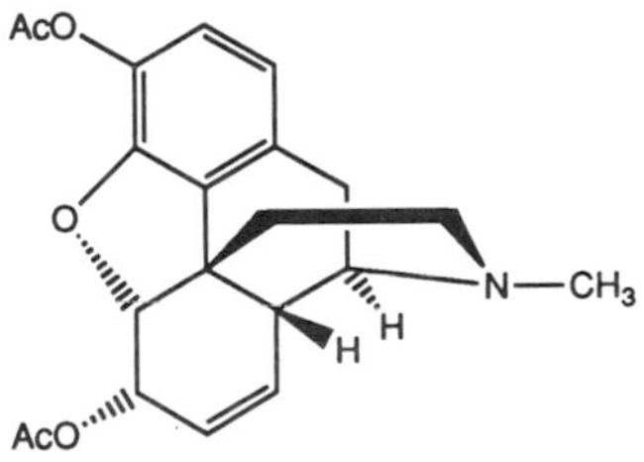

Fig. 12.5 Diamorphine (heroin).

Heroin has two polar groups which are masked and is therefore the most efficient compound of the three to cross the blood–brain barrier. However, before it can act at the receptor, the acetyl group on the phenolic group has to be removed by esterases in the brain. Therefore, it is more powerful than morphine because it enters the brain more easily, but it is less powerful than 6-acetylmorphine because the 3-acetyl group has to be removed before it can act.

Heroin and 6-acetylmorphine are both more potent analgesics than morphine. Unfortunately, they also have greater side-effects and have severe tolerance and dependence characteristics. Heroin is still used to treat terminally ill patients, such as those dying of cancer, but 6-acetylmorphine is so dangerous that its synthesis is banned in many countries.

To conclude, the 6-hydroxyl group is not required for analgesic activity and its removal can be beneficial to analgesic activity.

The double bond at 7–8

Several analogues including dihydromorphine (Fig. 12.6) have shown that the double bond is not necessary for analgesic activity.

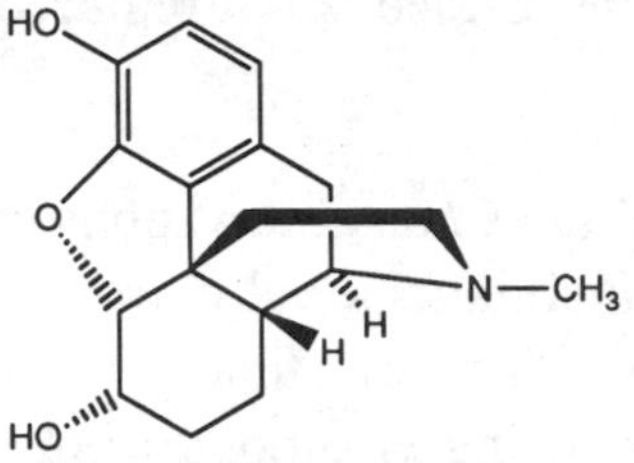

Fig. 12.6 Dihydromorphine.

The *N*-methyl group

The *N*-oxide and the *N*-methyl quaternary salts of morphine are both inactive, which might suggest that the introduction of charge destroys analgesic activity (Fig. 12.7). However, we have to remember that these experiments were done on animals and it is hardly surprising that no analgesia is observed, since a charged molecule has very little chance of crossing the blood–brain barrier. If these same compounds are injected directly into the brain, a totally different result is obtained and both these compounds are found to have similar analgesic activity to morphine. This fact, allied with the fact that neither compound can lose its charge, shows that the nitrogen atom of morphine is ionized when it binds to the receptor.

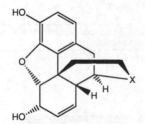

X		Analgesic Activity wrt morphine
NH	Normorphine	25%
$\overset{+}{N}\overset{Me}{\underset{O^-}{}}$	N-Oxide	0%
$\overset{+}{N}\overset{Me}{\underset{Me}{}}$	Quaternary salt	0%

Fig. 12.7 Effect of introduction of charge on analgesic activity.

The replacement of the *N*-methyl group with a proton reduces activity but does not eliminate it. The secondary NH group is more polar than the tertiary *N*-methyl group and therefore finds it more difficult to cross the blood–brain barrier, leading to a drop in activity. The fact that significant activity is retained despite this shows that the methyl substituent is not essential to activity.

However, the nitrogen itself is crucial. If it is removed completely, all analgesic activity is lost. To conclude, the nitrogen atom is essential to analgesic activity and interacts with the analgesic receptor in the ionized form.

The aromatic ring

The aromatic ring is essential. Compounds lacking it show no analgesic activity.

The ether bridge

As we shall see later, the ether bridge is not required for analgesic activity.

Stereochemistry

At this stage, it is worth making some observations on stereochemistry. Morphine is an asymmetric molecule containing several chiral centres, and exists naturally as a single enantiomer. When morphine was first synthesized, it was made as a racemic mixture of the naturally occurring enantiomer plus its mirror image. These were separated and the unnatural mirror image was tested for analgesic activity. It turned out to have no activity whatsoever.

This is not particularly surprising if we consider the interactions which must take place between morphine and its receptor. We have identified that there are at least three important interactions involving the phenol, the aromatic ring and the amine on

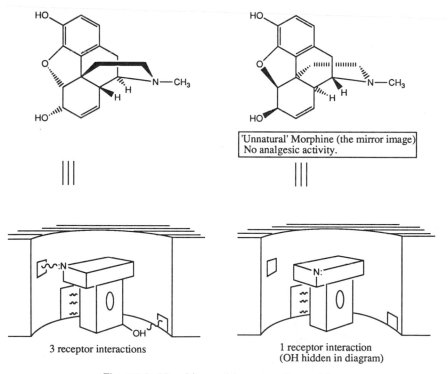

'Unnatural' Morphine (the mirror image) No analgesic activity.

3 receptor interactions

1 receptor interaction (OH hidden in diagram)

Fig. 12.8 Morphine and 'unnatural' morphine.

morphine. Let us consider a diagrammatic representation of morphine as a T-shaped block with the three groups marked as shown in Fig. 12.8. The receptor has complementary binding groups placed in such a way that they can interact with all three groups. If we now consider the mirror image of morphine, then we can see that it can interact with only one binding site at any one time.

Epimerization of a single chiral centre such as the 14-position (Fig. 12.9) is not beneficial either, since changing the stereochemistry at even one chiral centre can result in a drastic change of shape, making it impossible for the molecule to bind to the analgesic receptors.

To sum up, the important functional groups for analgesic activity in morphine are shown in Fig. 12.10.

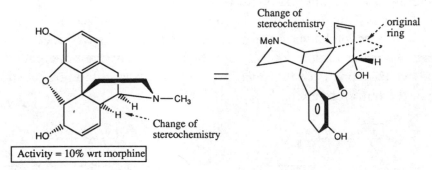

Fig. 12.9 Epimerization of a single chiral centre.

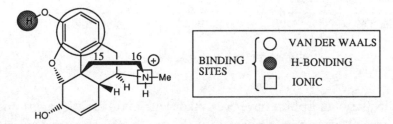

Fig. 12.10 Important functional groups for analgesic activity in morphine.

12.3 *Development of morphine analogues*

We now move on to consider the development of morphine analogues. As mentioned in Chapter 7, there are several strategies used in drug development.

We shall consider the following strategies in the development of morphine analogues.

- variation of substituents
- drug extension
- simplification
- rigidification

12.3.1 Variation of substituents

A series of alkyl chains on the phenolic group give compounds which are inactive or poorly active. We have already identified that the phenol group must be free for analgesic activity.

The removal of the *N*-methyl group to give normorphine allows a series of alkyl chains to be built on the basic centre. These results are discussed under drug extension since the results obtained are more relevant under that heading.

12.3.2 Drug extension

Drug extension is a strategy by which the molecule is 'extended' by the addition of extra 'binding groups'. The reasoning behind such a tactic is to probe for further binding sites which might be available on the receptor surface and which might improve the interaction between the drug and the receptor (Fig. 12.11).

This is a reasonable assumption since it is highly unlikely that a compound such as morphine (which is produced in a plant) would be the perfect binding substrate for a receptor in the human brain

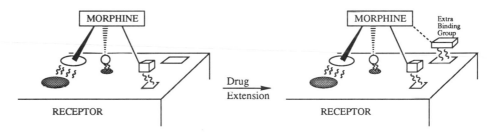

Fig. 12.11 Drug extension of morphine.

Many analogues of morphine have been made with extra functional groups attached. These have rarely shown any improvement. However, there are two exceptions. The introduction of a hydroxyl group at position 14 has been particularly useful (Fig. 12.12). This might be taken to suggest that there is a possible hydrogen bond interaction taking place between the 14-OH group and a suitable amino acid residue on the receptor. However, an alternative explanation is provided in Section 12.5.

The easiest position to add substituents (and the most advantageous) has been the nitrogen atom. The synthesis is easily achieved by removing the *N*-methyl group from morphine to give normorphine, then alkylating the amino group with an alkyl halide. Removal of the *N*-methyl group was originally achieved by a von Braun

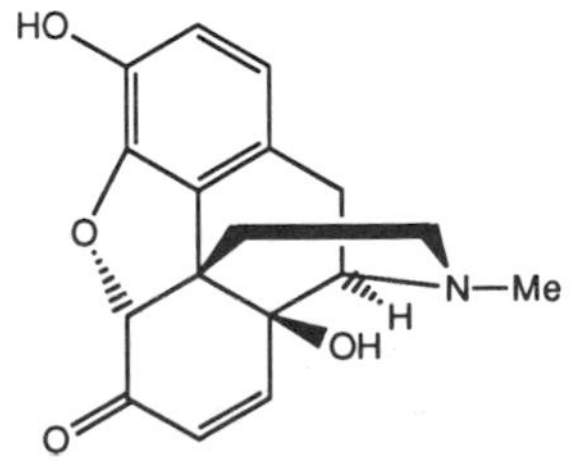

Fig. 12.12 Oxymorphine (2.5 × activity of morphine).

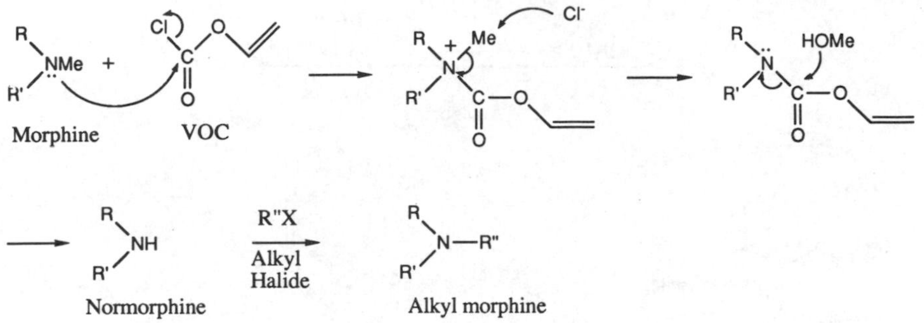

Fig. 12.13 Demethylation and alkylation of the basic centre.

R =	Me	Et	Pr	Bu	Amyl, Hexyl	CH$_2$CH$_2$Ph
	Agonism decreases Antagonism increases			Zero Activity	Agonists	14 x Activity wrt morphine

Fig. 12.14 Change in activity with respect to alkyl group size.

degradation with cyanogen bromide, but is now more conveniently carried out using a chloroformate reagent such as vinyloxycarbonyl chloride (Fig. 12.13). The alkylation step can sometimes be profitably replaced by a two-step process involving an acylation to give an amide, followed by reduction.

The results obtained from the alkylation studies are quite dramatic. As the alkyl group is increased in size from a methyl to a butyl group, the activity drops to zero (Fig. 12.14). However, with a larger group such as an amyl or a hexyl group, activity recovers slightly. None of this is particularly exciting, but when a phenethyl group is attached the activity increases 14 fold—a strong indication that a hydrophobic binding site has been located which interacts favourably with the new aromatic ring (Fig. 12.15).

To conclude, the size and nature of the group on the nitrogen is important to the activity spectrum. Drug extension can lead to better binding by making use of additional binding interactions.

Before leaving this subject, it is worth describing another series of important results arising from varying substituents on the nitrogen atom. Spectacular results were obtained when an allyl group or a cyclopropylmethylene group were attached (Fig. 12.16).

No increase in analgesic activity was observed

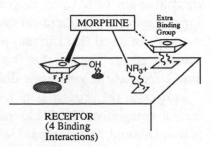

Fig. 12.15 Indication of fourth binding site.

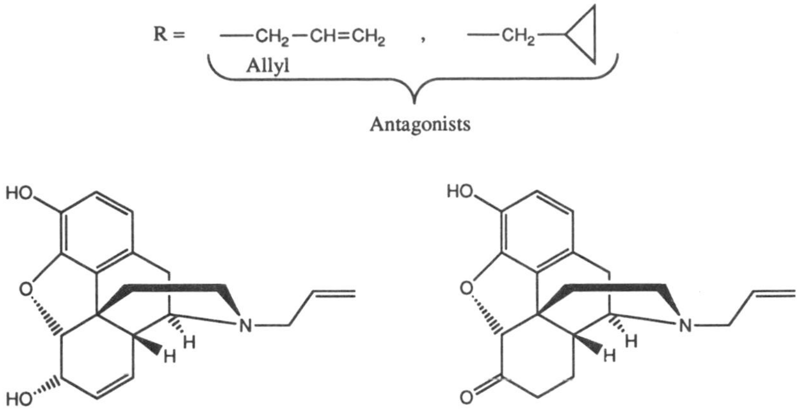

Fig. 12.16

and in fact the results were quite the opposite. Naloxone, for example, has no analgesic activity at all, whilst nalorphine retains only weak analgesic activity. However, the important feature about these molecules is that they act as antagonists to morphine. They do this by binding to the analgesic receptors without 'switching them on'. Once they have bound to the receptors, they block morphine from binding. As a result, morphine can no longer act as an analgesic. One might be hard pushed to see an advantage in this and with good reason. If we are just considering analgesia, there is none. However, the fact that morphine is blocked from all its receptors means that none of its side-effects are produced either, and it is the blocking of these effects which make antagonists extremely useful.

In particular, accident victims have sometimes been given an overdose of morphine. If this is not treated, then the casualty may die of suffocation.

By administering nalorphine, the antagonist displaces morphine from the receptor and binds more strongly, thus preventing morphine from continuing its action.

There is, however, a far more important observation arising from the biological results of these antagonists. For many years, chemists had been trying to find a morphine analogue with analgesic properties, but without the depressant effects on breathing, or the withdrawal symptoms. There had been so little success that many workers believed that the two properties were directly related, perhaps through the same receptor. The fact that the antagonist naloxone blocked morphine analgesia and side-effects at the same time did nothing to change that view.

However, the properties of nalorphine offered a glimmer of hope. Nalorphine is a strong antagonist and blocks morphine from its receptors. Therefore, no analgesic activity should be observed. However, a very weak analgesic activity *is* observed and what is more, this analgesia appears to be free of the undesired side-effects. This was the first sign that a non-addictive, safe analgesic might be possible.

But how can this be? How can a compound be an antagonist of morphine but also act as an agonist and produce analgesia. If it is acting as an agonist, why is the activity so weak and why is it free of the side-effects?

As we shall see later, there is not one single type of analgesic receptor, but several. Multiple receptors are common. We have already seen in Chapter 11 that there are two types of acetylcholine receptor—the nicotinic and muscarinic.

In the same way, there are at least three types of analgesic receptor. The differences between them are slight such that morphine cannot distinguish between them and activates them all, but in theory it should be possible to find compounds which would be selective for one type of analgesic receptor over another. However, this is not the way that nalorphine works.

Nalorphine binds to all three types of analgesic receptor and therefore blocks morphine from all three. Nalorphine itself is unable to switch on two of the receptors and is therefore a true antagonist at these receptors. However, at the third type of receptor, nalorphine is acting as a weak or partial agonist (see Section 5.8.). In other words, it has activated the receptor, but only weakly. We could imagine how this might occur if the third receptor is controlling something like an ion channel (Fig. 12.17).

Morphine is a strong agonist and interacts strongly with this receptor leading to a change in receptor conformation which fully opens the ion channel. Ions flow in or out of the cell, resulting in the activation or deactivation of enzymes. Naloxone is a pure antagonist. It binds strongly, but does not produce a change in the receptor conformation. Therefore, the ion channel remains closed. Nalorphine binds to the third receptor and changes the tertiary structure of the receptor very slightly, leading to a slight opening of the ion channel. It is therefore a weak agonist at this receptor, but it is also an antagonist since it blocks morphine from fully 'switching on' the receptor.

The results observed with nalorphine show that activation of this third type of analgesic receptor leads to analgesia without the undesirable side-effects associated with the other two analgesic receptors.

Unfortunately, nalorphine has hallucinogenic side-effects resulting from the activation of a non-analgesic receptor, and is therefore unsuitable as an analgesic, but for the first time a certain amount of analgesia had been obtained without the side-effects of respiratory depression and tolerance.

12.3.3 Simplification or drug dissection

We turn now to more drastic alterations of the morphine structure and ask whether the complete carbon skeleton is really necessary. After all, if we could simplify the molecule, it would be easier to make in the laboratory. This in turn would allow the chemist to make analogues much more easily, and any useful compounds could be made more efficiently and cheaply.

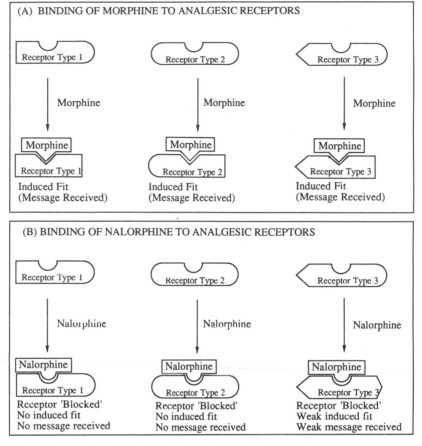

Fig. 12.17 Action of morphine and nalorphine at analgesic receptors.

There are five rings present in the structure of morphine (Fig. 12.18) and analogues were made to see which rings could be removed.

Removing ring E

Removing ring E leads to a complete loss of activity. This result emphasizes the importance of the basic nitrogen to analgesic activity.

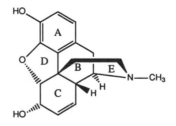

Fig. 12.18 Structure of morphine.

Removing ring D

Removing the oxygen bridge gives a series of compounds called the morphinans which have useful analgesic activity. This demonstrate that the oxygen bridge is not essential. Examples are shown in Fig. 12.19.

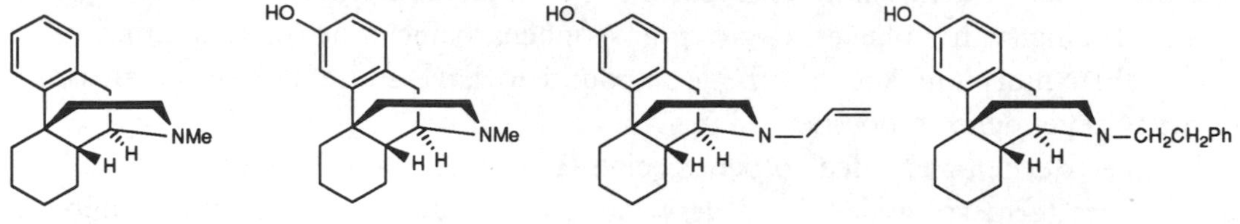

N-METHYL MORPHINAN
(20% activity of morphine)

LEVORPHANOL
(5x more potent than morphine)

LEVALLORPHAN
(Antagonist 5x more
potent than nalorphine)

(15x more potent than morphine)

Fig. 12.19 Examples of morphinans.

N-Methyl morphinan was the first such compound tested and is only 20 per cent as active as morphine, but since the phenolic group is missing, this is not surprising. The more relevant levorphanol structure is five times more active than morphine and, although side-effects are also increased, levorphanol has a massive advantage over morphine in that it can be taken orally and lasts much longer in the body. This is because levorphanol is not metabolized in the liver to the same extent as morphine.

As might be expected, the mirror image of levorphanol (dextrorphan) has insignificant analgesic activity.

The same strategy of drug extension already described for the morphine structures was also tried on the morphinans with similar results. For example, adding an allyl substituent on the nitrogen gives antagonists. Adding a phenethyl group to the nitrogen greatly increases potency. Adding a 14-OH group also increases activity.

Conclusions:
- Morphinans are more potent and longer acting than their morphine counterparts, but they also have higher toxicity and comparable dependence characteristics.
- The modifications carried out on morphine, when carried out on the morphinans, lead to the same biological results. This implies that both types of molecule are reacting with the same receptors in the same way.
- The morphinans are easier to synthesize since they are simpler molecules.

Removing rings C and D

Opening both rings C and D gives an interesting group of compounds called the benzomorphans (Fig. 12.20) which are found to retain analgesic activity. One of the simplest of these structures is metazocine which has the same analgesic activity as morphine. Notice that the two methyl groups in metazocine are *cis* with respect to each other and represent the 'stumps' of the C ring.

If the same type of chemical modifications are carried out on the benzomorphans as were described for the morphinans and morphine, then the same biological effects are observed. This suggests that the benzomorphans interact with the same receptors as the morphinans and morphine analogues. For example, replacing the *N*-methyl group of metazocine with a phenethyl group gives phenazocine which is four times more active than morphine and the first compound to have a useful level of analgesia without dependence properties.

Further developments led to pentazocine (Fig. 12.21) which has proved to be a useful long-term analgesic with a very low risk of addiction. A newer compound (bremazocine) has a longer duration, is 200 times the activity of morphine, appears to have no addictive properties, and does not depress breathing.

These compounds appear to be similar in their action to nalorphine in that they act as antagonists at two of the three types of analgesic receptors, but act as an agonist at the third. The big difference between nalorphine and compounds like pentazocine is that the latter are far stronger agonists, resulting in a more useful level of analgesia.

Unfortunately, many of these compounds have hallucinogenic side-effects due to interactions with a non-analgesic receptor.

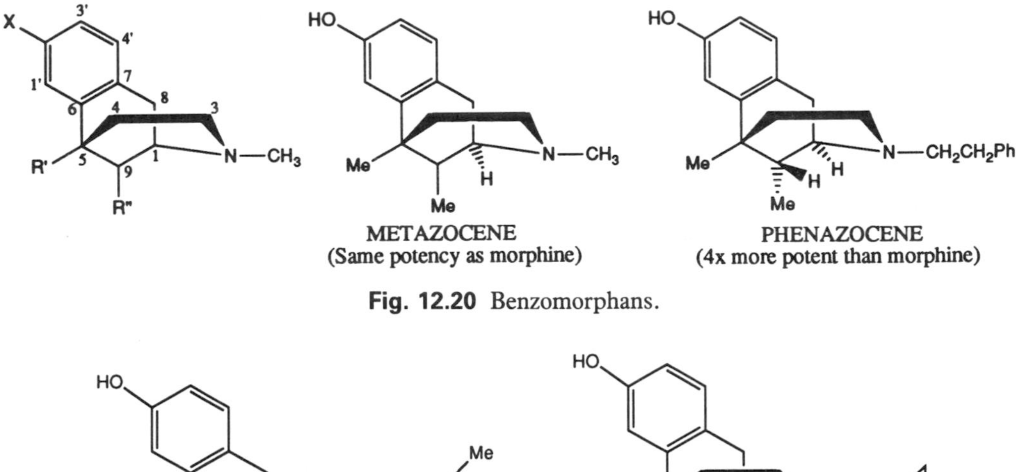

METAZOCENE
(Same potency as morphine)

PHENAZOCENE
(4x more potent than morphine)

Fig. 12.20 Benzomorphans.

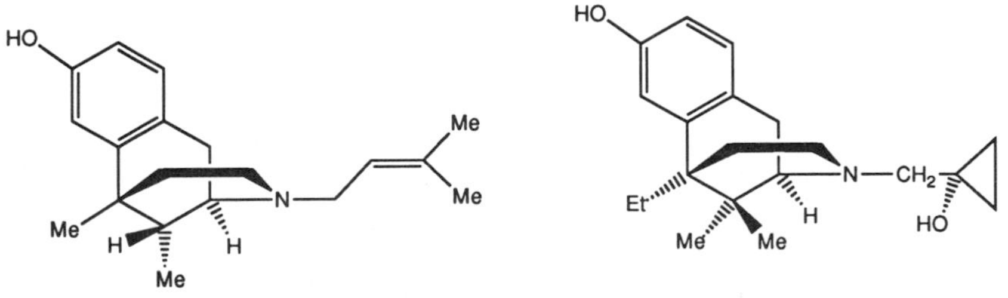

PENTAZOCINE
(33% activity of morphine,
short duration, low
addiction liability)

BREMAZOCINE

Fig. 12.21 Benzomorphans with low rates of dependency.

We shall come back to the interaction of benzomorphans with analgesic receptors later. For the moment, we can make the following conclusions about benzomorphans.

- Rings C and D are not essential to analgesic activity.
- Analgesia and addiction are not necessarily coexistent.
- 6,7-Benzomorphans are clinically useful compounds with reasonable analgesic activity, less addictive liability, and less tolerance.
- Benzomorphans are simpler to synthesize.

Removing rings B, C, and D

Removing rings B, C, and D gives a series of compounds known as 4-phenylpiperidines. The analgesic activity of these compounds was discovered by chance in the 1940s when chemists were studying analogues of cocaine for antispasmodic properties. Their structural relationship to morphine was only identified when they were found to be analgesics, and is evident if the structure is drawn as shown in Fig. 12.22. Activity can be increased sixfold by introducing the phenolic group and altering the ester to a ketone to give ketobemidone.

Meperidine (pethidine) is not as strong an analgesic as morphine and also shares the same undesirable side-effects. However, it has a rapid onset and a shorter duration and as a result has been used as an analgesic for difficult childbirths. The rapid onset and short duration of meperidine mean that there is less chance of it depressing the baby's breathing.

The piperidines are more easily synthesized than any of the above groups and a large number of analogues have been studied. There is some doubt as to whether they act in the same way as morphine at analgesic receptors since some of the chemical adaptations we have already described do not lead to comparable biological results. For example, adding allyl or cyclopropyl groups does not give antagonists. The replacement of the methyl group of meperidine with a cinnamic acid residue increases the activity by 30 times, whereas putting the same group on morphine eliminates activity (Fig. 12.23).

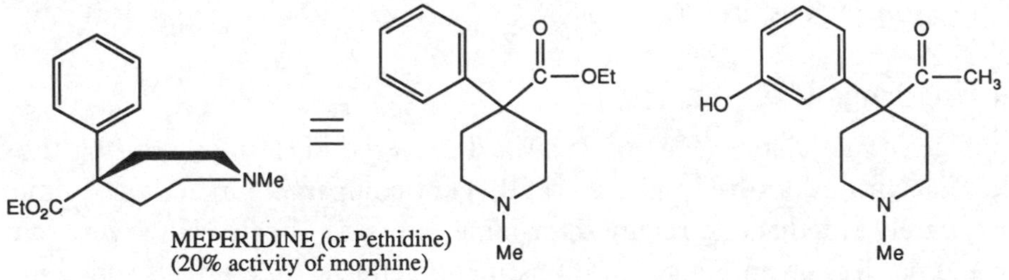

MEPERIDINE (or Pethidine)
(20% activity of morphine)

KETOBEMIDONE

Fig. 12.22 4-phenyl piperidines.

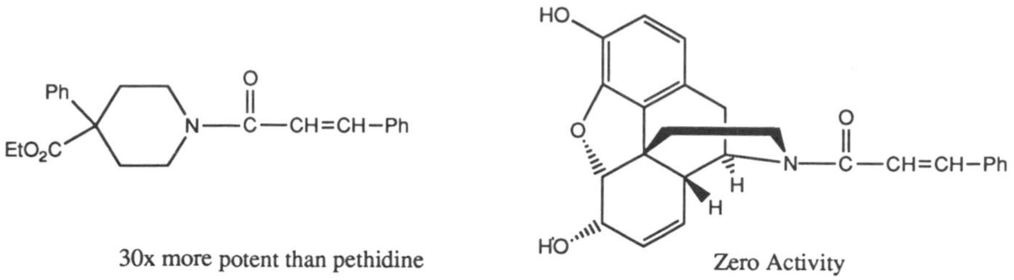

Fig. 12.23 Effect of addition of a cinnamic acid residue on meperidine and morphine.

These results might have something to do with the fact that the piperidines are far more flexible molecules than the previous structures and are thus more likely to interact with receptors in different ways.

One of the most successful piperidine derivatives is fentanyl (Fig. 12.24) which is up to 100 times more active than morphine. The drug lacks a phenolic group, but is very lipophilic. As a result, it can cross the blood–brain barrier efficiently.

Fig. 12.24 Fentanyl (no 2-C bridge).

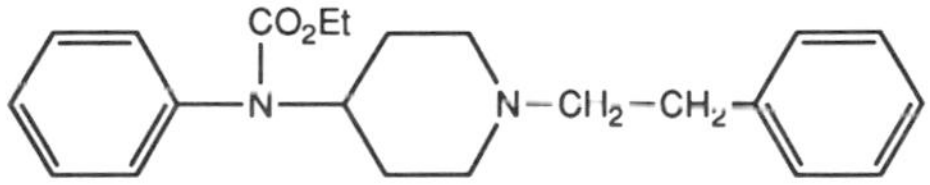

Conclusions:

- Rings C, D, and E are not essential for analgesic activity.
- Piperidines retain side-effects such as addiction and depression of the respiratory centre.
- Piperidine analgesics are faster acting and have shorter duration.
- The quaternary centre present in piperidines is usually necessary (fentanyl is an exception).
- The aromatic ring and basic nitrogen are essential to activity, but the phenol group is not.
- Piperidine analgesics appear to interact with analgesic receptors in a different manner to previous groups.

Removing rings B, C, D, and E

The analgesic methadone (Fig. 12.25) was discovered in Germany during the Second World War and has proved to be a useful agent comparable in activity to morphine. Unfortunately, methadone retains morphine-like side-effects. However, it is orally active and has less severe emetic and constipation effects. Side-effects such as sedation, euphoria, and withdrawal are also less severe and therefore the compound has been given to drug addicts as a substitute for morphine (or heroin) in order to wean them

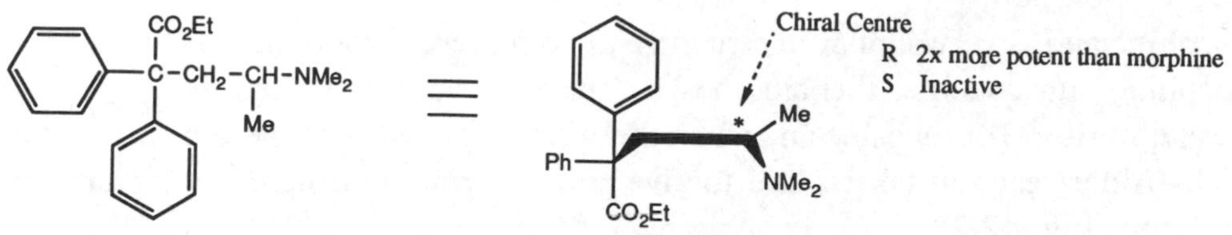

Fig. 12.25 Methadone.

off these drugs. This is not a complete cure since it merely swaps an addiction to heroin for an addiction to methadone. However, this is considered less dangerous.

The molecule has a single chiral centre and when the molecule is drawn in the same manner as morphine, we would expect the R enantiomer to be the more active enantiomer. This proves to be the case with the R enantiomer being twice as powerful as morphine, whereas the S enantiomer is inactive. This is quite a dramatic difference. Since the R and S enantiomers have identical physical properties and lipid solubility, they should both reach the receptor site to the same extent, and so the difference in activity is most probably due to receptor–substrate interactions.

Many analogues of methadone have been synthesized, but with little improvement over the parent drug.

12.3.4 Rigidification

Up till now, we have considered minor adjustments of functional groups on the periphery of the morphine skeleton or drastic simplification of the morphine skeleton.

A completely different strategy is to make the molecule more complicated or more rigid. This strategy is usually employed in an attempt to remove the side-effects of a drug or to increase activity.

It is usually assumed that the side-effects of a drug are due to interactions with additional receptors other than the one we are interested in. These interactions are probably due to the molecule taking up different conformations or shapes. If we make the molecule more rigid so that it takes up fewer conformations, we might eliminate the conformations which are recognized by undesirable receptors, and thus restrict the molecule to the specific conformation which fits the desired receptor. In this way, we would hope to eliminate such side-effects as dependence and respiratory depression. We might also expect increased activity since the molecule is more likely to be in the correct conformation to interact with the receptor.

The best example of this tactic in the analgesic field is provided by a group of compounds known as the oripavines. These structures often show remarkably high activity.

The oripavines are made from an alkaloid which we have not described so far—thebaine (Fig. 12.26). Thebaine can be extracted from opium along with codeine and morphine and is very similar in structure to both these compounds. However, unlike morphine and codeine, thebaine has no analgesic activity. There is a diene group present in ring C of thebaine and when thebaine is reacted with methyl vinyl ketone, a Diels–Alder reaction takes place to give an extra ring and increased rigidity to the structure (Fig. 12.26).

A comparison with morphine shows that the extra ring sticks out from what used to be the 'crossbar' of the T-shaped structure (Fig. 12.27).

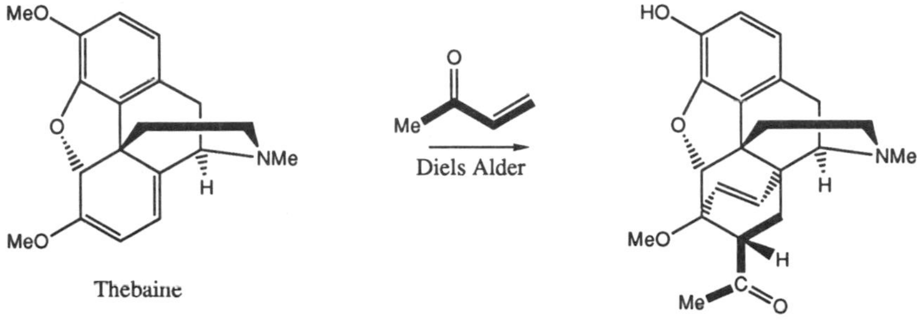

Fig. 12.26 Formation of oripavines.

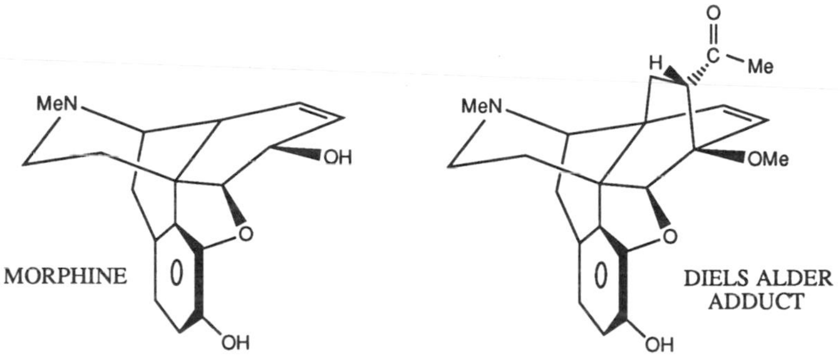

Fig. 12.27 Comparison of morphine and oripavine.

Since a ketone group has been introduced, it is now possible to try the strategy of drug extension, this time by adding various groups to the ketone via a Grignard reaction (Fig. 12.28).

It is noteworthy that the Grignard reaction is stereospecific. The Grignard reagent complexes to both the 6-methoxy group and the ketone, and is then delivered to the less-hindered face of the ketone to give an asymmetric centre (Fig. 12.29).

By varying the groups added by the Grignard reaction, some remarkably powerful compounds have been obtained. Etorphine (Fig. 12.30), for example, is 10 000 times more potent than morphine. This is a combination of the fact that it is a very hydrophobic molecule and can cross the blood–brain barrier 300 times more easily

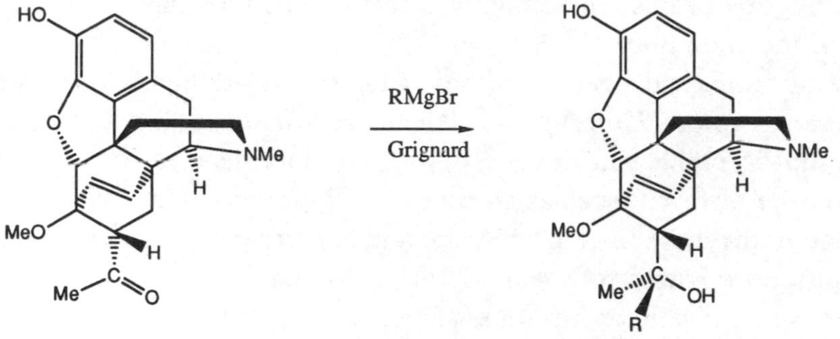

Fig. 12.28 Drug extension.

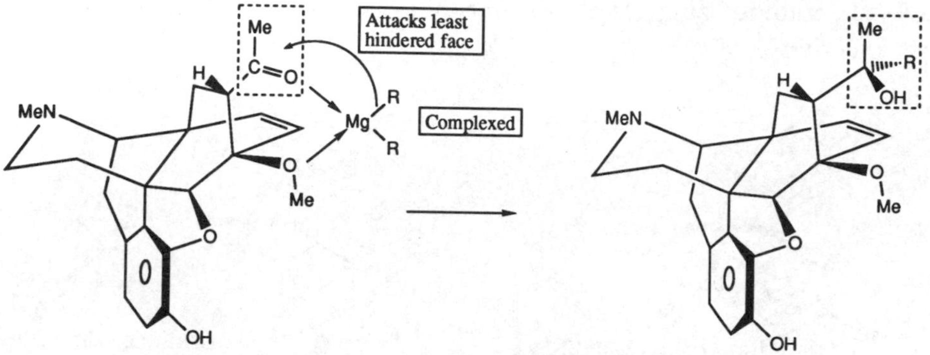

Fig. 12.29 Grignard reaction leads to an asymmetric centre.

than morphine, as well as the fact that it has 20 times more affinity for the analgesic receptor site due to better binding interactions.

At slightly higher doses than those required for analgesia, it can act as a 'knock-out' drug or sedative. The compound has a considerable margin of safety and is used to immobilize large animals such as elephants. Since the compound is so active, only very small doses are required and these can be dissolved in such small volumes (1 ml) that they can be placed in crossbow darts and fired into the hide of the animal.

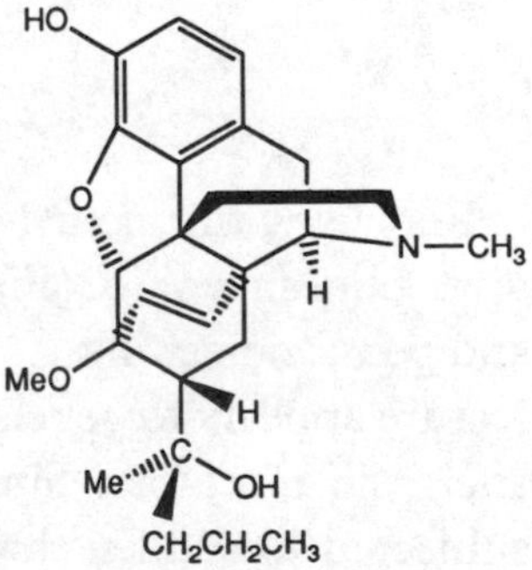

Fig. 12.30 Etorphine.

The addition of lipophilic groups (R) (Fig. 12.29) is found to improve activity dramatically, indicating the presence of a hydrophobic binding region close by on the receptor.[1] The group best able to interact with this region is a phenethyl substituent and the product containing this group is even more active than etorphine.

---

[1] It is believed that the phenylalanine aromatic ring on enkephalins (see later) interacts with this same binding site.

As one might imagine, these highly active compounds have to be handled very carefully in the laboratory.

Because of their rigid structures, these compounds are highly selective agents for the analgesic receptors. Unfortunately, the increased analgesic activity is also accompanied by unacceptable side-effects. It was therefore decided to see if putting substituents on the nitrogen, such as an allyl or cyclopropyl group, would give antagonists as found in the morphine, morphinan, and benzomorphan series of compounds. If so, it might be possible to obtain an oripavine equivalent of a pentazocine or a nalorphine—an antagonist with some agonist activity and with reduced side-effects.

Putting on a cyclopropyl group gives a very powerful antagonist called diprenorphine (Fig. 12.31), which is 100 times more potent than nalorphine and can be used to reverse the immobilizing effects of etorphine (see above). Diprenorphine has no analgesic activity.

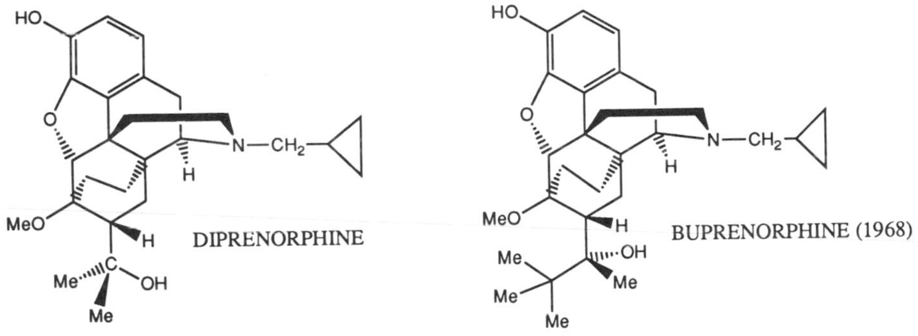

Fig. 12.31

Replacing the methyl group derived from the Grignard reagent with a *t*-butyl group gives buprenorphine (Fig. 12.31) which has similar properties to drugs like nalorphine and pentazocine, in that it has analgesic activity with a very low risk of addiction. This feature appears to be related to the slow onset and removal of buprenorphine from the analgesic receptors. Since these effects are so gradual, the receptor system is not subjected to sudden changes in transmitter levels.

Buprenorphine is the most lipophilic compound in the oripavine series of compounds and therefore enters the brain very easily. Usually, such a drug would react quickly with its receptor. The fact that it does not is therefore a feature of its interaction with the receptor rather than the ease with which it can reach the receptor. It is 100 times more active than morphine as an agonist and four times more active than nalorphine as an antagonist. It is a particularly safe drug since it has very little effect on respiration and what little effect it does have actually decreases at high doses. Therefore, the risks of suffocation from a drug overdose are much smaller than with morphine. Buprenorphine has been used in hospitals to treat patients suffering from cancer and also following surgery. Its drawbacks include side-effects such as nausea and vomiting as well as the fact that it cannot be taken orally. A further use for buprenorphine is as an alternative means to methadone for weaning addicts off heroin.

Buprenorphine binds slowly to analgesic receptors, but once it does bind, it binds very strongly. As a result, less buprenorphine is required to interact with a certain percentage of analgesic receptors than morphine.

On the other hand, buprenorphine is only a partial agonist. In other words, it is not very efficient at switching the analgesic receptor on. This means that it is unable to reach the maximum level of analgesia which can be acquired by morphine.

Overall, buprenorphine's stronger affinity for analgesic receptors outweighs its relatively weak action such that a lower dose of buprenorphine can produce analgesia, compared to morphine. However, if pain levels are high, buprenorphine is unable to counteract the pain and morphine has to be used.

Nevertheless, buprenorphine provides another example of an opiate analogue where analgesia has been separated from dangerous side-effects.

It is time to look more closely at the receptor theories relevant to the analgesics.

12.4 *Receptor theory of analgesics*

Although it has been assumed for many years that there are analgesic receptors, information about them has only been gained relatively recently (1973).

The present knowledge on the subject is that there are at least four different receptors with which morphine can interact, three of which are analgesic receptors. The initial theory on receptor binding assumed a single receptor site, but this does not invalidate many of the proposals which were made. Therefore, it is informative to look at the first theory—the Beckett–Casy hypothesis.

12.4.1 Beckett–Casy hypothesis

In this theory, it is assumed that there is a rigid receptor site and that morphine and its analogues fit into the site in a classic lock-and-key analogy.

Based on the results already described, the following features were proposed as being essential if an analgesic is to interact with its receptor.

● There must be a basic centre (nitrogen) which can be ionized at physiological pH to form a positively charged group. This group then forms an ionic bond with a comparable anionic group in the receptor. As a consequence of this, analgesics have to have a pK_a of 7.8–8.9 such that there is an approximately equal chance of the amine being ionized or un-ionized at physiological pH. This is necessary since the analgesic has to cross the blood–brain barrier as the free base, but once across has to be ionized in order to interact with the receptor.

The pK$_a$ values of useful analgesics all match this prediction.

- The aromatic ring in morphine has to be properly orientated with respect to the nitrogen atom to allow a van der Waals interaction with a suitable hydrophobic location on the receptor. The nature of this interaction suggests that there has to be a close spatial relationship between the aromatic ring and the surface of the receptor.

- The phenol group is probably hydrogen-bonded to a suitable residue at the receptor site.

- There might be a 'hollow' just large enough for the ethylene bridge of carbons 15 and 16 to fit. Such a fit would help to align the molecule and enhance the overall fit.

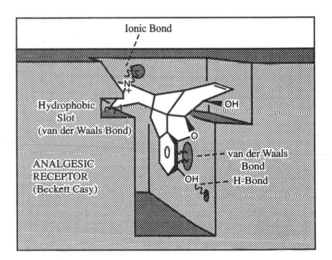

Fig. 12.32 Beckett–Casy hypothesis.

This was the first theory proposed and fitted in well with the majority of results. There can be no doubt that the aromatic ring, phenol, and the nitrogen groups are all important, but there is some doubt as to whether the ethylene bridge is important, since there are several analgesics which lack it (e.g. fentanyl).

The theory also fails to include the extra binding site which was discovered by drug extension. This fact can easily be fitted into the theory, but other anomalies exist which have already been discussed (e.g. the different results obtained for meperidine compared to morphine when a substituent such as the allyl group is attached to nitrogen).

Another anomaly was described earlier where the pethidine analogue containing a cinnamic acid residue is 30 times more active than pethidine itself, whereas the same group on morphine eliminates activity. Such results strongly suggest that a simple one-receptor theory is not applicable.

12.4.2 Multiple analgesic receptors

The previous theory tried to explain analgesic results based on a single analgesic receptor. It is now known that there are several different analgesic receptors which are associated with different types of side-effects. It is also known that several analgesics show preference for some of these receptors over others. This helps to explain the anomalies resulting from the previous Beckett–Casy hypothesis.

It is important to appreciate that the main points of the original theory still apply for each of the analgesic receptors now to be described. The important binding groups for each receptor are the phenol, the aromatic ring, and the ionized nitrogen centre. However, there are subtle differences between each receptor which can distinguish between the finer details of different analgesic molecules. As a result, some analgesics show preference for one analgesic receptor over another or interact in different ways.

There are three analgesic receptors to which the morphine molecule itself can bind and 'switch on'. These receptors have been tabbed with Greek letters.

The mu receptor (μ)

Morphine binds strongly to this receptor and produces analgesia. Receptor binding also leads to the undesired side-effects of respiratory depression, euphoria, and addiction. We can now see why it is so difficult to remove the side-effects of morphine and is analogues, since the receptor with which they bind most strongly is also inherently involved with these side-effects.

The kappa receptor (κ)

This is a different analgesic receptor to which morphine can bind and activate. However, the strength of binding is less than to the μ receptor.

The biological response is analgesia with sedation and none of the hazardous side-effects. It is this receptor which provides the best hope for the ultimate safe analgesic. The earlier results obtained from nalorphine, pentazocine, and buprenorphine can now be explained.

Nalorphine acts as an antagonist at the μ receptor, thus blocking morphine from acting there. However, it acts as a weak agonist at the κ receptor (as does morphine) and so the slight analgesia observed with nalorphine is due to the partial activation of the κ receptor. Unfortunately, nalorphine has hallucinogenic side-effects. This is caused by nalorphine also binding to a completely different, non-analgesic receptor in the brain called the sigma receptor (σ) (see Section 12.7.4.) where it acts as an agonist.

Pentazocine interacts with the μ and κ receptors in the same way, but is able to 'switch on' the κ receptor more strongly. It too suffers the drawback that it 'switches on' the σ receptor. Buprenorphine is slightly different. It binds strongly to all three analgesic receptors and acts as an antagonist at the Δ (see below) and κ receptors, but acts as a partial agonist at the μ receptor to produce its analgesic effect. This might suggest that buprenorphine should suffer the same side-effects as morphine. The fact that it does not is related in some way to the rate at which buprenorphine interacts with the receptor. It is slow to bind, but once it has bound, it is slow to leave.

The delta receptor (Δ)

The Δ receptor is where the brain's natural painkillers (the enkephalins, Section 12.6.) interact. Morphine can also bind quite strongly to this receptor.

A table showing the relative activities of morphine, nalorphine, pentazocine, enkephalins, pethidine, and naloxone is shown in Fig. 12.33. A plus sign indicates the compound is acting as an agonist. A minus sign means it acts as an antagonist. A zero sign means there is no activity or minor activity.

There is now a search going on for orally active opiate structures which can act as antagonists at the μ receptor, agonists at the κ receptor, and have no activity at the σ receptor. Some success has been obtained, especially with the compounds shown in Fig. 12.34, but even these compounds still suffer from certain side-effects, or lack the desired oral activity.

		Morphine	Nalorphine	Pentazocine	Enkephalins	Pethidine	Naloxone
Mu	Analgesia respiratory depression euphoria addiction	+++	−	−	+	+++	− − −
Kappa	Analgesia sedation	+	+	+	+	+	−
Sigma	Psychotomimetic	0	+	+	0	0	0
Delta	Analgesia	++	−	−	+++	+	−

+, Compounds acting as agonists; −, as antagonists. 0, No activity or minor activity.

Fig. 12.33 Relative activities of analgesics.

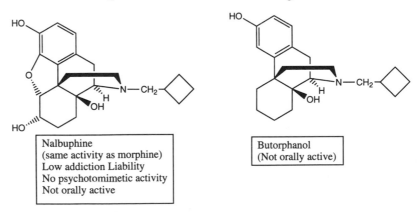

Nalbuphine
(same activity as morphine)
Low addiction Liability
No psychotomimetic activity
Not orally active

Butorphanol
(Not orally active)

Fig. 12.34 New analgesic structures.

12.5 *Agonists and antagonists*

We return now to look at a particularly interesting problem regarding the agonist/antagonist properties of morphine analogues. Why should such a small change as replacing an *N*-methyl group with an allyl group result in such a dramatic change in biological activity such that an agonist becomes an antagonist? Why should a molecule such as nalorphine act as an agonist at one analgesic receptor and an antagonist at another? How can different receptors distinguish between such subtle changes in a molecule?

We shall consider one theory which attempts to explain how these distinctions might take place, but it is important to realize that there are alternative theories. In this particular theory, it is suggested that there are two accessory hydrophobic binding sites present in an analgesic receptor. It is then proposed that a structure will act as an agonist or as an antagonist depending on which of these extra binding sites is used. In other words, one of the hydrophobic binding sites is an agonist binding site, whereas the other is an antagonist binding site.

The model was proposed by Snyder and co-workers and is shown in Figs. 12.35–12.37).[2] In the model, the agonist binding site is further away from the nitrogen and positioned axially with respect to it. The antagonist site is closer and positioned equatorially.

Let us now consider the morphine analogue containing a phenethyl substituent on the nitrogen (Fig. 12.35). It is proposed that this structure binds as already described, such that the phenol, aromatic ring, and basic centre are interacting with their respective binding sites. If the phenethyl group is in the axial position, the aromatic ring is in the correct position to interact with the agonist binding site. However, if the phenethyl group is in the equatorial position, the aromatic ring is placed beyond the antagonist binding site and cannot bind. The overall result is increased activity as an agonist.

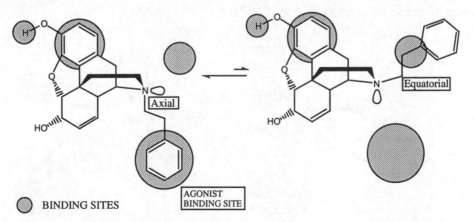

Fig. 12.35 Morphine analogue containing a phenethyl substituent on the nitrogen.

[2] Feinberg, A.P., Creese, I., and Snyder, S.H. (1976). *Proc. Natl. Acad. Sci. USA*, **73**, 4215.

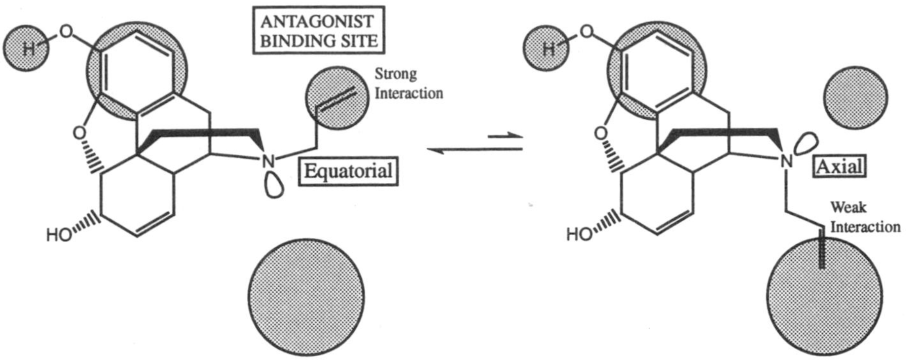

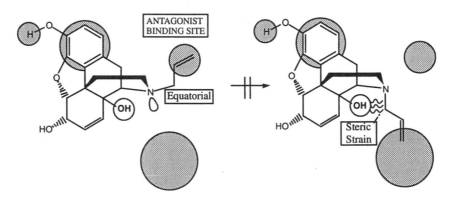

Fig. 12.36 Morphine analogue containing anallyl substituent.

Let us now consider what happens if the phenethyl group is replaced with an allyl group (Fig. 12.36). In the equatorial position, the allyl group is able to bind strongly to the antagonist binding site, whereas in the axial position it barely reaches the agonist binding site, resulting in a weak interaction.

In this theory, it is proposed that a molecule such as phenazocine (with a phenethyl group) acts as an agonist since it can only bind to the agonist binding site. A molecule such as nalorphine (with an allyl group) can bind to both agonist and antagonist sites and therefore acts as an agonist at one receptor and an antagonist at another. The ratio of these effects would depend on the relative equilibrium ratio of the axial and equatorial substituted isomers.

A compound which is a pure antagonist would be forced to have a suitable substituent in the equatorial position. It is believed that the presence of a 14-OH group sterically hinders the isomer with the axial substituent, and forces the substituent to remain equatorial (Fig. 12.37).

Fig. 12.37 Influence of 14-OH on binding interactions.

12.6 *Enkephalins and endorphins*

12.6.1 Naturally occurring enkaphalins and endorphins

Morphine, as we have already discussed, is an alkaloid which relieves pain and acts in the CNS. There are two conclusions which can be drawn from this. The first is that there must be analgesic receptors in the CNS. The second conclusion is that there must be chemicals produced in the body which interact with these receptors. Morphine itself is not produced by humans and therefore the body must be using a different chemical as its natural painkiller.

The search for this natural analgesic took many years, but ultimately led to the discovery of the enkephalins and the endorphins. The term enkephalin is derived from the Greek, meaning 'in the head', and that is exactly where the enkephalins are produced. The first enkephalins to be discovered were the pentapeptides Met-enkephalin and Leu-enkephalin.

$$\text{H–Tyr–Gly–Gly–Phe–Met–OH} \qquad \text{H–Tyr–Gly–Gly–Phe–Leu–OH}$$

At least 15 endogenous peptides have now been discovered, varying in length from 5 to 33 amino acids (the enkephalins and the endorphins). These compounds are thought to be neurotransmitters or neurohormones in the brain and operate as the body's natural painkillers as well as having a number of other roles. They are derived from three inactive precursor proteins—proenkephalin, prodynorphin, or pro-opiomelanocortin (Fig. 12.38).

Fig. 12.38 Production of the body's natural painkillers.

Proenkephalin
Prodynorphin ⟶ Endorphins + Enkephalins
Pro-opiomelanocortin

All 15 compounds are found to have either the Met- or the Leu-enkephalin skeleton at their N-terminus, which emphasizes the importance of this pentapeptide structure towards analgesic activity. It has also been shown conclusively that the tyrosine part of these molecules is essential to activity and much has been made of the fact that there is a tyrosine skeleton in the morphine skeleton (Fig. 12.39).

Enkephalins are thought to be responsible for the analgesia resulting from acupuncture.

12.6.2 Analogues of enkephalins

SAR studies on the enkephalins have shown the importance of the tyrosine phenol ring and the tyrosine amino group. Without either, activity is lost. If tyrosine is replaced with another amino acid, then activity is also lost (the only exception being D-serine). It has also been found that the enkephalins are easily inactivated by

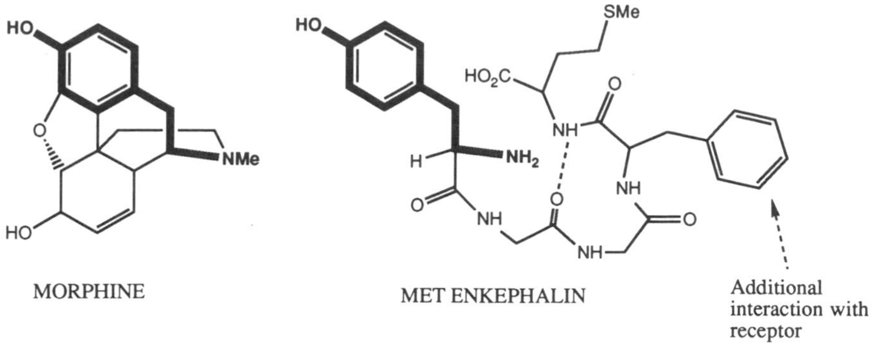

MORPHINE   MET ENKEPHALIN   Additional
                                                     interaction with
                                                     receptor

Fig. 12.39 The tyrosine section is essential to activity.

peptidase enzymes *in vivo*. The most labile peptide bond in the enkephalins is that between the tyrosine and glycine residues.

Much work has been done therefore, to try and stabilize this bond towards hydrolysis. It is possible to replace the amino acid glycine with an unnatural D-amino acid such as D-alanine. Since D-amino acids are not naturally occurring, peptidases do not recognize the structure and the peptide bond is not attacked. The alternative tactic of replacing L-tyrosine with D-tyrosine is not possible, since this completely alters the relative orientation of the tyrosine aromatic ring with respect to the rest of the molecule. As a result, the analogue is unable to bind to the analgesic receptor and is inactive.

Putting a methyl group on to the amide nitrogen can also block hydrolysis by peptidases. Another tactic is to use unusual amino acids which are either not recognized by peptidases or prevent the molecule from fitting the peptidase active site. Examples of these tactics at work are demonstrated in Fig. 12.40.

Unfortunately, the enkephalins also have some activity at the mu receptor and so the search for selective agents continues.

H—L-Tyr—Gly—Gly—L-Phe—L-Met—OH	Delta Agonist + a little mu activity
H—L-Tyr—D-AA—Gly—NMe-L-Phe—L-Met—OH	Resistant to peptidase. Orally active.
N,N-Diallyl-L-Tyr—aib—aib—L-Phe—L-Leu—OH	Antagonist to delta receptor. (aib = alpha-aminobutyric acid)
Longer enkaphalins/endorphins	Increase in kappa activity Slight increase in mu activity

Fig. 12.40 Tactics to stabilize the bond between the tyrosine and glycine residues.

12.7 *Receptor mechanisms*

Up until now we have discussed receptors very much as 'black boxes'. The substrate comes along, binds to the receptor, and switches it on. There is a biological response, be it analgesia, sedation, or whatever, but we have given no indication of how this response takes place. Why should morphine cause analgesia just by attaching itself to a receptor protein?

In general, all receptors in the body are situated on the surface of cells and act as communication centres for the various messages being sent from one part of the body to another. The message may be sent through nerves or via hormones, but ultimately the message has to be delivered from one cell to another by a chemical messenger. This chemical messenger has to 'dock' with the receptor which is waiting for it. When it does so, it forces the receptor to change shape. This change in shape of the receptor molecule may force a change in the shape of some neighbouring protein or perhaps an ion channel, resulting in an alteration of ion flows in and out of the cell. Such effects will ultimately have a biological effect, dependent on the cells affected.

We shall now look at the analgesic receptors in a little more detail.

12.7.1 The mu receptor (μ)

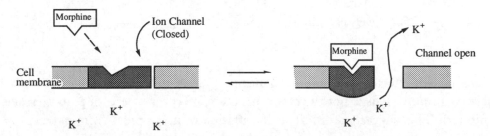

Fig. 12.41 Morphine binding to μ receptor.

As the diagram in Fig. 12.41 demonstrates, morphine binds to the μ receptor and induces a change in shape. This change in conformation opens up an ion channel in the cell membrane and as a result, potassium ions can flow out of the cell. This flow hyperpolarizes the membrane potential and makes it more difficult for an active potential to be reached (see Appendix 2).

Therefore, the frequency of action potential firing is decreased, which results in a decrease in neurone excitability.

This increase in potassium permeability has an indirect effect, since it also decreases the influx of calcium ions into the nerve terminal and this in turn reduces neurotransmitter release.

Both effects, therefore, 'shut down' the nerve and block the pain messages.

Unfortunately, this receptor is also associated with the hazardous side-effects of narcotic analgesics. There is still a search to see if there are possibly two slightly different mu receptors, one which is solely due to analgesia and one responsible for the side-effects.

12.7.2 The kappa receptor (κ)

The κ receptor is directly associated with a calcium channel (Fig. 12.42). When an agonist binds to the κ receptor, the receptor changes conformation and the calcium channel (normally open when the nerve is firing and passing on pain messages) is closed. Calcium is required for the production of the nerves neurotransmitters and therefore the nerve is shut down and cannot pass on pain messages.

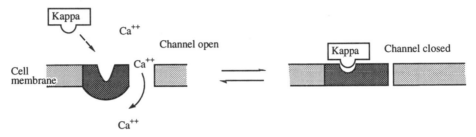

Fig. 12.42 The association of the κ receptor and calcium channels.

The nerves affected by the κ mechanism are those related to pain induced by non-thermal stimuli. This is not the case with the μ receptor where all pain messages are inhibited. This suggests a different distribution of κ receptors from μ receptors.

12.7.3 The delta receptor (Δ)

Like the μ receptor, the nerves containing the Δ receptor do not discriminate between pain from different sources.

In this case, there are no ion channels involved (Fig. 12.43). The substrate molecule binds to the Δ receptor and, in some way, the message is transmitted through the cell

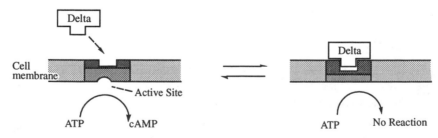

Fig. 12.43 The delta receptor.

membrane to a second membrane-bound protein. This protein then acts as an enzyme for the formation of cyclic AMP (Fig. 12.44). Normally, the active site is open when the nerve is receiving pain messages, such that cyclic AMP acts as a secondary messenger and passes on the pain messages. However, when the Δ receptor is activated it probably changes shape and as a result leads to a change in the shape of the cyclase enzyme to close down the active site by which it can make cyclic AMP (see also Appendix 3).

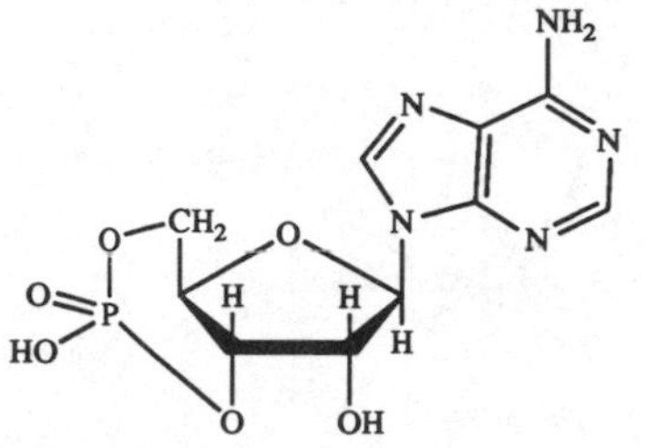

Fig. 12.44 Cyclic adenosine monophosphate (cAMP).

12.7.4 The sigma receptor (σ)

This receptor is not an analgesic receptor, but we have seen that it can be activated by certain opiate molecules such as nalorphine. When activated, it produces hallucinogenic effects. The σ receptor may be the one associated with the hallucinogenic and psychotomimetic effects of phencyclidine (PCP), otherwise known as 'angel dust' (Fig. 12.45).

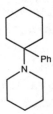

Fig. 12.45 Phencyclidine (PCP) or 'angel dust'.

12.8 *The future*

There is still a need for analgesic drugs with reduced side-effects. Four approaches are feasible in the field of opiates.

- κ Agonists.
  Such compounds should have much-reduced side-effects. However, a completely specific κ agonist has not yet been found and there may be a close link between the Δ receptor and the σ receptor. Examples of selective agonists are shown in Fig. 12.46.
- Selectivity between mu receptor subtypes.
  There might be two slightly different μ receptors, one of which is purely responsible for analgesia (μ_1) and the other solely responsible for unwanted side-effects such as respiratory depression (μ_2). An agent showing selectivity would prove such a theory and be very useful analgesic.

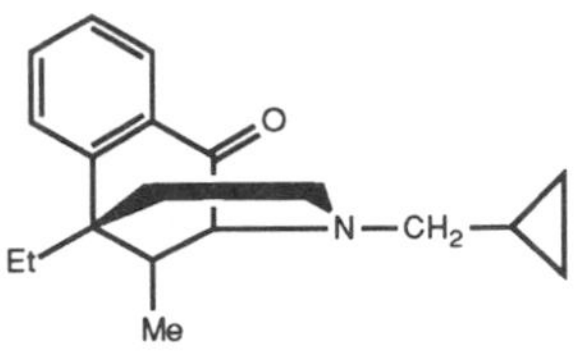

Ethylketocyclazocine
(Kappa=mu > delta)

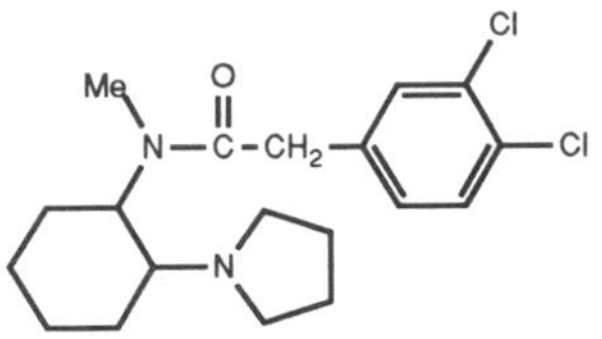

U 50488
(kappa > mu)

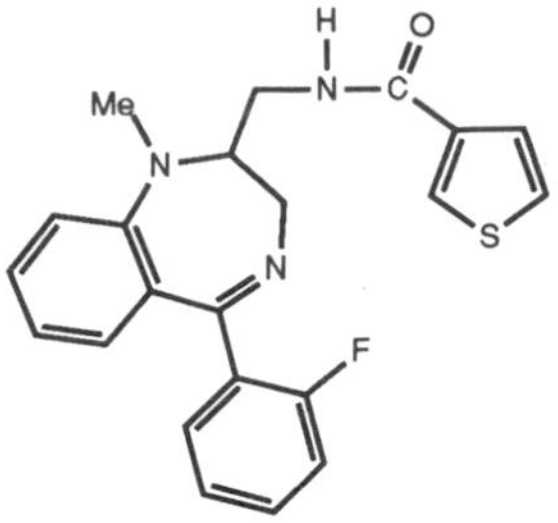

Tifluadom
(kappa > mu)

Fig. 12.46 Examples of selective agonists.

- Peripheral opiate receptors.
  Peripheral opiate receptors have been identified in the ileum and are responsible for the antidiarrhoeal activity of opiates. If peripheral sensory nerves also possess opiate receptors, drugs might be designed versus these sites and as a result would not need to cross the blood–brain barrier.
- Blocking postsynaptic receptors.
  Perhaps blocking the chemical messengers from transmitting pain information by blocking postsynaptic receptors with selective antagonists would be the best approach. This would involve non-opioids and non-opioid mechanisms, but might be the best way of eliminating side-effects.

  One promising lead is provided by the neurotransmitter GABA (Fig. 12.47) which appears to have a role in the regulation of enkephalinergic neurons and as such affects pain pathways.

  An undecapeptide called substance P is an excitatory neurotransmitter which appears to have a role in pain mediation and is worthy of further study.

Fig. 12.47 γ-aminobutanoic acid (GABA).

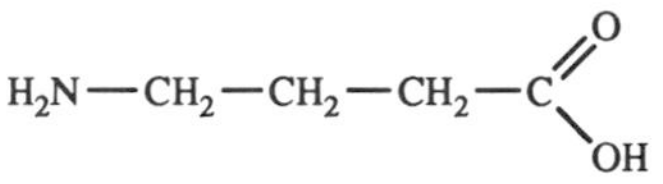
H₂N—CH₂—CH₂—CH₂—C
O
OH

13 · Cimetidine—a rational approach to drug design

13.1 Introduction

Many of the past successes in medicinal chemistry have involved the fortuitous discovery of useful pharmaceutical agents from natural sources such as plants or microorganisms. Analogues of these structures were then made in an effort to improve activity and/or to reduce side-effects, but often these variations were carried out on a trial-and-error basis. While this approach yielded a large range of medicinal compounds, it was wasteful with respect to the time and effort involved.

In the last twenty to thirty years, greater emphasis has been placed on rational drug design whereby drugs are designed to interact with a known biological system. For example, when looking for an enzyme inhibitor, a rational approach is to purify the enzyme and to study its tertiary structure by X-ray crystallography. If the enzyme can be crystallized along with a bound inhibitor, then the researcher can identify and study the binding site of the enzyme. The X-ray data can be read into a computer and the binding site studied to see whether new inhibitors can be designed to fit more strongly.

However, it is not often possible to isolate and purify enzymes, and when it comes to membrane-bound receptors, the difficulties become even greater. Nevertheless, rational drug design is still possible, even when the receptor cannot be studied directly. The strengths of the rational approach to drug design are amply demonstrated by the development of the antiulcer drug cimetidine (Tagamet) (Fig. 13.1), carried out by scientists at Smith, Kline, & French, (SK&F).

The remarkable aspect of the cimetidine story lies in the fact that at the onset of the

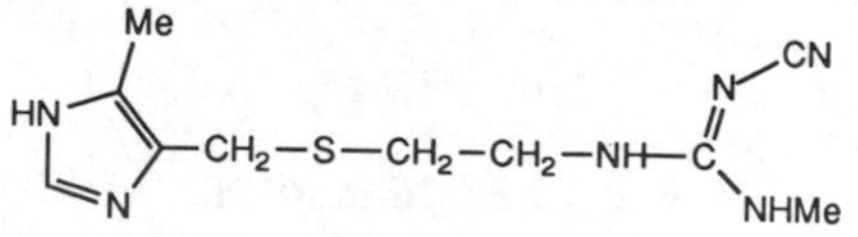

Fig. 13.1 Cimetidine.

project there were no lead compounds and it was not even known if the necessary receptor protein even existed!

13.2 *In the beginning—ulcer therapy in 1964*

When the cimetidine programme started in 1964, the methods available for treating peptic ulcers were few and generally unsatisfactory.

Ulcers are localized erosions of the mucous membranes of the stomach or duodenum. It is not known how these ulcers arise, but the presence of gastric acid aggravates the problem and delays recovery. In the early 1960s, the conventional treatment was to try and neutralize gastric acid in the stomach by administering bases such as sodium bicarbonate or calcium carbonate. However, the dose levels required for neutralization were large and caused unpleasant side-effects. It was reasoned that a better approach would be to inhibit the release of gastric acid at source.

Gastric acid (HCl) is released by cells known as parietal cells in the stomach (Fig. 13.2). These parietal cells are innervated with nerves (not shown on the diagram) from the automatic nervous system (see Chapter 11). When the autonomic nervous system is stimulated, a signal is sent to the parietal cells culminating in the release of the neurotransmitter acetylcholine at the nerve termini. Acetylcholine crosses the gap between nerve and parietal cell and activates the cholinergic receptors of the parietal cells leading to the release of gastric acid into the stomach. The trigger for this process is provided by the sight, smell, or even the thought, of food. Thus, gastric acid is released before food has even entered the stomach.

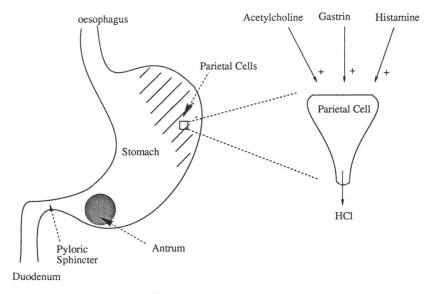

Fig. 13.2 The stomach.

Nerve signals also stimulate a region of the stomach known as the antrum which contains hormone-producing cells known as G cells. The hormone released is a peptide called gastrin (Fig. 13.3) which is also released when food is present in the stomach. The gastrin moves into the blood supply and travels to the parietal cells further stimulating the release of gastric acid. Release of gastric acid should therefore be inhibited by antagonists blocking either the acetylcholine receptor or the receptor for gastrin.

Fig. 13.3 Gastrin.   p-Glu-Gly-Pro-Trp-Leu-[Glu]$_5$-Ala-Tyr-Gly-Trp-Met-Asp-Phe-NH$_2$

Agents which block the acetylcholine receptor are known as anticholinergic drugs (see Chapter 11). These agents certainly block the cholinergic receptor in parietal cells and inhibit release of gastric acid. Unfortunately, they also inhibit acetylcholine receptors at other parts of the body and cause unwanted side-effects.

Therefore, in 1964, the best hope of achieving an antiulcer agent appeared to be in finding a drug which would block the hormone gastrin. Several research teams were active in this field, but the research team at SK&F decided to follow a different tack altogether.

It was known that histamine (Fig. 13.4) could also stimulate gastric acid release, and it was proposed by the SK&F team that an antihistamine agent might also be effective in treating ulcers. At the time, this was a highly speculative proposal. Although histamine had been shown experimentally to stimulate gastric acid release, it was by no means certain that it played any significant role *in vivo*. Many workers at the time discounted the importance of histamine, especially when it was found that conventional antihistamines failed to inhibit gastric acid release. This result appeared to suggest the absence of histamine receptors in the parietal cells. The fact that histamine did have a stimulatory effect could be explained away by suggesting that histamine coincidentally switched on the gastrin or acetylcholine receptors.

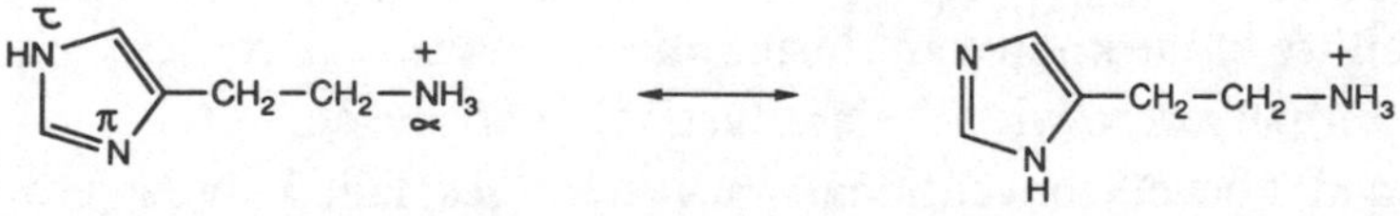

Fig. 13.4 Histamine.

Why then did the SK&F team persevere in their search for an effective antihistamine? What was their reasoning? Before answering that, let us look at histamine itself and the antihistamines available at that time.

13.3 *Histamine*

Histamine is made up of an imidazole ring which can exist in two tautomeric forms as shown in Fig. 13.4. Attached to the imidazole ring there is a two-carbon chain with a terminal α-amino group. The pK_a of this amino group is 9.80, which means that at a plasma pH of 7.4, the side-chain of histamine is 99.6 per cent ionized.

The pK_a of the imidazole ring is 5.74 and so the ring is mostly un-ionized at pH 7.4.

Whenever cell damage occurs, histamine is released and stimulates the dilation and increased permeability of small blood vessels. The advantage of this to the body is that defensive cells (e.g. white blood cells) are released from the blood supply into an area of tissue damage and are able to combat any potential infection. Unfortunately, the release of histamine can also be a problem. For example, when an allergic reaction or irritation is experienced, histamine is released and produces the same effects when they are not really needed.

The early antihistamine drugs were therefore designed to treat conditions such as hay fever, rashes, insect bites, or asthma.

Two examples of these early antihistamines are mepyramine (Fig. 13.5) and diphenhydramine ('Benadryl') (Fig. 13.6).

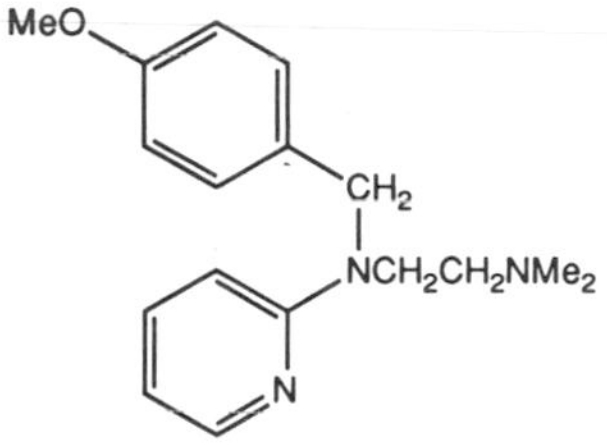

Fig. 13.5 Mepyramine. 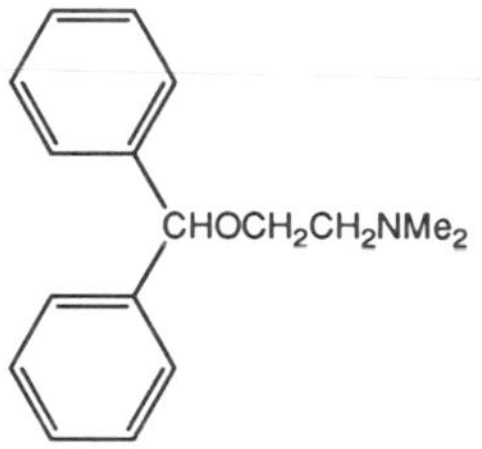   **Fig. 13.6** Diphenhydramine.

13.4 *The theory—two histamine receptors?*

We are now able to return to the question we asked at the end of Section 13.2. Bearing in mind the failure of the known antihistamines to inhibit gastric acid release, why did the SK&F team persevere with the antihistamine approach?

As mentioned above, conventional antihistamines failed to have any effect on gastric acid release. However, they also failed to inhibit other actions of histamine. For example, they failed to fully inhibit the dilation of blood vessels induced by histamine. The SK&F scientists therefore proposed that there might be two different types of histamine receptor, analogous to the two types of acetylcholine receptor mentioned in Chapter 11. Histamine—the natural messenger—would switch both on equally effectively and would not distinguish between them. However, suitably designed antagonists should in theory be capable of making that distinction. By implication, this meant that the conventional antihistamines known in the early sixties were already selective in that they were able to inhibit the histamine receptors involved in the inflammation process (classified as H1 receptors), and were unable to inhibit the proposed histamine receptors responsible for gastric acid secretion (classified as H2 receptors).

It was an interesting theory, but the fact remained that there was no known antagonist for the proposed H2 receptors. Until such a compound was found, it could not be certain that the H2 receptors even existed, and yet without a receptor to study, how could one design an antagonist to act with it?

13.5 *Searching for a lead—histamine*

The SK&F team obviously had a problem. They had a theory but no lead compound. How could they make a start?

Their answer was to start from histamine itself. If histamine was stimulating the release of gastric acid by binding to a hypothetical H2 receptor, then clearly histamine was being 'recognized' by the receptor. The task then was to vary the structure of histamine in such a way that it would still be recognized by the receptor, but bind in such a way that it acted as an antagonist rather than an agonist.

It was necessary then to find out how histamine itself was binding to its receptors. Structure activity studies on histamine and histamine analogues revealed that the binding requirements for histamine to the H1 and the proposed H2 receptors were slightly different.

At the H1 receptor, the essential requirements were as follows:

- The side-chain had to have a positively charged nitrogen atom with at least one attached proton. Quaternary ammonium salts which lacked such a proton were extremely weak in activity.
- There had to be a flexible chain between the above cation and a heteroaromatic ring.
- The heteroaromatic ring did not have to be imidazole, but it did have to contain a nitrogen atom with a lone pair of electrons, *ortho* to the side-chain.

For the proposed H2 receptor, structure–activity studies were carried out to determine whether histamine analogues could bring about the physiological effects proposed for this receptor (e.g. stimulating gastric acid release).

The essential structure–activity requirements were the same as for the H1 receptor except that the heteroaromatic ring had to contain an amidine unit (HN–CH–N:). These results are summarized in Fig. 13.7.

From these results, it appeared that the terminal α-amino group was involved in a binding interaction with both types of receptor via ionic or hydrogen bonding, while

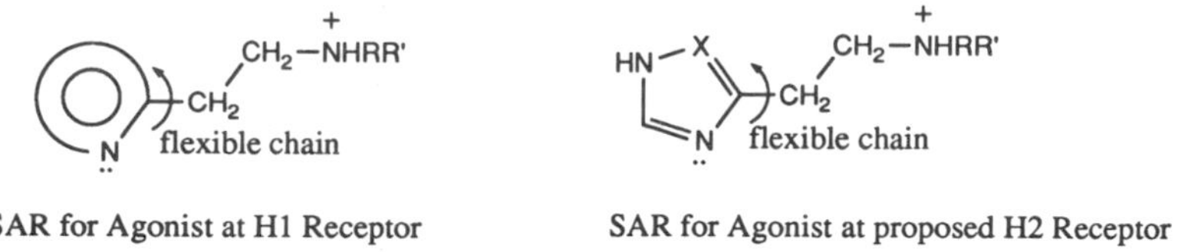

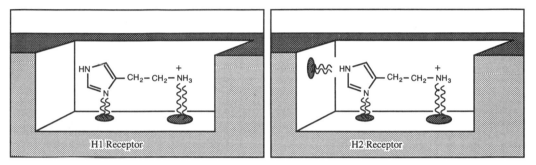

Fig. 13.7 Summary of SAR results.

Fig. 13.8 Binding interactions for H1 and H2.

the nitrogen atom(s) in the heteroaromatic ring bound via hydrogen bonding, as shown in Fig. 13.8.

13.6 *Searching for a lead—N^α-guanylhistamine*

Having gained a knowledge of the structure–activity relationships for histamine, the task was now to design a molecule which would be recognized by the H2 receptor, but which would not activate it. In other words, an agonist had to be converted to an antagonist. In order to do that, it would be necessary to alter the way in which the molecule was bound to the receptor.

Pictorially, one can imagine histamine fitting into its receptor site and inducing a change in shape which 'switches the receptor on' (Fig. 13.9). An antagonist might be found by adding a functional group which would bind to another binding site on the receptor and prevent the change in shape required for activation.

This was one of several strategies tried out by the SK&F workers. To begin with, a study of known agonists and antagonists in other fields of medicinal chemistry was carried out. The structural differences between agonists and antagonists for a particular receptor were identified and then similar alterations were tried on histamine.

For example, fusing an aromatic ring on to noradrenaline had been a successful tactic used in the design of antagonists for the noradrenaline receptor (see Section 7.5.5.). This same tactic was tried with histamine to give analogues such as the one shown in Fig. 13.10, but none of the compounds synthesized proved to be an antagonist.

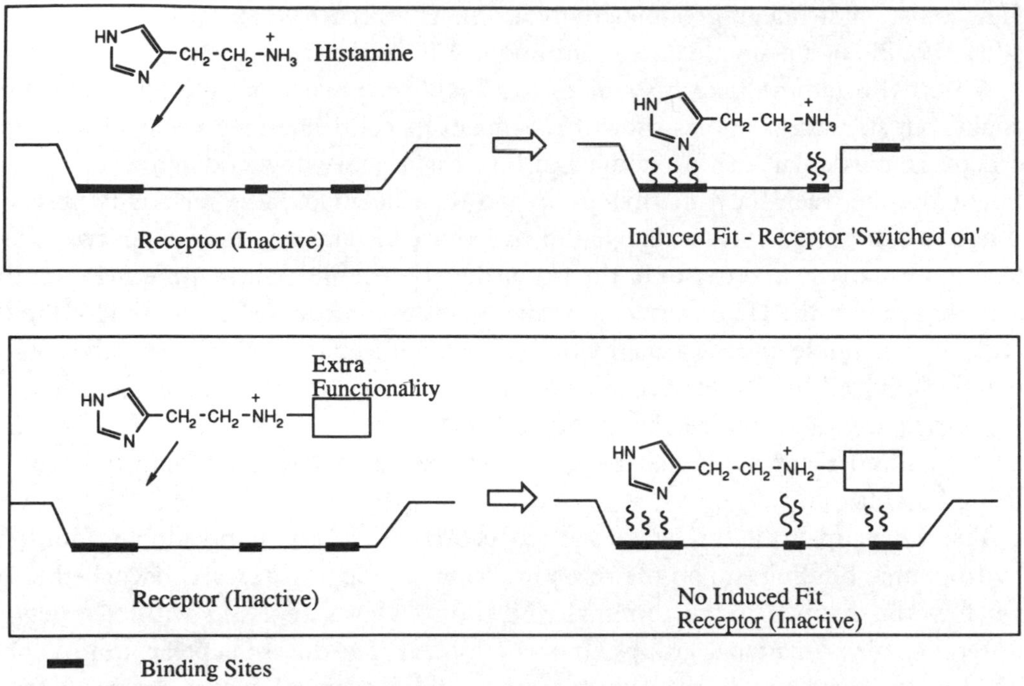

Fig. 13.9 Possible receptor interactions of histamine and an antagonist.

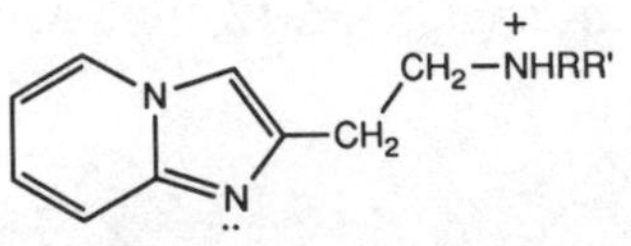

Fig. 13.10 Histamine analogue—not an antagonist.

Another approach which had been used successfully in the development of anti-cholinergic agents (section 11.11.2.) had been the addition of non-polar, hydrophobic substituents. This approach was tried with histamine by attaching various alkyl and arylalkyl groups to different locations on the histamine skeleton. Unfortunately, none of these analogues proved to be antagonists.

However, one interesting result was obtained which was to be relevant to later studies. It was discovered that 4-methylhistamine (Fig 13.11) was a highly selective

Fig. 13.11 4-Methylhistamine.

H2 agonist, showing far greater activity for the H2 receptor than for the H1 receptor. Why should such a simple alteration produce this selectivity?

4-Methylhistamine (like histamine) is a highly flexible molecule due to its side-chain, but structural studies show that some of its conformations are less stable than others. In particular, conformation I in Fig. 13.11 is disallowed due to a large steric interaction between the 4-methyl group and the side-chain. The selectivity observed suggests that 4-methylhistamine (and by inference histamine) has to adopt two different conformations in order to fit the H1 or the H2 receptor. Since 4-methylhistamine is more active at the H2 receptor, it implies that the conformation required for the H2 receptor is a stable one for 4-methylhistamine (conformation II), whereas the conformation required for the H1 receptor is an unstable one (conformation I).

Despite this interesting result, the SK&F workers were no closer to an H2 antagonist. Two hundred compounds had been synthesized and not one had shown a hint of being an antagonist.

The work up until this stage had concentrated on searching for an additional hydrophobic binding site on the receptor. Now the focus of research switched to see what would happen if the terminal αNH_3^+ group was replaced with a variety of different polar functional groups. It was reasoned that different polar groups could bond to the same site on the receptor as the NH_3^+ group, but that the geometry of bonding might be altered sufficiently to produce an antagonist. It was from this study that the first crucial breakthrough was achieved with the discovery that N^α-guanylhistamine (Fig. 13.12) was acting very weakly as an antagonist.

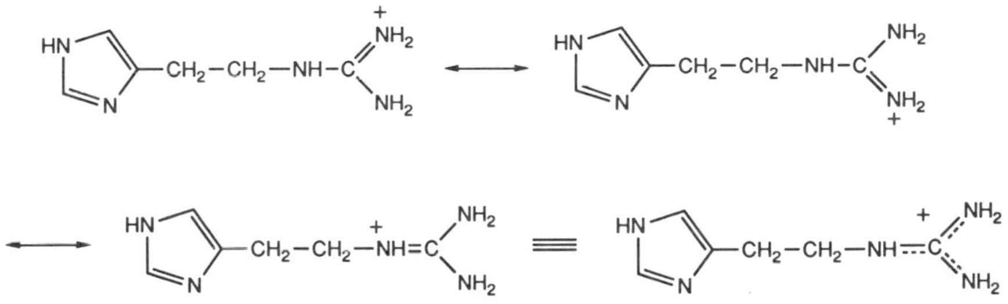

Fig. 13.12 N^α-Guanylhistamine.

This structure had in fact been synthesized early on in the project, but had not been recognized as an antagonist. This is not too surprising since it acts as an agonist! It was not until later pharmacological studies were carried out that it was realized that N^α-guanylhistamine was also acting as an antagonist to histamine. In other words, it was a partial agonist (see Section 5.8.).

N^α-guanylhistamine activates the H2 receptor, but not to the same extent as histamine. As a result, the amount of gastric acid released is lower. More importantly, as long as N^α-guanylhistamine is bound to the receptor, it prevents histamine from binding and thus prevents complete receptor activation.

This was the first indication of any sort of antagonism to histamine.

The question now arose as to which part or parts of the N^α-guanylhistamine skeleton were really necessary for this effect. Perhaps the guanidine group itself could act as an antagonist?

Various guanidine structures were synthesized which lacked the imidazole ring, but none had the desired antagonist activity, demonstrating that both the imidazole ring and the guanidine group were required.

The structures of N^α-guanylhistamine and histamine were now compared. Both structures contain an imidazole ring and a positively charged group linked by a two-carbon bridge. The guanidine group is basic and protonated at pH 7.4 so that the analogue has a positive charge similar to histamine. However, the charge on the guanidine group can be spread around a planar arrangement of three nitrogens and can potentially be further away from the imidazole ring (Fig. 13.12). This leads to the possibility that the analogue could be interacting with another binding group on the receptor which is 'out of reach' of histamine. This is demonstrated in Fig. 13.13 and 13.14. Two alternative binding sites might be available for the cationic group—an agonist site where binding leads to activation of the receptor and an antagonist site where binding does not activate the receptor. In Fig. 13.14, histamine is only able to

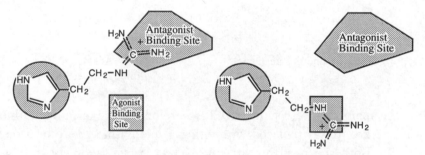

Fig. 13.13 Possible binding modes for N^α-guanylhistamine.

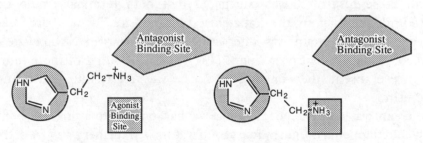

Fig. 13.14 Binding of histamine: agonist mode only.

reach the agonist site. However, the analogue with its extended functionality is capable of reaching either site (Fig. 13.13).

If most of the analogue molecules bind to the agonist site and the remainder bind to the antagonist site, then this could explain the partial agonist activity. Regardless of the mode of binding, histamine would be prevented from binding and an antagonism would be observed due to the percentage of N^α-guanylhistamine bound to the antagonist site.

13.7 *Developing the lead—a chelation bonding theory*

Variations were now necessary to see if an analogue could be made which would only bind to the antagonist site.

The synthesis of the isothiourea (Fig. 13.15) gave a structure where the nitrogen nearest to the imidazole ring was replaced with a sulfur atom.

The positive charge in this molecule is now restricted to the terminal portion of the chain and should interact more strongly with the proposed antagonist binding site if it is indeed further away.

Antagonist activity did increase, but the compound was still a partial agonist, showing that binding was still possible to the agonist site.

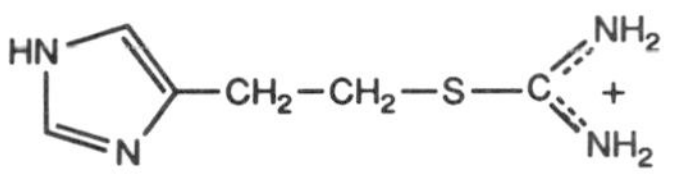

Fig. 13.15 Isothiourea analogue.

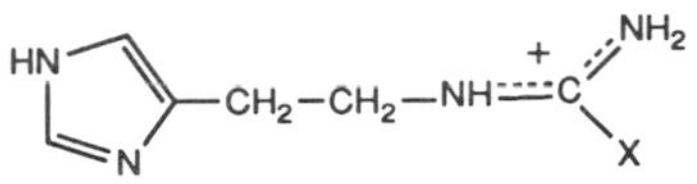

Fig. 13.16 Analogue, where X is a methylthio group or methyl group.

Two other analogues were synthesized, where one of the terminal amino groups in the guanidine group was replaced with either a methylthio group or a methyl group (Fig. 13.16). Both the resulting structures was partial agonists, but with poorer antagonist activity.

From these results, it was concluded that both terminal amino groups were required for binding to the antagonist binding site. It was proposed that the charged guanidine group was interacting with a charged carboxylate residue on the receptor via two hydrogen bonds (Fig. 13.17). If either of these terminal amino groups were absent, then binding would be weaker, resulting in a lower level of antagonism.

The chain was now extended from a two-carbon unit to a three-carbon unit to see what would happen if the guanidine group was moved further away from the imidazole ring. The antagonist activity increased for the guanidine structure (Fig. 13.18), but

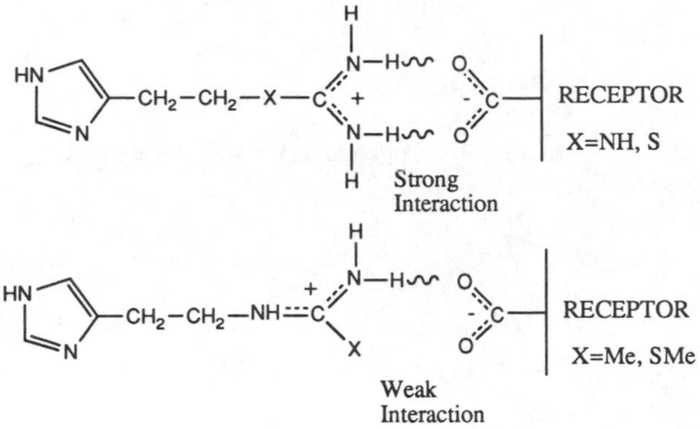

Fig. 13.17 Proposed interaction of the charged guanidine group.

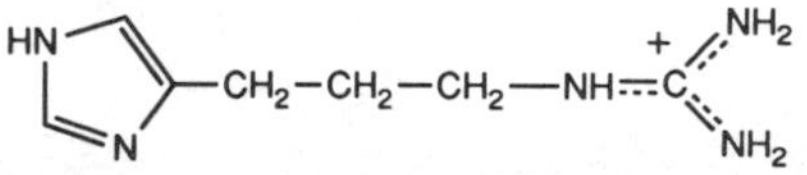

Fig. 13.18 Guanidine structure.

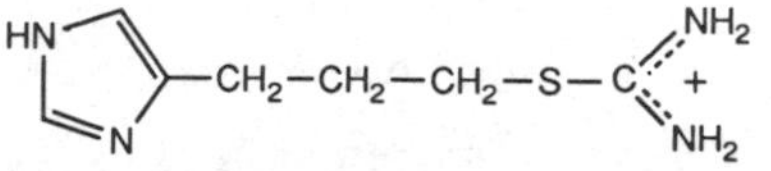

Fig. 13.19 Isothiourea structure.

strangely enough, decreased for the isothiourea structure (Fig. 13.19). It was therefore proposed that with a chain length of two carbon units, hydrogen bonding to the receptor involved the terminal NH_2 groups, but with a chain length of three carbon units, hydrogen bonding involved one terminal NH_2 group along with the NH group within the chain (Fig. 13.20). Support for this theory was provided by the fact that replacing one of the terminal NH_2 groups in the guanidine analogue (Fig. 13.18) with SMe or Me (Fig. 13.21) did not adversely affect the antagonist activity. This was completely different from the results obtained when similar changes were carried out on the two-carbon bridged compound.

These bonding interactions are represented pictorially in Figs. 13.22 and 13.23.

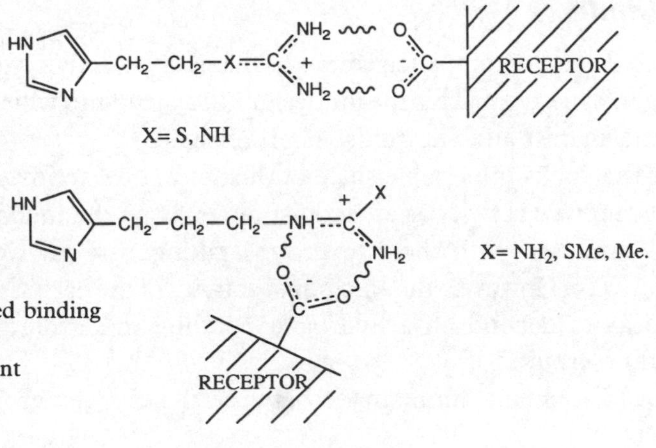

Fig. 13.20 Proposed binding interactions for analogues of different chain length.

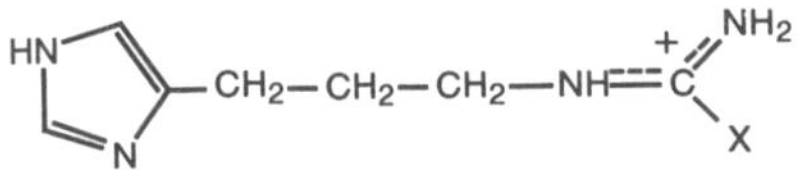

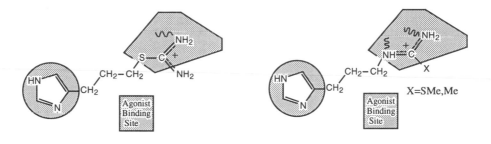

Fig. 13.21 Guanidine analogue with SMe or Me.

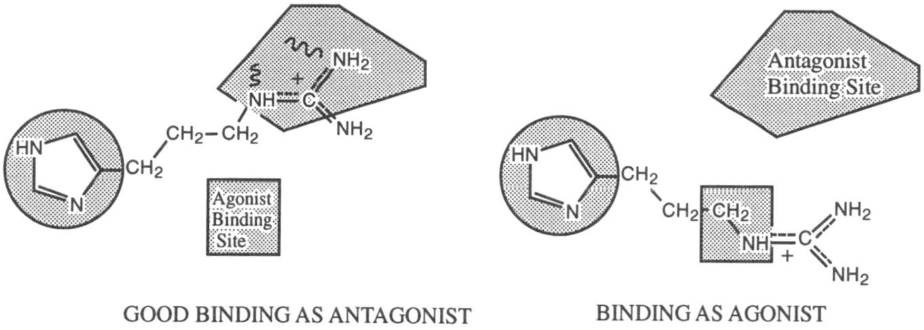

GOOD BINDING AS ANTAGONIST   BINDING AS AGONIST

Fig. 13.22 Binding interactions for 3C bridged analogue.

POOR BINDING AS ANTAGONIST   GOOD BINDING AS ANTAGONIST

Fig. 13.23 Effect of varying the guanidine group on binding to the antagonist site.

13.8 *From partial agonist to antagonist—the development of burimamide*

The problem now was to completely remove the agonist activity to get compounds with pure antagonist activity. This meant designing a structure which would differentiate between the agonist and antagonist binding sites.

At first sight this looks impossible since both sites appear to involve the same type of bonding. Histamine's activity as an agonist depends on the imidazole ring and the charged amino function, with the two groups taking part in hydrogen and ionic bonding, respectively. However, the antagonist activity of the partial agonists described so far also appear to depend on a hydrogen bonding imidazole ring and an ionic bonding guanidine group.

Fortunately, a distinction can be made between the charged groups.

The structures which show antagonist activity are all capable of forming a chelated bonding structure as previously shown in Fig. 13.20. This interaction involves two hydrogen bonds between two charged species, but is it really necessary for the chelating group to be charged? Could a neutral group also chelate to the antagonist site by hydrogen bonding alone? If so, it might be possible to distinguish between the agonist and antagonist sites, especially since ionic bonding appears necessary for the agonist site.

It was therefore decided to see what would happen if the strongly basic guanidine group was replaced with a neutral group capable of interacting with the receptor by two hydrogen bonds. There are a large variety of such groups, but the SK&F workers limited the options by adhering to a principle which they followed throughout their research programme. Whenever they wished to alter any specific physical or chemical property, they strove to ensure that other properties were changed as little as possible. Only in this way could they rationalize any observed improvement in activity.

Thus, in order to study the effect of replacing the basic guanidine group with a neutral group, it was necessary to ensure that the new group was as similar as possible to guanidine in terms of size, shape, and hydrophobicity.

Several functional groups were tried, but success was ultimately achieved by using a thiourea group. The thiourea derivative SK&F 91581 (Fig. 13.24) proved to be a weak antagonist with no agonist activity.

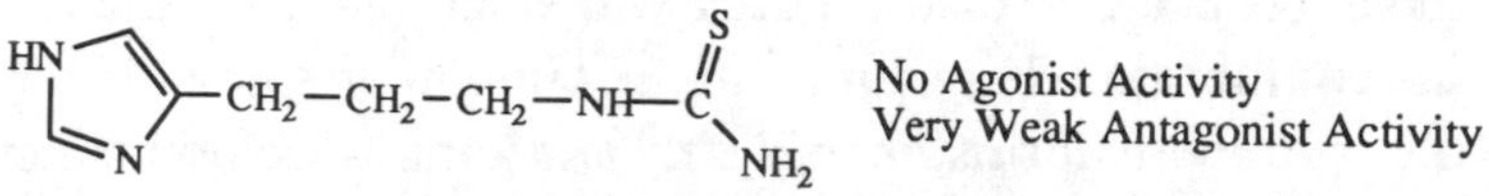

No Agonist Activity
Very Weak Antagonist Activity

Fig. 13.24 SK&F 91581.

Apart from basicity, the properties of the thiourea group are very similar to the guanidine group. Both groups are planar, similar in size, and can take part in hydrogen bonding. Thus, the alteration in biological activity can reasonably be attributed to the differences in basicity between the two groups.

Unlike guanidine, the thiourea group is neutral. This is due to the $C{=}S$ group which has an electron withdrawing effect on the neighbouring nitrogens, making them non-basic and more like amide nitrogens.

The fact that a neutral group could bind to the antagonist site and not to the agonist site could be taken to imply that the agonist binding site involves ionic bonding, whereas the antagonist site involves hydrogen bonding.

Further chain extension and the addition of an N-methyl group led to burimamide (Fig. 13.25) which was found to have enhanced activity.

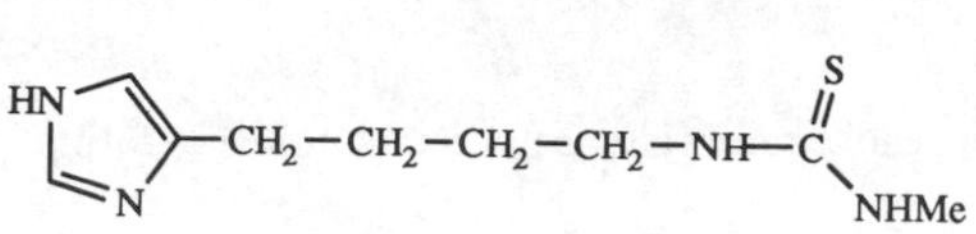

Fig. 13.25 Burimamide.

These results suggest that chain extension has moved the thiourea group closer to the antagonist binding site, and that the addition of the *N*-methyl group has resulted in a beneficial increase in hydrophobicity. A possible explanation for this latter result will be described in Section 13.12.2.

Burimamide is a highly specific competitive antagonist of histamine at H2 receptors, and is 100 times more potent than N^α-guanylhistamine. Its discovery finally proved the existence of the H2 receptors.

13.9 *Development of metiamide*

Despite this success, burimamide was not suitable for clinical trials since its antagonist activity was still too low for oral administration. Further developments were needed. Attention was now directed to the imidazole ring of burimamide and, in particular, to the various possible tautomeric forms of this ring. It was argued that if one particular tautomer was preferred for binding with the H2 receptor, then activity might be enhanced by modifying the burimamide structure to favour that tautomer.

At pH 7.4, it is possible for the imidazole ring to equilibrate between the two tautomeric forms (I) and (II) via the protonated intermediate (III) (Fig. 13.26). The necessary proton for this process has to be supplied by water or by an exchangeable proton on a suitable amino acid residue in the binding site. If the exchange is slow, then it is possible that the drug will enter and leave the receptor at a faster rate than the equilibration between the three tautomeric forms. If bonding involves only one of the tautomeric forms, then clearly antagonism would be increased if the structure was varied to prefer that tautomeric form over the others. Our model hypothesis for receptor binding shows that the imidazole ring is important for the binding of both agonists and antagonists. Therefore, it is reasonable to assume that the preferred imidazole tautomer is the same for both agonists and antagonists. If this is so, then the preferred tautomer for a strong agonist such as histamine should also be the preferred tautomer for a strong antagonist.

Figure 13.26 shows that the imidazole ring can exist as one ionized tautomer and two unionized tautomers. Let us first consider whether the preferred tautomer is likely to be ionized or not.

We have already seen that the pK_a for the imidazole ring in histamine is 5.74,

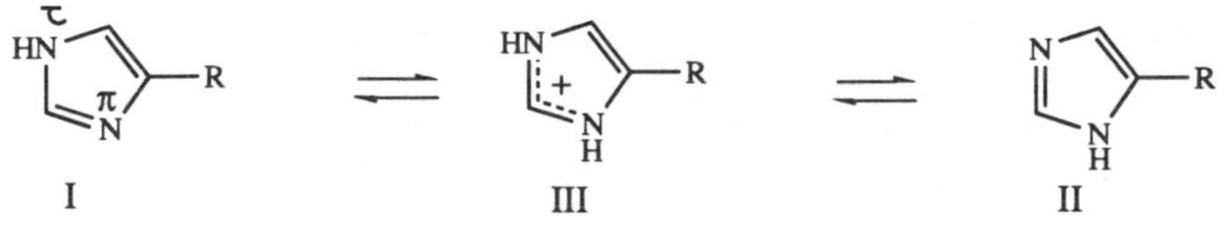

Fig. 13.26 Imidazole ring can equilibrate between tautomeric forms (I and II) via the protonated intermediate (III).

meaning that the ring is a weak base and mostly un-ionized. The pK_a value for imidazole itself is 6.80 and for burimamide 7.25. These values show that these imidazole rings are more basic than histamine and more likely to be ionized. Why should this be so?

The explanation must be that the side-chain has an electronic effect on the imidazole ring. If the side-chain is electron withdrawing or electron donating, then it will affect the basicity of the ring. A measure of the side-chain's electronic effect can be worked out by the Hammett equation (see Chapter 9):

$$pK_{a(R)} = pK_{a(H)} + \rho\sigma_R$$

where $pK_{a(R)}$ is the pK_a of the imidazole ring bearing a side-chain R, $pK_{a(H)}$ is the pK_a of the unsubstituted imidazole ring, ρ is a constant, and σ_R is the Hammett substituent constant for the side-chain R.

From the pK_a values, the value of the Hammett substituent constant can be calculated to show whether the side-chain R is electron withdrawing or electron donating.

In burimamide, the side-chain is calculated to be slightly electron donating (of the same order as a methyl group). Therefore, the imidazole ring in burimamide is more likely to be ionized than in histamine, where the side-chain is electron withdrawing. At pH 7.4, 40 per cent of burimamide is ionized in the imidazole ring compared to approximately 3 per cent of histamine. This represents quite a difference between the two structures and since the binding of the imidazole ring is important for antagonist activity as well as agonist activity, it suggests that a pK_a value closer to that of histamine might lead to better binding and to better antagonist activity.

It was necessary, therefore, to make the side-chain electron withdrawing rather than electron donating. This can be done by inserting an electronegative aom into the side-chain—preferably one which has a minimum disturbance on the rest of the molecule. In other words, an isostere for a methylene group is required—one which has an electronic effect, but which has approximately the same size and properties as the methylene group.

The first isostere to be tried was a sulfur atom. Sulfur is quite a good isostere for the methylene unit in that both groups have similar van der Waals radii and similar bond angles. However, the C–S bond length is slightly longer than a C–C bond, leading to a slight extension (15 per cent) of the structure.

The methylene group replaced was next but one to the imidazole ring. This site was chosen, not for any strategic reasons, but because a synthetic route was readily available to carry out that particular transformation.

As hoped, the resulting compound, thiaburimamide (Fig. 13.27), had a significantly lower pK_a of 6.25 and was found to have enhanced antagonistic activity. This result supported the theory that a reduction in the proportion of ionized tautomer was beneficial to receptor binding and activity.

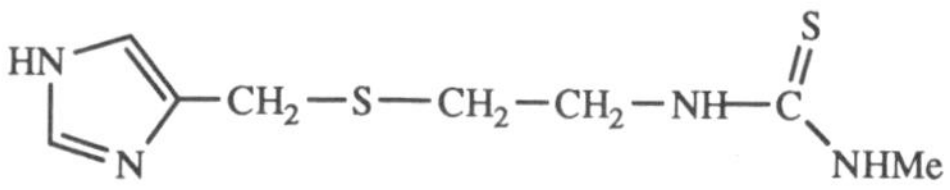

Fig. 13.27 Thiaburimamide.

Thiaburimamide had been synthesized in order to favour the un-ionized imidazole ring over the ionized ring. However, as we have seen, there are two possible un-ionized tautomers. The next question is whether either of these are preferred for receptor binding.

Let us return to histamine. If one of the un-ionized tautomers is preferred over the other in histamine, then it would be reasonable to assume that this is the favoured tautomer for receptor binding. The preferred tautomer for histamine is tautomer I (Fig. 13.26).

Why is tautomer I favoured? The answer lies in the fact that the side-chain on histamine is electron withdrawing. This electron withdrawing effect on the imidazole ring is inductive and therefore the strength of the effect will decrease with distance round the ring. This implies that the nitrogen atom on the imidazole ring closest to the side-chain ($N\pi$) will experience a greater electron withdrawing effect than the one further away ($N\tau$). As a result, the closer nitrogen is less basic, which in turn means that it is less likely to bond to hydrogen (tautomer I).

Since the side-chain in thiaburimamide is electron withdrawing, then it too will favour tautomer I.

It was now argued that this tautomer could be further enhanced if an electron **donating** group was placed at position 4 in the ring. At this position, the inductive effect would be felt most at the neighbouring nitrogen (N^τ), further enhancing its basic character and increasing the population of tautomer I. However, it was important to choose a group which would not interfere with the normal receptor binding interaction. For example, a large substituent might be too bulky and prevent the analogue fitting the receptor. A methyl group was chosen since it was known that 4-methylhistamine was an agonist and also highly selective for the H2 receptor (see Section 13.6.).

The compound obtained was metiamide (Fig. 13.28) which was found to have enhanced activity as an antagonist, supporting the previous theory.

It is interesting to note that the above effect outweighs an undesirable rise in pK_a. By adding an electron donating methyl group, there has been a rise in the pK_a of the imidazole ring to 6.80 compared to 6.25 for thiaburimamide. (Coincidentally, this is

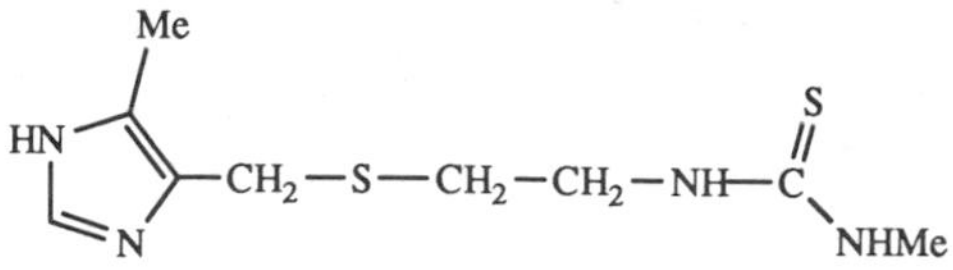

Fig. 13.28 Metiamide.

the same pK_a as for imidazole itself, which shows that the electronic effects of the methyl group and the side-chain are cancelling each other out as far as pK_a is concerned.) A pK_a of 6.80 means that 20 per cent of metiamide is ionized in the imidazole ring. However, this is still significantly lower than the corresponding 40 per cent for burimamide.

Compared to burimamide, the percentage of ionized imidazole ring has been lowered in metiamide and the ratio of the two possible un-ionized imidazole tautomers reversed. The fact that activity is increased with respect to **thiaburimamide** suggests that the increase in the population of tautomer (I) outweighs the increase in population of the ionized tautomer (III).

4-Methylburimamide (Fig. 13.29) was also synthesized for comparison. Here, the introduction of the 4-methyl group does not lead to an increase in activity. The pK_a is increased to 7.80, resulting in the population of ionized imidazole ring rising to 72 per cent. This demonstrates how important it is to rationalize structural changes. Adding the 4-methyl group to thiaburimamide is advantageous, but adding it to burimamide is not.

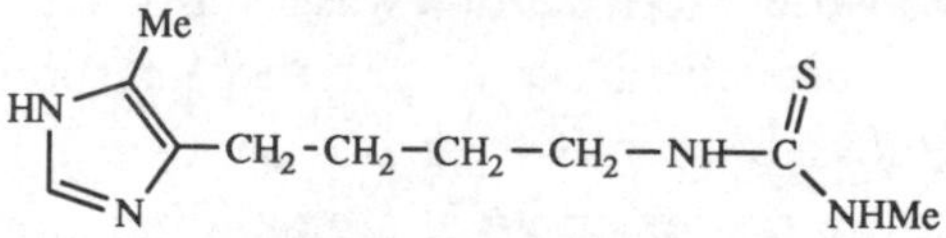

Fig. 13.29 4-Methylburimamide.

The design and synthesis of metiamide followed a rational approach aimed at favouring one specific tautomer. Such a study is known as a dynamic structure–activity analysis.

Strangely enough, it has since transpired that the improvement in antagonism may have resulted from conformational effects. X-ray crystallography studies have indicated that the longer thioether linkage in the chain increases the flexibility of the side-chain and that the 4-methyl substituent in the imidazole ring may help to orientate the imidazole ring correctly for receptor binding. It is significant that the oxygen analogue oxaburimamide (Fig. 13.30) is less potent than burimamide despite the fact that the electron withdrawing effect of the oxygen-containing chain on the ring is similar to the sulfur-containing chain. The bond lengths and angles of the ether link are similar to the methylene unit and in this respect it is a better isostere than sulfur. However, the oxygen atom is substantially smaller. It is also significantly more basic and more hydrophilic than either sulfur or methylene. Oxaburimamide's lower activity might be due to a variety of reasons. For example, the oxygen may not allow the same flexibility permitted by the sulfur atom. Alternatively, the oxygen may be involved in a hydrogen bonding interaction either with the receptor or with its own imidazole ring, resulting in a change in receptor binding interaction.

Metiamide is ten times more active than burimamide and showed promise as an

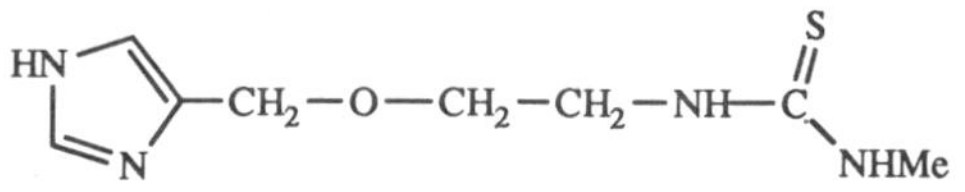

Fig. 13.30 Oxaburimamide.

antiulcer agent. Unfortunately, a number of patients suffered from kidney damage and granulocytopenia—a condition which results in the reduction of circulating white blood cells and which makes patients susceptible to infection. Further developments were now required to find an improved drug lacking these side-effects.

13.10 *Development of cimetidine*

It was proposed that metiamide's side-effects were associated with the thiourea group—a group which is not particularly common in the body's biochemistry. There-fore, consideration was given to replacing this group with a group which was similar in property but would be more acceptable in a biochemical context. The urea analogue (Fig. 13.31) was tried, but found to be less active. The guanidine analogue (Fig. 13.32) was also less active, but it was interesting to note that this compound had no agonist activity. This contrasts with the three-carbon bridged guanidine (Fig. 13.18) which we have already seen is a partial agonist. Therefore, the guanidine analogue (Fig 13.32) was the first example of a guanidine having pure antagonist activity.

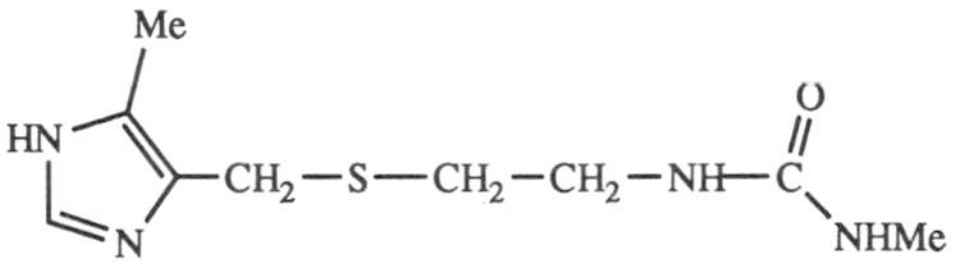

Fig. 13.31 Urea analogue.

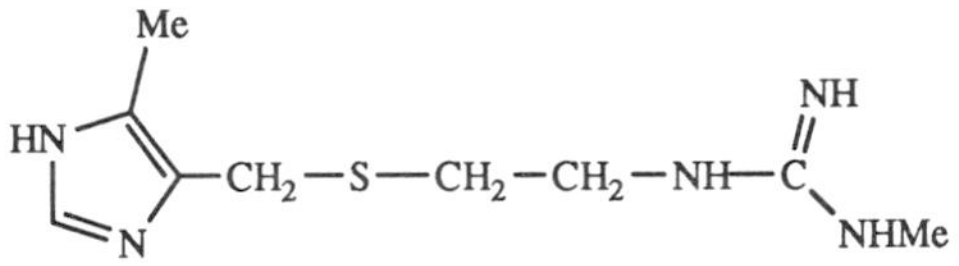

Fig. 13.32 Guanidine analogue.

One possible explanation for this is that the longer four-unit chain extends the guanidine binding group beyond the reach of the agonist binding site (Fig. 13.33), whereas the shorter three-unit chain still allows binding to both agonist and antagonist site (Fig. 13.34).

The antagonist activity for the guanidine analogue (Fig. 13.32) is weak, but it was decided to look more closely at this compound since it was thought that the guanidine unit would be less likely to have toxic side-effects than the thiourea. This is a reasonable assumption since the guanidine unit is present naturally in the amino acid arginine (Fig. 13.35). The problem now was to retain the guanidine unit, but to increase activity. It seemed likely that the low activity was due to the fact that the basic guanidine group would be ionized at pH 7.4. The problem was how to make this group neutral—no easy task, considering that guanidine is one of the strongest bases in organic chemistry.

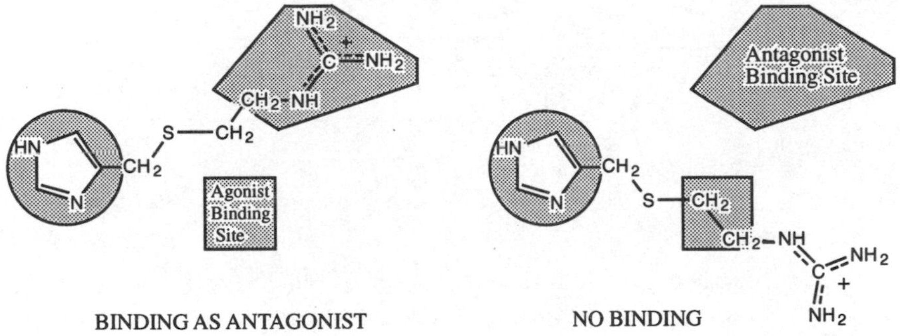

Fig. 13.33 Four-carbon unit chain.

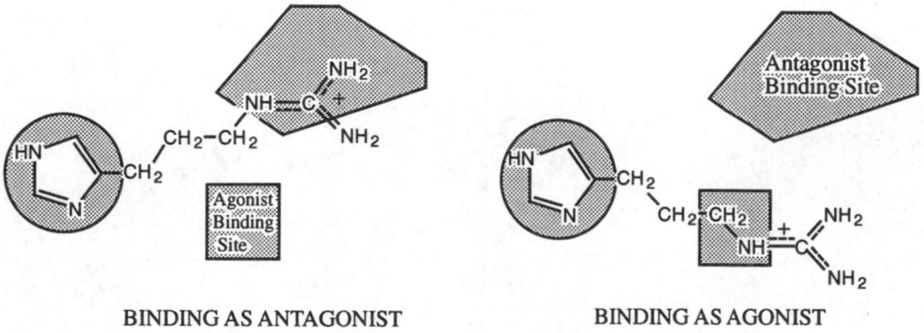

Fig. 13.34 Three-carbon unit chain.

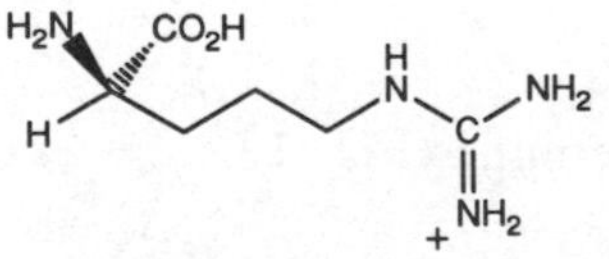

Fig. 13.35 Arginine.

Nevertheless, a search of the literature revealed a useful study on the ionization of monosubstituted guanidines (Fig. 13.36). A comparison of the pK_a values of these compounds with the inductive substituent constants σ_I for the substituents X gave a straight line as shown in Fig. 13.37, showing that pK_a is inversely proportional to the electron withdrawing power of the substituent. Thus, strongly electron withdrawing substituents make the guanidine group less basic and less ionized. The nitro and cyano groups are particularly strong electron withdrawing groups. The ionization constants for cyanoguanidine and nitroguanidine are 0.4 and 0.9 respectively (Fig. 13.37)—similar values to the ionization constant for thiourea itself (-1.2).

Both the nitroguanidine and cyanoguanidine analogues of metiamide were synthesized and found to have comparable antagonist activities to metiamide. The

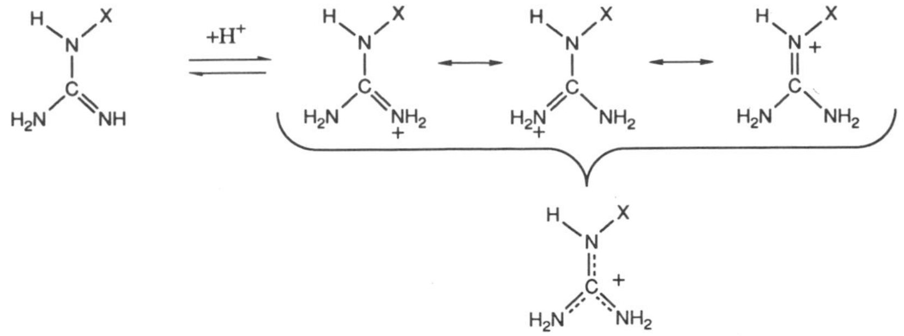

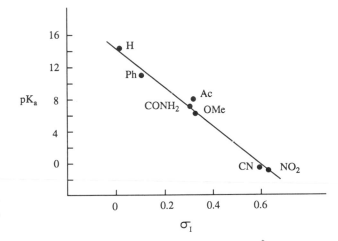

Fig. 13.36 Ionization of monosubstituted guanidines.

Fig. 13.37 PK_a vs. inductive substituent constants (σ_I) for the substituents X in Fig. 13.36.

cyanoguanidine analogue (cimetidine (Fig. 13.1)) was the more potent analogue and was chosen for clinical studies.

13.11 *Cimetidine*

13.11.1 Biological activity of cimetidine

Cimetidine inhibits H2 receptors and thus inhibits gastric acid release. The drug does not show the toxic side-effects observed for metiamide and has been shown to be slightly more active. It has also been found to inhibit pentagastrin (Fig. 13.38) from stimulating release of gastric acid. Pentagastrin is an analogue of gastrin and the fact that cimetidine blocks its stimulatory activity suggests some relationship between histamine and gastrin in the release of gastric acid.

Cimetidine was first marketed in the UK in 1976 and for several years was the

Fig. 13.38 Pentagastrin.   N-t-BOC-β-Ala-Trp-Met-Asp-Phe-NH₂

world's biggest selling prescription product until it was pushed into second place in 1988 by ranitidine (Section 13.13.).

13.11.2 Structure and activity of cimetidine

The finding that metiamide and cimetidine are both good H2 antagonists of similar activity shows that the cyanoguanidine group is a good bioisostere for the thiourea group.

This is despite the fact that three tautomeric forms (Fig. 13.39) are possible for the guanidine group compared to only one for the isothiourea group. In fact, this is more apparent than real, since the imino tautomer (II) is the preferred tautomeric form for the guanidine unit. Tautomer II is favoured since the cyano group has a stronger electron withdrawing effect on the neighbouring nitrogen compared to the two nitrogens further away. This makes the neighbouring nitrogen less basic and therefore less likely to be protonated.

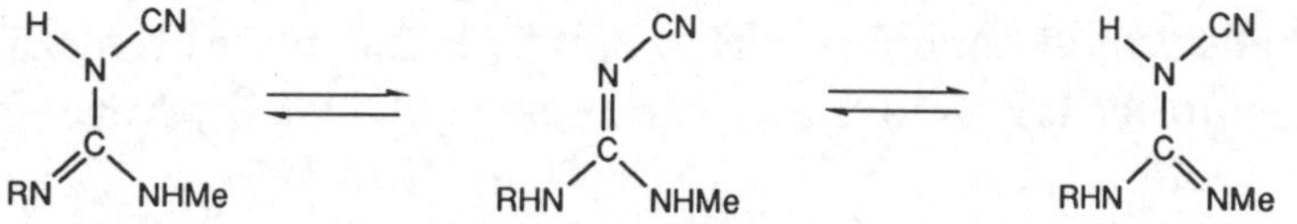

I Amino Tautomer — II Imino Tautomer — III Amino Tautomer

Fig. 13.39 Three tautomeric forms of guanidine unit.

Since tautomer II is favoured, the guanidine group does in fact bear a close structural similarity to the thiourea group. Both groups have a planar pi electron system with similar geometries (equal C–N distances and angles). They are polar and hydrophilic with high dipole moments and low partition coefficients. They are weakly basic and also weakly acidic such that they are un-ionized at pH 7.4.

13.11.3 Metabolism of cimetidine

It is important to study the metabolism of any new drug in case the metabolites have biological activity in their own right. Any such activity might lead to side-effects. Alternatively, a metabolite might have enhanced activity of the type desired and give clues to further development.

Cimetidine itself is metabolically stable and is excreted largely unchanged. The only metabolites which have been identified are due to oxidation of the sulfur link (Fig. 13.40) or oxidation of the ring methyl group (Fig. 13.41).

It has been found that cimetidine inhibits the P-450 cytochrome oxidase system in the liver. This is an important enzyme system in the metabolism of drugs and care must be taken if other drugs are being taken at the same time as cimetidine, since

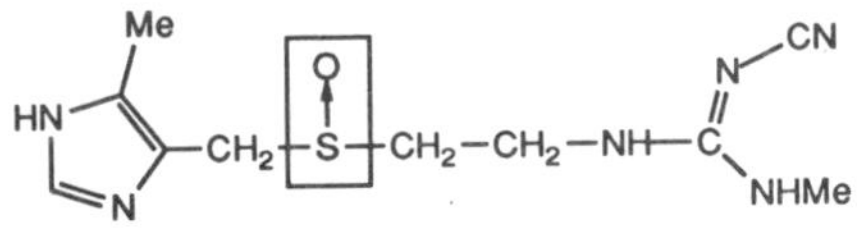

Fig. 13.40 Oxidation of the sulfur link in cimetidine.

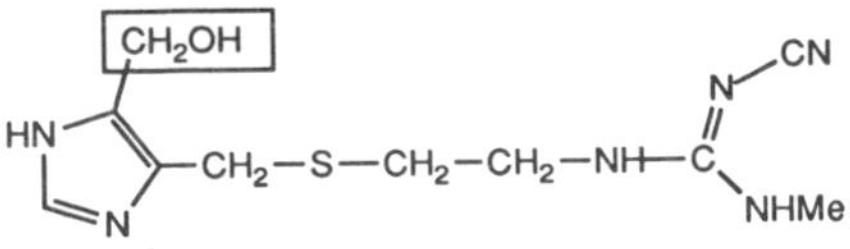

Fig. 13.41 Oxidation of the ring methyl group in cimetidine.

cimetidine may inhibit the metabolism of these drugs, leading to higher blood levels and toxic side-effects. In particular, care must be taken when cimetidine is taken with drugs such as diazepam, lidocaine, warfarin, or theophylline.

13.12 *Further studies—cimetidine analogues*

13.12.1 Conformational isomers

A study of the various stable conformations of the guanidine group in cimetidine led to a rethink of the type of bonding which might be taking place at the antagonist site. Up until this point, the favoured theory had been a bidentate hydrogen interaction as shown in Fig. 13.17.

In order to achieve this kind of bonding, the guanidine group in cimetidine would have to adopt the *Z,Z* conformation shown in Fig. 13.42.

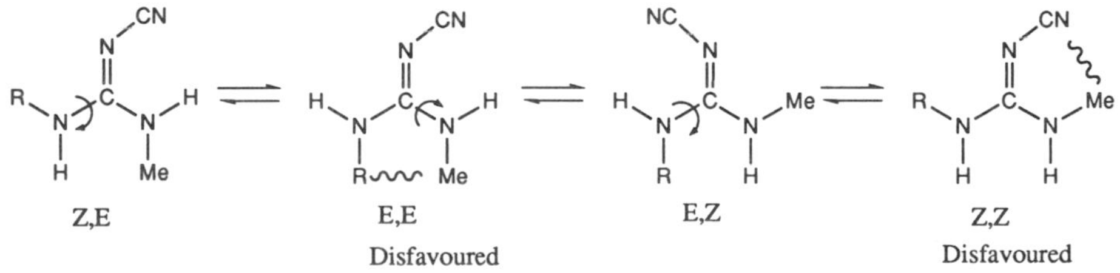

Fig. 13.42 Conformations of the guanidine group in cimetidine.

However, X-ray and NMR studies have shown that cimetidine exists as an equilibrium mixture of the *E,Z* and *Z,E* conformations. The *Z,Z* form is not favoured since the cyano group is forced too close to the *N*-methyl group. If either the *E,Z* or *Z,E* form is the active conformation, then it implies that the chelation type of hydrogen bonding described previously is not essential. An alternative possibility is that the guanidine unit is hydrogen bonding to two distinct hydrogen bonding sites rather than to a single carboxylate group (Fig. 13.43).

Further support for this theory is provided by the weak activity observed for the urea analogue (Fig. 13.32). This compound is known to prefer the *Z,Z* conformation over the *Z,E* or *E,Z* and would therefore be unable to bind to both hydrogen bonding sites.

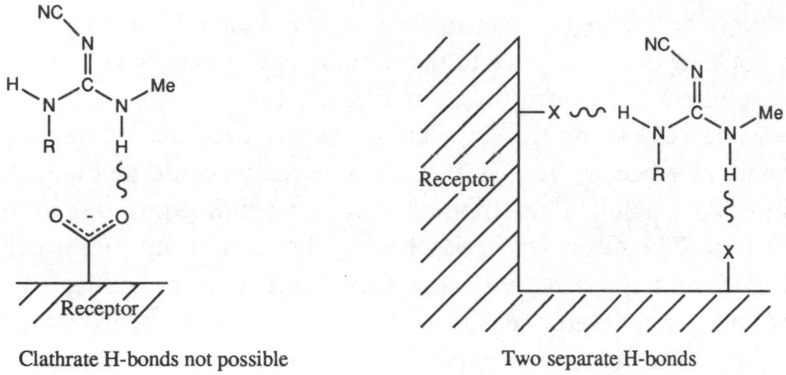

Clathrate H-bonds not possible   Two separate H-bonds

Fig. 13.43 Alternative theory for cimetidine bonding at the agonist site.

If this bonding theory is correct and the active conformation is the E,Z or Z,E form, then restricting the group to adopt one or other of these forms may lead to more active compounds and an identification of the active conformation. This can be achieved by incorporating part of the guanidine unit within a ring—a strategy of rigidification.

For example, the nitropyrrole derivative (Fig. 13.44) has been shown to be the strongest antagonist in the cimetidine series, implying that the E,Z conformation is the active conformation.

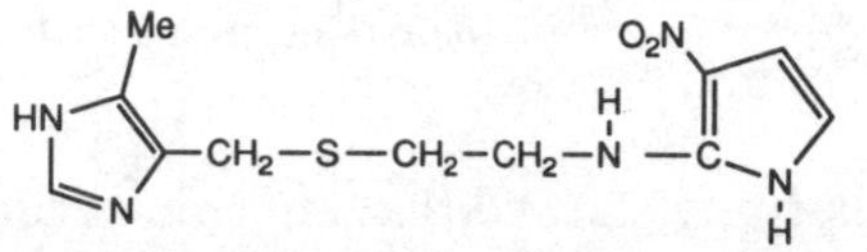

Fig. 13.44 Nitropyrrole derivative of cimetidine.

The isocytosine ring (Fig. 13.45) has also been used to 'lock' the guanidine group, limiting the number of conformations available. The ring allows further substitution and development as seen below (Section 13.12.2.).

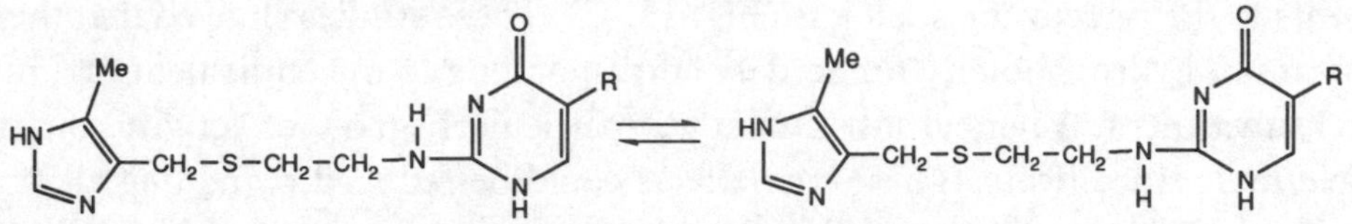

Fig. 13.45 Isocytosine ring.

13.12.2 Desolvation

Development of oxmetidine

It has already been stated that the guanidine and thiourea groups, used so successfully in the development of H2 antagonists, are polar and hydrophilic. This implies that they are likely to be highly solvated (i.e. surrounded by a 'water coat'). Before hydrogen bonding can take place to the receptor, this 'water coat' has to be removed. The more solvated the group, the more difficult that will be.

One possible reason for the low activity of the urea derivative (Fig. 13.32) has already been described above. Another possible reason could be the fact that the urea group is more hydrophilic than thiourea or cyanoguanidine groups and therefore more highly solvated. The difficulty in desolvating the urea group might explain why the urea analogue has a lower activity than cimetidine, despite having a lower partition coefficient and greater water solubility.

Leading on from this, if the ease of desolvation is a factor in antagonist activity, then reducing the solvation of the polar group should increase activity. One way of achieving this would be to increase the hydrophobic character of the polar binding group.

A study was carried out on a range of cimetidine analogues containing different planar aminal systems (Z) (Fig. 13.46) to see whether there was any relationship between antagonist activity and the hydrophobic character of the aminal system (HZ).

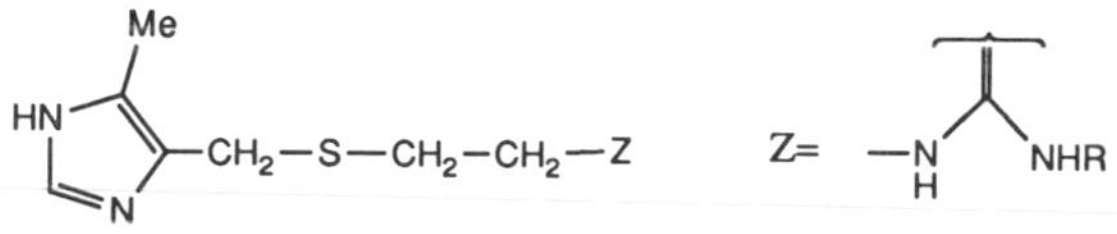

Fig. 13.46 Cimetidine analogue with planar animal system (Z).

This study showed that antagonist activity was proportional to the hydrophobicity of the aminal unit Z (Fig. 13.47) and supported the desolvation theory. The relationship could be quantified as follows:

$$\log(\text{activity}) = 2.0 \log P + 7.4$$

Further studies on hydrophobicity were carried out by adding hydrophobic substituents to the isocytosine analogue (Fig. 13.45). These studies showed that there was an optimum hydrophobicity for activity corresponding to the equivalent of a butyl or pentyl substituent. A benzyl substituent was particularly good for activity, but proved to have toxic side-effects. These side-effects could be reduced by adding alkoxy substituents to the aromatic ring and this led to the synthesis of oxmetidine (Fig. 13.48) which had enhanced activity over cimetidine. Oxmetidine was considered for clinical use, but was eventually withdrawn since it still retained undesirable side-effects.

The development of the nitroketeneaminal binding group

As we have seen, antagonist activity increases with the hydrophobicity of the polar binding group. It was therefore decided to see what would happen if the polar imino

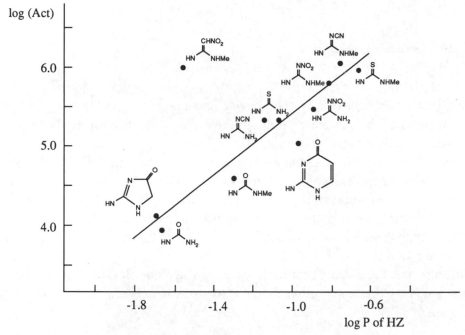

Fig. 13.47 Antagonist activity is proportional to the hydrophobicity of the animal unit Z.

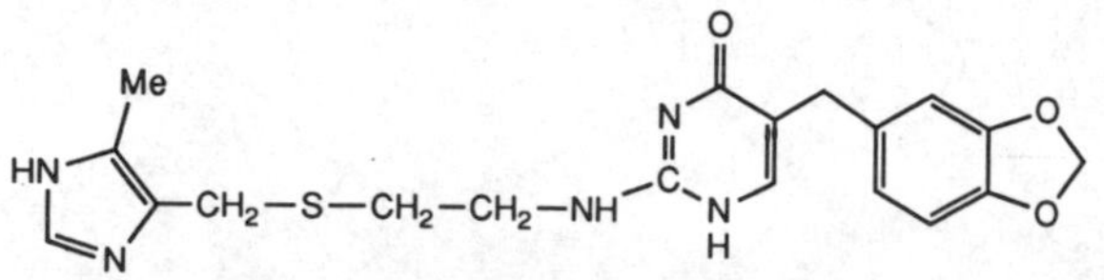

Fig. 13.48 Oxmetidine.

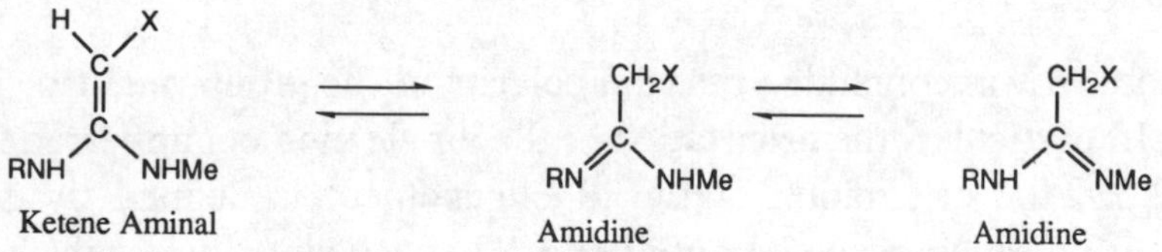

Ketene Aminal   Amidine   Amidine

Fig. 13.49

nitrogen of cimetidine was replaced with a non-polar carbon atom. This would result in a keteneaminal group as shown in Fig. 13.49. Unfortunately, keteneaminals are more likely to exist as their amidine tautomers unless a strongly electronegative group (e.g. NO_2) is attached to the carbon atom.

Therefore, a nitroketeneaminal group was used to give the structure shown in Fig. 13.50. Surprisingly, there was no great improvement in activity, but when the structure was studied in detail, it was discovered that it was far more hydrophilic then

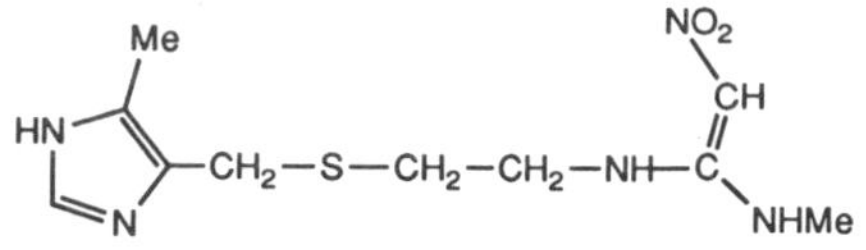

Fig. 13.50 Cimetidine analogue with a nitroketeneaminal group.

expected. This explained why the activity had not increased, but it highlighted a different puzzle. The compound was *too* active. Based on its hydrophilicity, it should have been a weak antagonist (Fig. 13.47).

It was clear that this compound did not fit the pattern followed by previous compounds since the antagonist activity was 30 times higher than would have been predicted by the equation above. Nor was the nitroketeneaminal the only analogue to deviate from the normal pattern. The imidazolinone analogue (Fig. 13.51), which is relatively hydrophobic, had a much lower activity than would have been predicted from the equation.

Findings like these are particularly exciting since any deviation from the normal pattern suggests that some other factor is at work which may give a clue to future development.

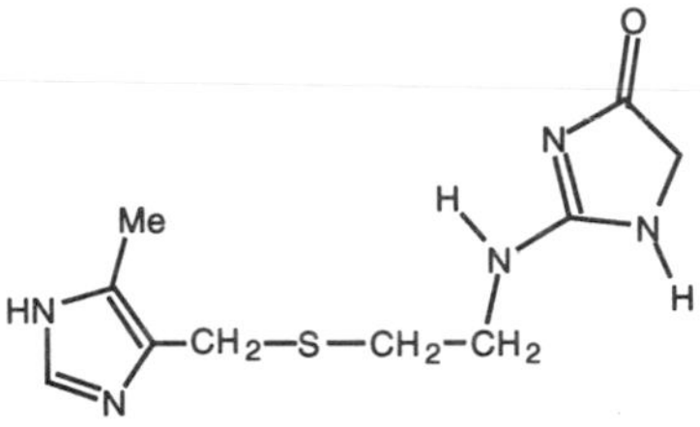

Fig. 13.51 Imidazolinone analogue.

Fig. 13.52 Orientation of dipole moment.

In this case, it was concluded that the polarity of the group might be important in some way. In particular, the **orientation** of the dipole moment appeared to be crucial. In Fig. 13.52, the orientation of the dipole moment is defined by ϕ—the angle between the dipole moment and the NR bond. The cyanoguanidine, nitroketeneaminal, and nitropyrrole groups all have high antagonist activity and have dipole moment orientations of 13, 33, and 27° respectively (Fig. 13.53). The isocytosine and imidazolinone groups result in lower activity and have dipole orientations of 2 and −6°, respectively. The strength of the dipole moment (μ) does not appear to be crucial.

Why should the orientation of a dipole moment be important? One possible explanation is as follows. As the drug approaches the receptor, its dipole interacts with a dipole on the receptor surface such that the dipole moments are aligned. This orientates the drug in a specific way before hydrogen bonding takes place and will determine how strong the subsequent hydrogen bonding will be (Fig. 13.54). If the

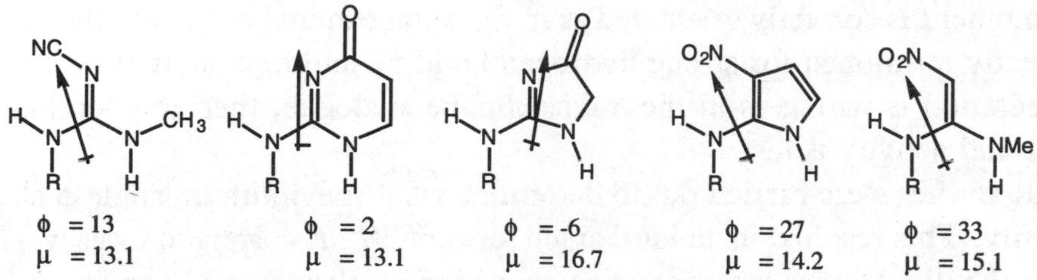

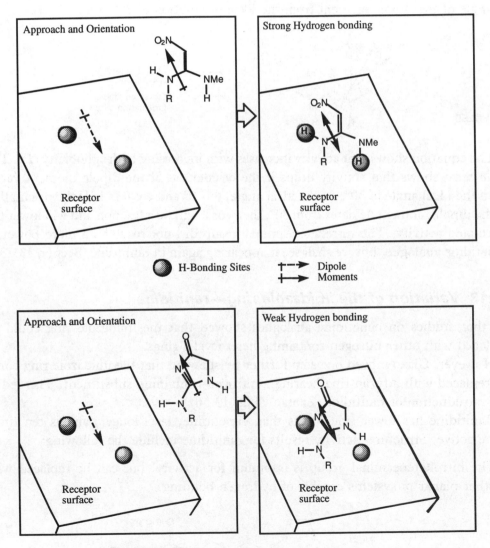

Fig. 13.53 Dipole moments of various antagonistic groups.

$\phi = 13$
$\mu = 13.1$

$\phi = 2$
$\mu = 13.1$

$\phi = -6$
$\mu = 16.7$

$\phi = 27$
$\mu = 14.2$

$\phi = 33$
$\mu = 15.1$

Approach and Orientation

Strong Hydrogen bonding

Receptor surface

Receptor surface

H-Bonding Sites

Dipole Moments

Approach and Orientation

Weak Hydrogen bonding

Receptor surface

Receptor surface

Fig. 13.54 Orientation effects on receptor bonding.

dipole moment is correctly orientated as in the keteneaminal analogue, the group will be correctly positioned for strong hydrogen bonding and high activity will result. If the orientation is wrong as in the imidazolinone analogue, then the bonding is less efficient and activity is lost.

QSAR studies were carried out to determine what the optimum angle ϕ should be for activity. This resulted in an ideal angle for ϕ of 30°. A correlation was worked out between the dipole moment orientation, partition coefficient, and activity as follows:

$$\log A = 9.12 \cos \theta + 0.6 \log P - 2.71$$

where A is the antagonist activity, P is the partition coefficient, and θ is the deviation in angle of the dipole moment from the ideal orientation of 30° (Fig. 13.55).

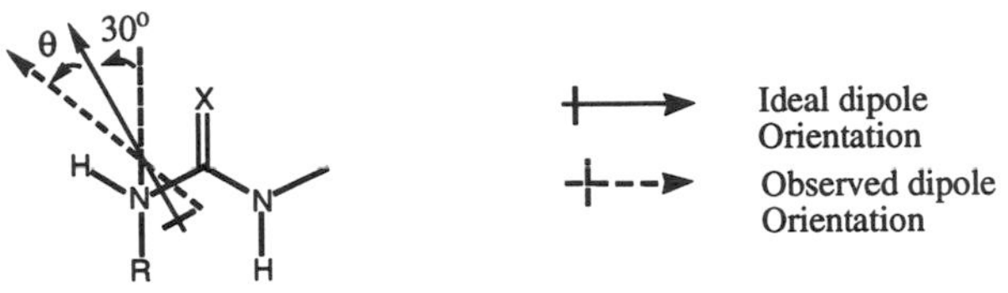

Fig. 13.55

The equation shows that activity increases with increasing hydrophobicity (P). The $\cos \theta$ term shows that activity drops if the orientation of the dipole moment varies from the ideal angle of 30°. At the ideal angle, θ is 0° and $\cos \theta$ is 1. If the orientation of the dipole moment deviates from 30°, then $\cos \theta$ will be a fraction and will lower the calculated activity. The nitroketeneaminal group did not result in a more powerful cimetidine analogue, but we shall see it appearing again in ranitidine (Section 13.13.).

13.13 *Variation of the imidazole ring—ranitidine*

Further studies on cimetidine analogues showed that the imidazole ring could be replaced with other nitrogen-containing heterocyclic rings.

However, Glaxo moved one step further by showing that the imidazole ring could be replaced with a furan ring bearing a nitrogen-containing substituent. This led to the introduction of ranitidine (Zantac) (Fig. 13.56).

Ranitidine has fewer side-effects than cimetidine, lasts longer, and is ten times more active. Structure–activity results for ranitidine include the following:

• The nitroketeneaminal group is optimum for activity, but can be replaced with other planar pi systems capable of hydrogen bonding.

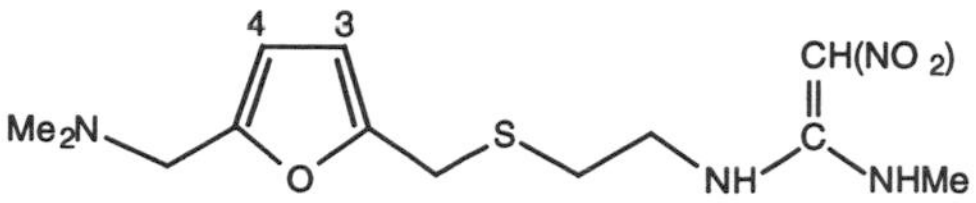

Fig. 13.56 Ranitidine.

- Replacing the sulfur atom with a methylene unit leads to a drop in activity.
- Placing the sulfur next to the ring lowers activity.
- Replacing the furan ring with more hydrophobic rings such as phenyl or thiophene reduces activity.
- 2,5-Disubstitution is the best substitution pattern for the furan ring.
- Substitution on the dimethylamino group can be varied, showing that the basicity and hydrophobicity of this group are not crucial to activity.
- Methyl substitution at carbon-3 of the furan ring eliminates activity, whereas the equivalent substitution in the imidazole series increases activity.
- Methyl substitution at carbon-4 of the furan ring increases activity.

The latter two results imply that the heterocyclic rings for cimetidine and ranitidine are not interacting in the same way with the H2 receptor. This is supported by the fact that a corresponding dimethylaminomethylene group attached to cimetidine leads to a drop in activity.

Ranitidine was introduced to the market in 1981 and by 1988 was the world's biggest selling prescription drug.

13.14 *Famotidine and nizatidine*

During 1985/87 two new antiulcer drugs were introduced to the market—famotidine and nizatidine.

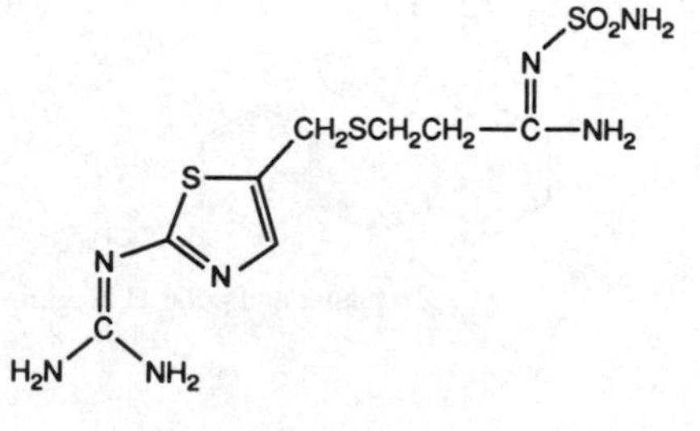

Fig. 13.57 Famotidine. 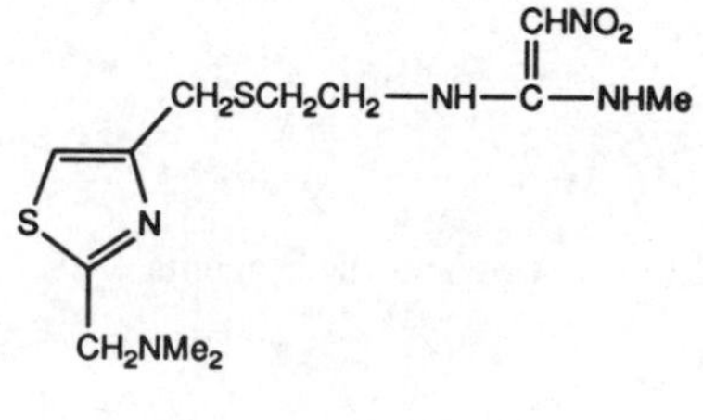　　　**Fig. 13.58** Nizatidine.

Famotidine (Pepcid) (Fig. 13.57) is 30 times more active than cimetidine in vitro. The side-chain contains a sulfonylamidine group while the heterocyclic imidazole ring of cimetidine has been replaced with a 2-guanidinothiazole ring. Structure–activity studies gave the following results:

- The sulfonylamidine binding group is not essential and can be replaced with a variety of structures as long as they are planar, have a dipole moment, and are capable of interacting with the receptor by hydrogen bonding. A low pK_a is not essential, which allows a larger variety of planar groups to be used than is possible for cimetidine.

- Activity is optimum for a chain length of four or five units.
- Replacement of sulfur with a CH_2 group *increases* activity.
- Modification of the chain is possible with, for example, inclusion of an aromatic ring.
- A methyl substituent *ortho* to the chain leads to a drop in activity (unlike the cimetidine series).
- Three of the four hydrogens in the two NH_2 groups are required for activity.

There are several results here which are markedly different from cimetidine, implying that famotidine and cimetidine are not interacting in the same way with the H2 receptor. Further evidence for this is the fact that guanidine substitution at the equivalent position of cimetidine analogues leads to very low activity.

Nizatidine (Fig. 13.58) was introduced into the UK in 1987 by the Lilly Corporation and is equipotent with ranitidine. The furan ring in ranitidine is replaced with a thiazole ring.

13.15 *H2 antagonists with prolonged activity*

There is presently a need for longer lasting antiulcer agents which require once daily doses. Glaxo carried out further development on ranitidine by placing the oxygen of the furan ring exocyclic to a phenyl ring and replacing the dimethylamino group with a piperidine ring to give a series of novel structures (Fig. 13.59).

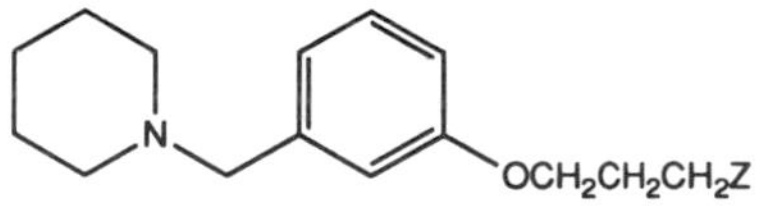

Fig. 13.59 Long lasting anti-ulcer agents.

Z= planar and polar H bonding group

The most promising of these compounds were lamitidine and loxtidine (Fig. 13.60) which were five to ten times more potent than ranitidine and three times longer lasting.

Unfortunately, these compounds showed toxicity in long-term animal studies with

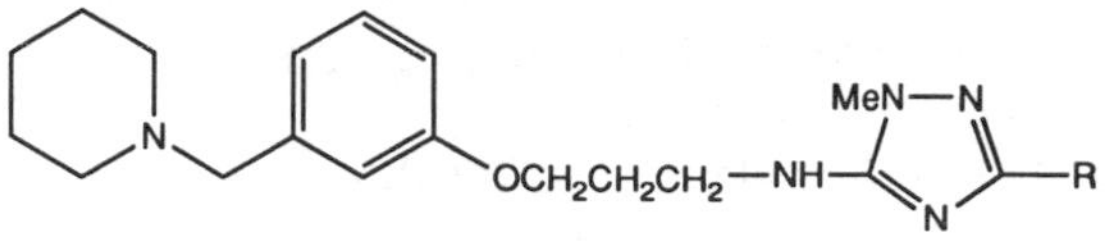

Fig. 13.60 Lamitidine and loxtidine.

R= NH₂   LAMITIDINE
R= CH₂OH LOXTIDINE

the possibility that they caused gastric cancer, and they were subsequently withdrawn from clinical study. However, the relevance of these results has been disputed.

13.16 *Comparison of H1 and H2 antagonists*

The structures of the H2 antagonists are markedly different to the classical H1 antagonists and so there can be little surprise that these original antihistamines failed to antagonize the H2 receptor.

H1 antagonists, like H1 agonists, possess an ionic amino group at the end of a flexible chain. Unlike the agonists, they possess two aryl or heteroaryl rings in place of the imidazole ring. (Fig. 13.61). Because of the aryl rings, H1 antagonists are hydrophobic molecules having high partition coefficients.

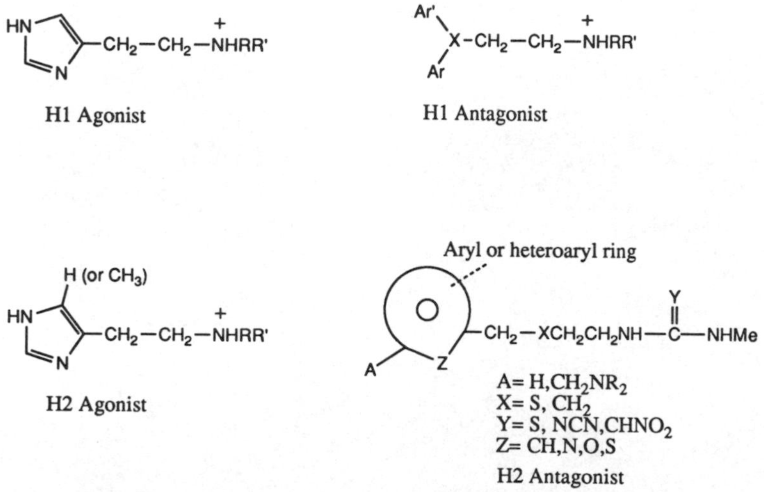

Fig. 13.61 Comparison between H1 and H2 agonists and antagonists.

In contrast, H2 antagonists are polar, hydrophilic molecules having high dipole moments and low partition coefficients. At the end of the flexible chain they have a polar, pi electron system which is weakly amphoteric and un-ionized at pH 7.4. This binding group appears to be the key feature leading to antagonism of H2 receptors (Fig. 13.61). The five-membered heterocycle generally contains a nitrogen atom or, in the case of furan or phenyl, a nitrogen-containing side-chain. The hydrophilic character of H2 antagonists helps to explain why H2 antagonists are less likely to have the CNS side-effects often associated with H1 antagonists.

13.17 *The H2 receptor and H2 antagonists*

H2 receptors are present in a variety of organs and tissues, but their main role is in acid secretion. As a result, H2 antagonists are remarkably safe and mostly free of side-effects. The four most used agents on the market are cimetidine, ranitidine, famotidine, and nizatidine. They inhibit all aspects of gastric secretion and are rapidly absorbed from the gastrointestinal tract with half-lives of 1–2 h. About 80 per cent of ulcers are healed after 4–6 weeks.

Attention must be given to possible drug interactions when using cimetidine due to inhibition of drug metabolism (Section 13.11.). The other three H2 antagonists mentioned do not inhibit the P-450 cytochrome oxidase system and are less prone to such interactions.

IONISED

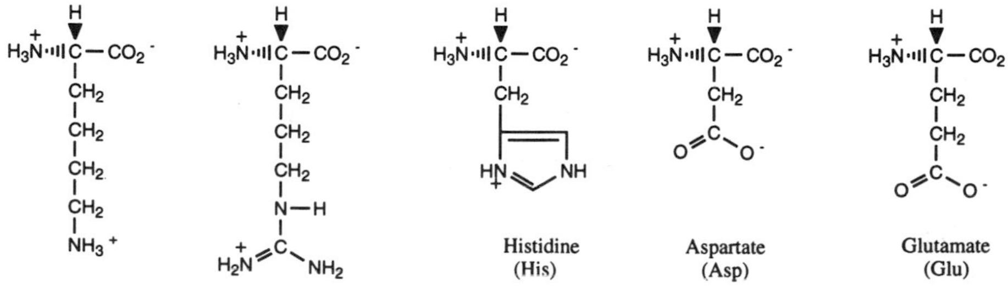

Lysine
(Lys)

Arginine
(Arg)

Histidine
(His)

Aspartate
(Asp)

Glutamate
(Glu)

Appendix 1 ▪ Essential amino acids

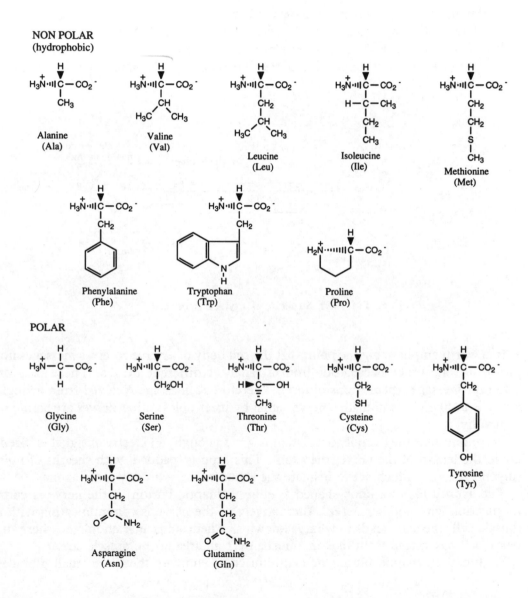

Appendix 2 ▪ The action of nerves

The structure of a typical nerve is shown in Fig. A2.1. The nucleus of the cell is found in the large cell body situated at one end of the nerve cell. Small arms (dendrites) radiate from the cell body and receive messages from other nerves. These messages either stimulate or destimulate the nerve. The cell body 'collects' the sum total of these messages.

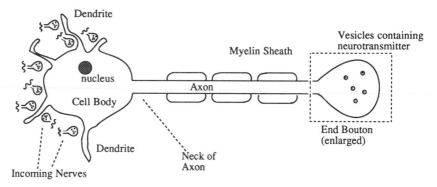

Fig. A2.1 Structure of a typical nerve cell.

It is worth emphasizing the point that the cell body of a nerve receives messages not just from one other nerve, but from a range of different nerves. These pass on different messages (neurotransmitters). Therefore, a message received from a single nerve is unlikely to stimulate a nerve signal by itself unless other nerves are acting in sympathy.

Assuming that the overall stimulation is great enough, an electrical signal is 'fired' down the length of the nerve (the axon). This axon is 'padded' with sheaths of lipid (myelin sheaths) which act to insulate the signal as it passes down the axon.

The axon leads to a knob-shaped swelling (synaptic button) if the nerve is communicating with another nerve. Alternatively, if the nerve is communicating with a muscle cell, the axon leads to what is known as a neuromuscular endplate, where the nerve cell has spread itself like an amoeba over an area of the muscle cell.

Within the synaptic button or neuromuscular endplate there are small globules (vesicles) containing the neurotransmitter chemical. When a signal is received from the axon, the vesicles merge with the cell membrane and release their neurotransmitter into the gap between the nerve and the target cell (synaptic gap). The neurotransmitter binds to the receptor as described in Chapter 5 and passes on its message. Once the message has been received, the neurotransmitter leaves the receptor and is either broken down enzymatically (e.g. acetylcholine) or taken up intact by the nerve cell (e.g. noradrenaline). Either way, the neurotransmitter is removed from the synaptic gap and is unable to bind with its receptor a second time.

To date, we have talked about nerves 'firing' and the generation of 'electrical signals' without really considering the mechanism of these processes. The secret behind nerve transmission lies in the movement of ions across cell membranes, but there is an important difference in what happens in the cell body of a nerve compared to the axon. We shall consider what happens in the cell body first.

All cells contain sodium, potassium, calcium, and chloride ions and it is found that the concentration of these ions is different inside the cell compared to the outside. The concentration of potassium inside the cell is larger than the surrounding medium, whereas the concentration of sodium and chloride ions is smaller. Thus, a concentration gradient exists across the membrane.

Potassium is able to move down its concentration gradient (i.e. out of the cell) since it can pass through the potassium ion channels (Fig. A2.2). However, if potassium

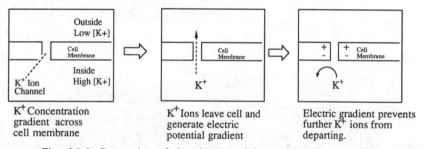

Fig. A2.2 Generation of electric potential across a cell membrane.

can move out of the cell, why does the potassium concentration inside the cell not fall to equal that of the outside? The answer lies in the fact that potassium is a positively charged ion and as it leaves the cell an electric potential is set up across the cell membrane. This would not happen if a negatively charged counterion could leave with the potassium ion. However, the counterions in question are large proteins which cannot pass through the cell membrane. As a result, a few potassium ions are able to escape down the ion channels out of the cell and an electric potential builds up across the cell membrane such that the inside of the cell membrane is more negative than the outside. This electric potential (50–80 mV) opposes and eventually prevents the flow of potassium ions

But what about the sodium ions? Could they flow into the cell along their concentration gradient to balance the charged potassium ions which are departing? The answer is that they cannot because they are too big for the potassium ion channels. This appears to be a strange argument since sodium ions are smaller than potassium ions. However, it has to be remembered that we are dealing with an aqueous environment where the ions are solvated (i.e. they have a 'coat' of water molecules). Sodium, being a smaller ion than potassium, has a greater localization of charge and is able to bind its solvating water molecules more strongly. As a result, sodium along with its water coat is bigger than a potassium ion with or without its water coat.

Ion channels for sodium do exist and these channels are capable of removing the water coat around sodium and letting it through. However, the sodium ion channels are mostly closed when the nerve is in the resting state. As a result, the flow of sodium ions across the membrane is very small compared to potassium. Nevertheless, the presence of sodium ion channels is crucial to the transmission of a nerve signal.

To conclude, the movement of potassium across the cell membrane sets up an electric potential across the cell membrane which opposes this same flow. Charged protein structures are unable to move across the membrane, while sodium ions cross very slowly, and so an equilibrium is established. The cell is polarized and the electric potential at equilibrium is known as the resting potential.

The number of potassium ions required to establish that potential is of the order of

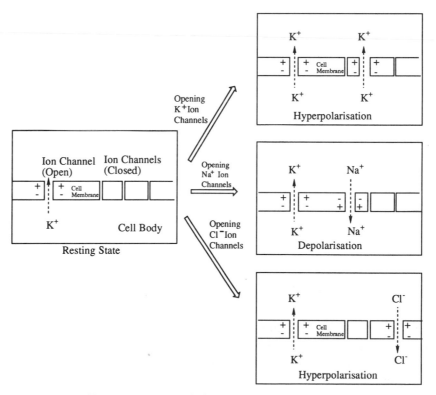

Fig. A2.3 Hyperpolarization and depolarization.

a few million compared to the several hundred billion present in the cell. Therefore the effect on concentration is negligible.

As mentioned above, potassium ions are able to flow out of potassium ion channels. However, not all of these channels are open in the resting state. What would happen if more were to open? The answer is that more potassium ions would flow out of the cell and the electric potential across the cell membrane would become more negative to counter this increased flow. This is known as hyperpolarization and the effect is to destimulate the nerve (Fig. A2.3).

Suppose instead that a few sodium ion channels were to open up. In this case, sodium ions would flow into the cell and as a result the electric potential would become less negative. This is known as depolarization and results in a stimulation of the nerve.

If chloride ions channels are opened, chloride ions flow into the cell and the cell membrane becomes hyperpolarized, destimulating the nerve.

Ion channels do not open or close by chance. They are controlled by the neurotransmitters released by communicating nerves. The neurotransmitters bind with their receptors and lead to the opening or closing of ion channels. For example, acetylcholine controls the sodium ion channel, whereas GABA and glycine control chloride ion channels. The resulting flow of ions leads to a localized hyperpolarization or depolarization in the area of the receptor. The cell body collects and sums all this information such that the neck of the axon experiences an overall depolarization or hyperpolarization depending on the sum total of the various excitatory or inhibitory signals received.

We shall now consider what happens in the axon of the nerve (Fig. A2.4). The cell membrane of the axon also has sodium and potassium ion channels but they are different in character from those in the cell body. The axon ion channels are not controlled by neurotransmitters, but by the electric potential of the cell membrane.

The sodium ion channels located at the junction of the nerve axon with the cell body are the crucial channels since they are the first channels to experience whether the cell body has been depolarized or hyperpolarized.

If the cell body is strongly depolarized then a signal is fired along the nerve. However, a specific threshold value has to be reached before this happens. If the depolarization from the cell body is weak, only a few sodium channels open up and the depolarization at the neck of the axon does not reach that threshold value. The sodium channels then reclose and no signal is sent.

With stronger depolarization, more sodium channels open up until the flow of sodium ions entering the axon becomes greater than the flow of potassium ions leaving it. This results in a rapid increase in depolarization, which in turn opens up more sodium channels, resulting in very strong depolarization at the neck of the axon. The flow of sodium ions into the cell increases dramatically, such that it is far greater than the flow of potassium ions out of the axon, and the electric potential across the membrane is reversed, such that it is positive inside the cell and negative outside the cell. This process lasts less than a millisecond before the sodium channels reclose and sodium permeability returns to its normal state. More potassium channels then open

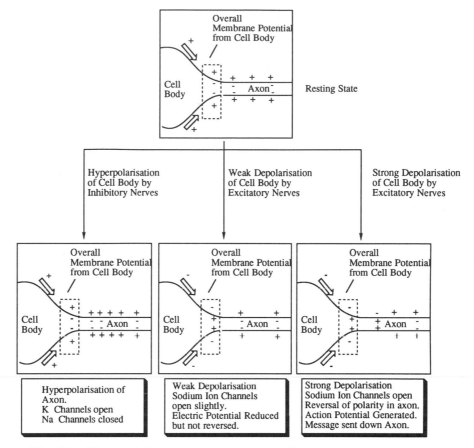

Fig. A2.4 Hyperpolarization and depolarization effects at the neck of the axon.

and permeability to potassium ions increases for a while to speed up the return to the resting state.

The process is known as an action potential and can only take place in the axon of the nerve. The cell membrane of the axon is said to be excitable, unlike the membrane of the cell body.

The important point to note is that once an action potential has fired at the neck of the axon, it has reversed the polarity of the membrane at the point. This in turn has an effect on the neighbouring area of the axon and depolarizes it beyond the critical threshold level. It too fires an action potential and so the process continues along the whole length of the axon (see Fig. A2.5).

The number of ions involved in this process is minute, such that the concentrations are unaffected.

Once the action potential reaches the synaptic button or the neuromuscular endplate, it causes an influx of calcium ions into the cell and an associated release of neurotransmitter into the synaptic gap. The mechanism of this is not well understood.

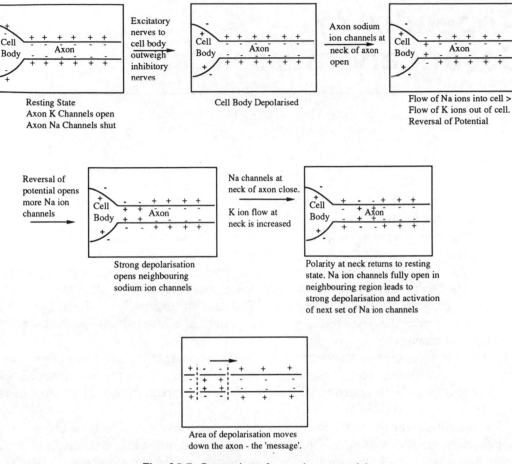

Fig. A2.5 Generation of an action potential.

Appendix 3 ▪ Secondary messengers

In Chapter 5, we discussed how simple chemicals interact with receptor proteins to change the tertiary structure of these proteins. This change of structure can result in the opening or closing of ion channels, e.g. the acetylcholine receptor (see chapter 11).

Alternatively, the neurotransmitter/receptor interaction leads to the activation or deactivation of enzymes. In Chapter 5, this was represented as a direct process whereby the receptor and target enzyme are closely associated. In reality, the receptor and target enzyme are not directly associated and the interaction of receptor and neurotransmitter is the first step in a complex chain of events which involves several proteins and enzymes.

In fact, the receptor is not thought to be *directly* associated with any sort of protein, enzyme or otherwise. The sequence of events leading from the combination of receptor and ligand (the neurotransmitter) to the final activation of a target enzyme is thought to be as shown in Fig. A3.1.

The neurotransmitter binds to the receptor as already described, resulting in a change in the tertiary structure. This receptor/ligand complex is then thought to drift through the cell membrane until it meets a protein called the G_s protein. (The G_s protein acquires the label G since it binds the guanyl nucleotide GDP. The subscript S refers to the fact that the G_s protein stimulates the enzymic synthesis of a secondary messenger.) The G_s protein is situated at the inner surface of the cell membrane and is made up of three protein subunits, possibly linked by disulfide bridges.

The receptor ligand complex interacts in some way with the G_s protein and leads to the exchange of GDP with GTP (guanosine triphosphate). This in turn results in the fragmentation of the G_s protein giving off the $alpha_s$ subunit with GTP attached. The receptor ligand complex has now done its job and it dissociates.

The message is now carried by the $alpha_s$ subunit which interacts with a membrane-bound enzyme called adenylate cyclase and 'switches' it on. This enzyme then catalyses the synthesis of the secondary messenger—cyclic AMP—which moves into the cell's cytoplasm and proceeds to activate enzymes called protein kinases.

The role of the kinases is to phosphorylate and thus activate enzymes with functions specific to the particular cell or organ in question. For example, a protein kinase would activate lipase enzymes in fat cells, whereas in muscle and liver cells enzymes involved in glycogenolysis and glycogen synthesis are regulated through an even

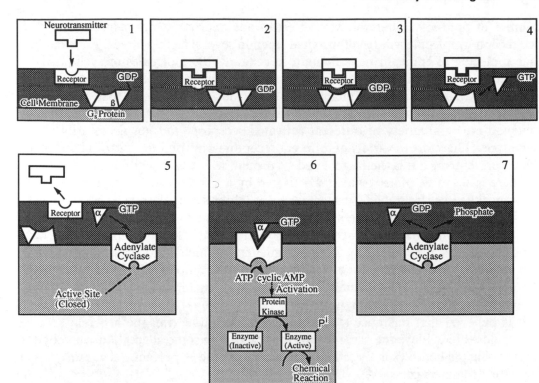

Fig. A3.1 Processes between the combination of receptor and ligand to activation of a target enzyme.

longer sequence of enzymes. How does phosphorylation activate enzymes? Phosphorylation is known to occur on the phenolic group of a tyrosine residue. If this phenolic group is involved in hydrogen bonding, then the addition of a bulky phosphate group disrupts this bonding and also introduces a couple of negatively charged oxygens. These charged groups can now form strong ionic bonds to acidic residues such as aspartate or glutamate and result in the enzyme changing its tertiary structure in order to optimize these new interactions. This change in shape results in the exposure of the active site.

While this is in progress, the G_s protein can be regenerated by the hydrolysis of a phosphate unit from GTP and recombination of the three subunits. The complete process is summarized in Fig. A3.1.

There are several points worth noting about this complicated looking procedure: First of all, the generation of a secondary messenger explains how the original message delivered to the outside of the cell surface can be transmitted to enzymes within the cell and which have no association with the cell membrane.

Secondly, the process involves the switching on of at least three different enzymes —the enzyme synthesizing the secondary messenger, the phosphorylating enzyme (protein kinase A), and the final target enzyme. At each of these three stages, the action of one activating molecule will result in the production of a much larger number of products. Therefore, the effect of one neurotransmitter interacting with one receptor molecule will result in a final reaction several factors larger. For example, each molecule of epinephrine is thought to generate 100 molecules of cyclic AMP.

Thirdly, there is an advantage in having the receptor, the G_s protein, and adenylate cyclase as separate entities, although this might not be obvious at first sight. Since adenylate cyclase is not intimately associated with a single receptor, the enzyme can be switched on by a variety of different activated receptors and not necessarily of the same type. Therefore, a variety of different receptors can combine with and dissociate the G_s protein since it is the fragmented G_s protein which activates adenylate cyclase.

There is also a G_i protein which is activated by a different set of receptors and has the opposite effect from the G_s protein in that it inhibits adenylate cyclase. As a result, the secondary messenger process is under the dual control of 'brake and accelerator' and this explains the process by which two different neurotransmitters can have opposing effects at a target cell. A neurotransmitter which stimulates the production of a secondary messenger will form a receptor/ligand complex which activates the G_s protein, whereas a neurotransmitter which acts as an inhibitor will fit a different receptor which activates the G_i protein.

It is believed that tolerance to drugs such as morphine (i.e. the necessity to take higher doses to achieve the same result) may be due to some adaptation whereby the receptor/ligand complex is less efficient in activating the G_i protein. As a result, larger levels of drug are necessary.

It has also been proposed that dependence might result from the action of a drug acting through the G_i protein to inhibit the synthesis of cyclic AMP. If this action was prolonged, the system might compensate by increasing the levels of acetylcholine messenger reaching the target cell. This in turn would shift the balance back towards the normal level of cyclic AMP. If the original drug was now removed, there would be an excess of acetylcholine and hence too much cyclic AMP produced, resulting in the observed withdrawal symptoms.

Cyclic AMP is not the only secondary messenger existing in cells. To date, three other secondary messengers have been identified. Cyclic GMP is a secondary messenger synthesized by the enzyme guanylate cyclase by the same mechanism described for cyclic AMP. Two other secondary messengers which have been identified are inositol triphosphate (Fig. A3.2) and diacylglycerol (Fig. A3.3). These are both generated by

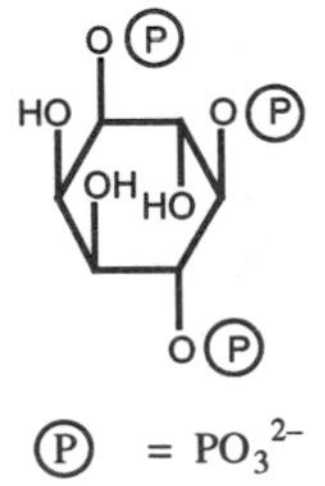

Fig. A3.2 Inositol triphosphate.

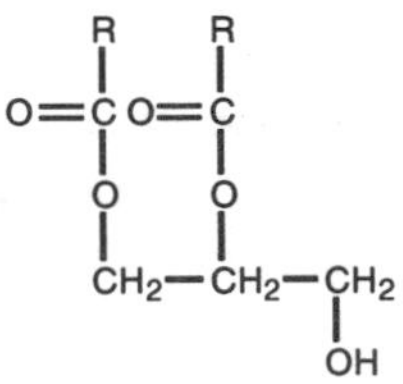

Fig. A3.3 Diacylglycerol.

the process shown in Fig. A3.5. The first part of the process parallels that of cyclic AMP. A neurotransmitter combines with a membrane-bound receptor to form a receptor/ligand complex. This receptor/ligand complex drifts through the cell membrane until it meets a G_s protein. Binding of the receptor/ligand complex with the G_s protein leads to splitting off of the alpha subunit as before. This time however, the alpha subunit goes on to activate an enzyme called phospholipase C. This enzyme catalyses the hydrolysis of phosphatidylinositol diphosphate (PIP$_2$) (Fig. A3.4) to generate the two secondary messengers inositol triphosphate (IP$_3$) and diacylglycerol (DG).

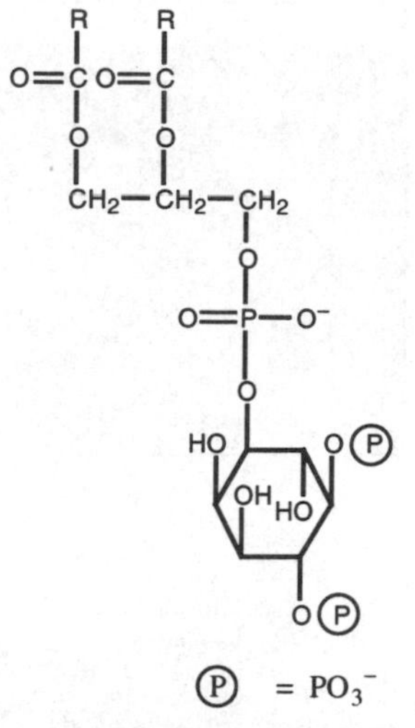

The neurotransmitter, receptor, and G_s protein have now completed their part of the process and return to their original states. We now go on to see what the secondary messengers do. Inositol triphosphate is a hydrophilic molecule and moves into the cytoplasm, whereas diacylglycerol is a hydrophobic molecule and remains in the cell membrane (Fig. A3.6). There, it activates a membrane enzyme called protein kinase C (PKC) which catalyses the phosphorylation of enzymes within the cell. Once phosphorylated, these enzymes are activated and catalyse specific reactions within the cell.

Meanwhile, inositol triphosphate is at work within the cell (Fig. A3.7). This messenger works by mobilizing calcium from calcium stores in microsomes or the

Fig. A3.4 Phosphatidylinositol diphosphate (PIP$_2$).

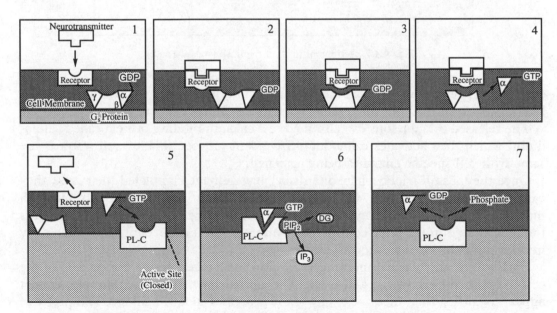

Fig. A3.5 Generation process of inositol triphosphate and diacylglycerol.

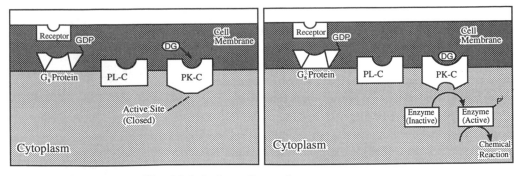

Fig. A3.6 Actions of secondary messengers.

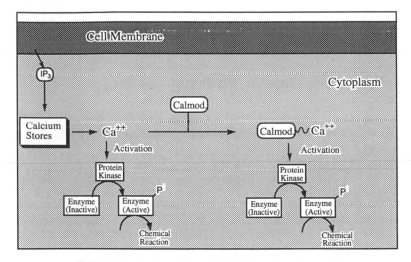

Fig. A3.7 Actions of inositol triphosphate in a cell.

endoplasmic reticulum. Once released, the calcium ions activate calcium-dependent kinases which in turn phosphorylate and activate cell specific enzymes.

The released calcium ions can also bind to a calcium binding protein called calmodulin, which then activates calmodulin-dependent kinases. The overall result is the same with cell specific enzymes being activated.

Once the inositol triphosphate and diacylglycerol have completed their task, they can be linked together again. Oddly enough, they cannot be linked directly and both molecules have to undergo several metabolic steps before resynthesis can occur. It is thought that lithium salts control the symptoms of manic depressive illness by interfering with this involved resynthesis.

Finally, not all chemical messengers bind to their receptors on the outside of the cell, pass on their message and leave. The steroid hormones, for example, do not appear to entrust their message to any of the processes so far mentioned and prefer to deliver their message into the cell in person. The process is believed to be as follows.

The steroid molecule is hydrophobic and is able to diffuse through the cell membrane and finds its receptor waiting for it in the cytoplasm of the cell. Receptor and steroid combine and the complex travels across the nuclear membrane into the nucleus whereupon it binds (still as the receptor/steroid complex) to an acceptor site on the cell's DNA. This binding then switches on transcription and the synthesis of mRNA.

Appendix 4 • Bacteria and bacterial nomenclature

Bacterial nomenclature

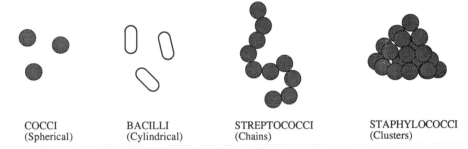

Fig. A4.1 Bacterial nomenclature.

Some clinically important bacteria

Organism	Gram	Infections
Staphylococcus aureus	Positive	Skin and tissue infections, septicaemia, endocarditis, accounts for about 25 per cent of all hospital infections
Streptococcus	Positive	Several types—commonly cause sore throats, upper respiratory tract infections, and pneumonia
Escherichia coli	Negative	Urinary tract and wound infections, common in the gastrointestinal tract and often causes problems after surgery, accounts for about 25 per cent of hospital infections
Proteus species	Negative	Urinary tract infections
Salmonella species	Negative	Food poisoning and typhoid
Shigella species	Negative	Dysentery
Enterobacter species	Negative	Urinary tract and respiratory tract infections, septicaemia
Pseudomonas aeruginosa	Negative	An 'opportunist' pathogen, can cause very severe infections in burn victims and other compromised

patients, i.e. cancer patients, commonly causes chest infections in patients with cystic fibrosis.

Haemophilus influenzae   Negative   Chest and ear infections, occasionally meningitis in young children

Bacteroides fragilis   Negative   Septicaemia following gastrointestinal surgery

The Gram stain

A staining procedure of great value in the identification of bacteria. The procedure is as follows:

1. Stain the cells purple.
2. Decolourize with organic solvent.
3. Stain the cells red.

This test will discriminate between two types of bacterial cell:

(A) Gram-negative bacteria —these bacteria cells are easily decolourized at stage two and will therefore be coloured red;

(B) Gram-positive bacteria —these resist the decolourization at stage 2 and will therefore remain purple.

The different result in the Gram test observed between these two types of bacteria is due to differences in their cell wall structure. These differences in cell wall structure have important consequences in the sensitivity of the two types of bacteria to certain types of antibacterial agents. It is believed that Gram-negative bacteria have an extra outer layer.

The cells of Gram-positive bacteria have an outer covering or membrane containing teichoic acids, whereas the walls of Gram-negative bacteria are covered with a smooth, soft lipopolysaccharide which also contains phospholipids, lipoproteins, and proteins. This layer acts a barrier and penicillins have to negotiate a limited number of protein channels in order to reach the cell.

Glossary

ADDICTION

Addiction can be defined as a habitual form of behaviour. It need not be harmful. For example, one can be addicted to eating chocolate or watching television without suffering more than a bad case of toothache or a surplus of soap operas.

AGONIST

A drug producing a response at a certain receptor.

ANTAGONIST

A drug which interacts with a defined receptor to block an agonist.

ANTIBACTERIAL AGENT

A synthetic or naturally occurring agent which can kill or inhibit the growth of bacterial cells.

ANTIBIOTIC

An antibacterial agent derived from a natural source (e.g. penicillin from penicillium mould).

BACTERIOSTATIC

Bacteriostatic drugs inhibit the growth and multiplication of bacteria, but do not directly kill them (e.g. sulfonamides, tetracyclines, chloramphenicol).

BACTERICIDAL

Bactericidal drugs irreversibly damage and kill bacteria, usually by attacking the cell wall or plasma membrane (e.g. penicillins, cephalosporins, polymyxins).

BIOISOSTERE

A chemical group which can replace another chemical group in a drug without affecting the biological activity of the drug.

DEPENDENCE

A compulsive urge to take a drug for psychological or physical needs. The psychological need is usually why the drug was taken in the first place (to change one's mood) but physical needs are often associated with this. This shows up when the drug is no longer taken leading to psychological withdrawal symptoms (feeling miserable) and physical withdrawal symptoms (headaches, shivering, etc.) Dependence need not be a serious matter if it is mild and the drug is non-toxic (e.g.

dependence on coffee). However, it is a serious matter if the drug is toxic and/or shows tolerance. Examples: opiates, alcohol, barbiturates, diazepams.

ED50

The ED50 is the mean effective dose of a drug necessary to produce a therapeutic effect in 50 per cent of the test sample.

ISOSTERE

A chemical group which can be considered to be equivalent in size and behaviour to another chemical group; for example, replacing a methylene group with an ether bridge (CH_2 for O).

LD50

The LD50 is the mean lethal dose of a drug required to kill 50 per cent of the test sample.

PARTIAL AGONIST

A drug which acts like an antagonist by blocking an agonist, but retains some agonist activity of itself.

RECEPTOR

A protein in the cell membrane of a nerve or target organ with which a transmitter substance or drug can interact to produce a biological response. Example: cholinergic receptors at nerve synapses.

THERAPEUTIC INDEX (OR RATIO)

The therapeutic index is the ratio of a drug's undesirable effects with respect to its desirable effects and is therefore a measure of how safe that drug is. Usually this involves comparing the dose levels leading to a toxic effect with respect to the dose levels leading to a therapeutic effect. The larger the therapeutic index, the safer the drug.

To be more precise, the therapeutic index compares the drug dose levels which lead to toxic effects in 50 per cent of cases studied, with respect to the dose levels leading to maximum therapeutic effects in 50 per cent of cases studied. This is a more reliable method of measuring the index since it eliminates any peculiar individual results.

TOLERANCE

Repeat doses of a drug may result in smaller biological results. The drug may block or antagonize its own action and larger doses are needed for the same pharmacological effect.

Alternatively, the body may 'learn' how to metabolize the drug more efficiently. Again larger doses are needed for the same pharmacological effect, increasing the chances of toxic side-effects.

Examples: morphine, hexamethonium.

Further reading

Albert, A. (1987. *Xenobiosis*. Chapman and Hall.

Bowman, W. C. and Rand, M. J. (1980). *Textbook of pharmacology*. Blackwell Scientific.

Mann, J. (1992). *Murder, magic, and medicine*. Oxford University Press.

Nogrady, T. (1988). *Medicinal chemistry: a biochemical approach*. Oxford University Press, New York.

Roberts, S. M. and Price, B. J. (ed.) (1985). *Medicinal chemistry—the role of organic chemistry in drug research*. Academic Press.

Sammes, P. G. (ed.) (1990). *Comprehensive medicinal chemistry*. Pergamon Press.

Silverman, R. (1992). *The organic chemistry of drug design and drug action*. Academic Press.

Smith, C. M. and Reynard, A. M. (1992). *Textbook of pharmacology*. W. B. Saunders.

Sneader, W. (1985). *Drug discovery: the evolution of modern medicine*. Wiley.

Stenlake, J. B. (1979). *Foundations of molecular pharmacology*. Volumes 1 and 3. Athlone Press.

Suckling, K. E. and Suckling, C. J. (1980). *Biological chemistry*. Cambridge University Press.

Wolff, M. E. *Burger's medicinal chemistry and drug discovery*. Volumes 1–3. Wiley. (In press.)

Index